ENDANGERED SPECIES

PROTECTING BIODIVERSITY

ISSN 1930-3319

ENDANGERED SPECIES

PROTECTING BIODIVERSITY

Kim Masters Evans

INFORMATION PLUS® REFERENCE SERIES
Formerly Published by Information Plus, Wylie, Texas

Detroit • New York • San Francisco • New Haven, Conn • Waterville, Maine • London

Endangered Species: Protecting Biodiversity

Kim Masters Evans

Kepos Media, Inc.: Paula Kepos and Janice Jorgensen, Series Editors

Project Editors: Kimberley McGrath, Kathleen J. Edgar, Elizabeth Manar

Rights Acquisition and Management: Kimberly Potvin, Robyn V. Young

Composition: Evi Abou-El-Seoud, Mary Beth Trimper

Manufacturing: Cynde Lentz

Gale
27500 Drake Rd.
Farmington Hills, MI 48331-3535

ISBN-13: 978-0-7876-5103-9 (set)
ISBN-13: 978-1-4144-8138-8
ISBN-10: 0-7876-5103-6 (set)
ISBN-10: 1-4144-8138-1

ISSN 1930-3319

This title is also available as an e-book.
ISBN-13: 978-1-4144-9726-6 (set)
ISBN-10: 1-4144-9726-1 (set)
Contact your Gale sales representative for ordering information.

Printed in the United States of America
1 2 3 4 5 16 15 14 13 12

FD289

THE IMPORTANCE OF BIODIVERSITY

The term *biodiversity* is short for biological diversity. It refers to the richness and variety of living organisms across the planet. Biodiversity is important at levels within the taxonomic table and at the genetic level. For example, all humans are members of one species (*Homo sapiens*), but humans can vary widely in their personal characteristics, such as race, hair color, and eye color. These differences are due to slight variations in genetic material from person to person. Genetic biodiversity results in different individual properties within a species. It also helps ensure that deformities or disorders in genetic material do not become concentrated in a population.

Inbreeding is mating between closely related individuals with extremely similar genetic material. It is almost certain that if one of these individuals has a gene disorder, the other individual will also have it. This disorder might not cause any notable problems in the parents, but it could become concentrated in the offspring and cause serious health problems for them. Thus, there is a certain lower limit to the population of some species, particularly those that are isolated in location. If the population falls too low, the remaining individuals will be so closely related that any inherent gene problems can kill off the resulting offspring and ultimately wipe out the entire species.

EXTINCT SPECIES

A species is described as extinct when no living members remain. Scientists know from the study of fossils that dinosaurs, mammoths, saber-toothed cats, and countless other animal and plant species that once lived on the earth no longer exist. These species have "died out," or become extinct.

Mass Extinctions

In the billions of years since life began on the earth, species have formed, existed, and then become extinct. Scientists call the natural extinction of a few species per million years a background, or normal, rate. When the extinction rate doubles for many different groups of plants and animals at the same time, this is described as a mass extinction. Mass extinctions have occurred infrequently and, in general, have been attributed to major cataclysmic geological or astronomical events. Five mass extinctions have occurred during the last 600 million years. These episodes, known as the Big Five, occurred at the end of five geologic periods:

- Ordovician (505 million to 440 million years ago)
- Devonian (410 million to 360 million years ago)
- Permian (286 million to 245 million years ago)
- Triassic (245 million to 208 million years ago)
- Cretaceous (146 million to 65 million years ago)

After each mass extinction the floral (plant) and faunal (animal) composition of the earth changed drastically. The largest mass extinction on record occurred at the end of the Permian, when an estimated 90% to 95% of all species became extinct. The Cretaceous extinction, which is hypothesized to have resulted from the collision of an asteroid with the earth, saw the demise of many species of dinosaurs.

The Sixth Mass Extinction?

Scientists estimate that hundreds, or even thousands, of species are being lost annually worldwide. This suggests that another mass extinction is taking place. However, the current extinction is not associated with a cataclysmic physical event. Rather, the heightened extinction rate has coincided with the success and spread of human beings. Researchers predict that as humans continue to destroy and alter natural habitats, create pollution, introduce nonnative species, and contribute to global climate change, the extinction rate will continue to rise. The United Nations (UN) concludes in *Global Biodiversity Outlook 3* (2010, http://www.cbd.int/doc/publications/gbo/gbo3-final-en.pdf) that "the news is not good. We continue to lose biodiversity at a rate never before seen in history—extinction rates may be up to 1,000 times higher than the historical background rate." The UN warns that continuing this trend could "catastrophically reduce the capacity of ecosystems" to provide humans with food, fresh water, and other essential services.

Resurrecting Extinct Species?

In the 1993 film *Jurassic Park* scientists use ancient deoxyribonucleic acid (DNA) that was preserved in amber to clone dozens of prehistoric dinosaur species. With the advent of modern genetic science the possibility of resurrecting extinct species has left the realm of science fiction and is fast becoming a reality. According to Charles Q. Choi, in "First Extinct-Animal Clone Created" (*National Geographic News*, February 10, 2009), in 2003 scientists conducted the first-known cloning of an extinct subspecies: *Capra pyrenaica pyrenaica*—commonly called the Pyrenean ibex. Ibex are wild goats. The Pyrenean ibex went extinct in 2000. Choi indicates that hundreds of clone embryos were created using frozen skin cells that were taken from the last living Pyrenean ibex. They were implanted in Spanish ibex and ibex-goat hybrids. Only seven of the embryos developed into fetuses, and only one of the fetuses survived to full term, but it died "immediately after birth" due to lung abnormalities. Choi notes that "such abnormalities are common in cloning" due to irregularities that arise during DNA transfer.

Cloning as a means of resurrecting extinct species is fraught with challenges and is highly controversial. Scientists note that many nonclosely related members of a species would have to be reproduced for the species as a whole to survive. The resurrection of long-extinct species, such as the wooly mammoth, is even more daunting, because of DNA deterioration over time and the lack of suitable surrogate mothers. Cloning also raises ethical questions, for example, the morality of treating animals as objects for scientific manipulation and experimentation.

U.S. HISTORY: SOME EXTINCTIONS AND SOME CLOSE CALLS

The colonization of the North American continent by European settlers severely depleted the ranks of some native wild species. The introduction of livestock brought new animal diseases that devastated some native animals. Widespread hunting and trapping led to the demise of other species. During the early 1800s the United States was home to millions, perhaps billions, of passenger pigeons. These migratory birds traveled in enormous flocks and were extremely popular with hunters. By the beginning of the 20th century the species was virtually exterminated. The last known passenger pigeon died in the Cincinnati Zoo in 1914. The heath hen, a small wild fowl native to the United States and once very abundant, was wiped out of existence by 1932. Stocks of other animals—beaver, elk, and bison (American buffalo; see Figure 1.1)—were driven to the brink of extinction, but saved by conservation efforts.

HOW MANY SPECIES ARE ENDANGERED?

As noted earlier, the IUCN compiles the annual *Red List of Threatened Species*, in which it examines the status of species worldwide. The so-called Red List categorizes species based on the level of risk of their extinction in the wild as follows:

- Critically endangered—extremely high risk
- Endangered—very high risk
- Vulnerable—high risk
- Near threatened—likely to qualify for a risk category soon

FIGURE 1.1

Bison (or American buffaloes) are the largest terrestrial animals in North America. (© *Steve Degenhardt/Shutterstock.com.*)

The IUCN refers to species in all these categories as threatened species.

Determining how many species of plants and animals are imperiled is difficult. In fact, only a small fraction of the species in existence have even been identified and named, let alone studied in detail. In 2011 the IUCN designated 19,570 species as threatened. Nearly 62,000 species were examined out of the 1.7 million species that the IUCN considers "described species." Thus, only 4% of all known species were evaluated. Further study will likely result in many more species being added to the Red List.

In the United States imperiled species are identified and listed as either endangered or threatened in accordance with the Endangered Species Act (ESA), which is described in detail in Chapter 2. Endangered species are at risk of extinction through all or a significant portion of their natural habitats. Threatened species are likely to become endangered in the future. Management at the federal level of listed species is handled by two agencies: the U.S. Fish and Wildlife Service (USFWS) and the National Marine Fisheries Service (NMFS). The USFWS is an agency of the U.S. Department of the Interior (DOI) and oversees terrestrial (land-based) and freshwater species. The NMFS is an agency of the National Oceanic and Atmospheric Administration under the U.S. Department of Commerce and has jurisdiction for marine (ocean-dwelling) species and those that are anadromous (migrate between the ocean and freshwater).

Table 1.2 tabulates the number of ESA-listed species as of November 2011. Of the 1,193 animal species, 589 were found in the United States. Among these, 422 were endangered and 167 were threatened. Among animals, the greatest numbers of endangered and threatened species occurred among fish, birds, and mammals. Of the 797 plant species, 794 were found in the United States. Among these, 644 were endangered and 150 were threatened. Nearly all the endangered plants were flowering plants.

Figure 1.2 shows the number of U.S. species listed per calendar year between 1967 and 2011. The peak year was 1994, when 127 species were listed. As of November 2011, 30 species had been added to the list for the year.

Within the United States, endangered and threatened species are not evenly distributed but are clustered in specific geographical areas. Table 1.3 shows the number of listed species in each state as of November 2011. Four states (Hawaii, California, Alabama, and Florida) each had over 100 listed species. Despite its small size, Hawaii harbored 380 listed species, more than any other state. This is because a significant proportion of Hawaiian plant and animal life is endemic—that is, found nowhere else on the

TABLE 1.2

Count of endangered and threatened species and U.S. species with recovery plans, November 1, 2011

	United States			Foreign				
Group	Endangered	Threatened	Total listings	Endangered	Threatened	Total listings	Total listings (U.S. and foreign)	U.S. listings with active recovery plans
Mammals	70	14	84	256	20	276	360	60
Birds	76	16	92	204	14	218	310	85
Reptiles	13	24	37	66	16	82	119	37
Amphibians	15	10	25	8	1	9	34	17
Fishes	77	68	145	11	1	12	157	101
Clams	64	8	72	2	0	2	74	70
Snails	25	12	37	1	0	1	38	29
Insects	51	10	61	4	0	4	65	40
Arachnids	12	0	12	0	0	0	12	12
Crustaceans	19	3	22	0	0	0	22	18
Corals	0	2	2	0	0	0	2	0
Animal totals	**422**	**167**	**589**	**552**	**52**	**604**	**1,193**	**469**
Flowering plants	613	147	760	1	0	1	761	638
Conifers and cycads	2	1	3	0	2	2	5	3
Ferns and allies	27	2	29	0	0	0	29	26
Lichens	2	0	2	0	0	0	2	2
Plant totals	**644**	**150**	**794**	**1**	**2**	**3**	**797**	**669**
Grand totals	**1,066**	**317**	**1,383**	**553**	**54**	**607**	**1,990**	**1,138**

The U.S. species counted more than once are:

- Plover, piping (Charadrius melodus)
- Salamander, California tiger (Ambystoma californiense)
- Salmon, chinook (Oncorhynchus (=Salmo) tshawytscha)
- Salmon, chum (Oncorhynchus (=Salmo) keta)
- Salmon, coho (Oncorhynchus (=Salmo) kisutch)
- Salmon, sockeye (Oncorhynchus (=Salmo) nerka)
- Sea-lion, Steller (Eumetopias jubatus)
- Sea turtle, green (Chelonia mydas)
- Steelhead (Oncorhynchus (=Salmo) mykiss)
- Tern, roseate (Sterna dougallii dougallii)
- Wolf, gray (Canis lupus)

The foreign species counted more than once are:

- Argali (Ovis ammon)
- Chimpanzee (Pan troglodytes)
- Crocodile, saltwater (Crocodylus porosus)
- Dugong (Dugong dugon)
- Leopard (Panthera pardus)

Notes: Sixteen animal species (11 in the United States and 5 foreign) are counted more than once in the above table, primarily because these animals have distinct population segments (each with its own individual listing status). There are a total of 591 distinct active (draft and final) recovery plans. Some recovery plans cover more than one species, and a few species have separate plans covering different parts of their ranges. This count includes only plans generated by the United States Fish and Wildlife Services (USFWS) (or jointly by the USFWS and National Marine Fisheries Service), and only listed species that occur in the United States. United States listings include those populations in which the United States shares jurisdiction with another nation.

SOURCE: "Summary of Listed Species Listed Populations and Recovery Plans As of Tue, 1 Nov 2011 14:29:49 UTC," in *Species Reports*, U.S. Department of the Interior, U.S. Fish and Wildlife Service, November 1, 2011, http://ecos.fws.gov/tess_public/pub/Boxscore.do (accessed November 1, 2011)

earth. Endemism is dangerous for imperiled species for a variety of reasons. A single calamitous event, such as a hurricane, earthquake, or disease epidemic, could wipe out the entire population at one time. The likelihood of interbreeding and resulting genetic problems is also higher for species that are geographically limited.

SPECIES LOSS: CRISIS OR FALSE ALARM?

Environmental issues, which tend to pit conservation against business or economic development, are often hotly debated. With respect to biodiversity loss, some critics argue that the scale of loss is not as great as it seems. They point to the uncertainty regarding the total number of species and to the geographic distributions of species. Other challengers claim that loss of habitat and disruption by human activity are not powerful enough to cause the massive extinction being documented. Still other challengers contend that extinction is inevitable and that the earth has experienced, and recovered from, mass extinctions before. They conclude that the current biodiversity loss, while huge, is not disastrous.

In addition, opponents of the ESA frequently argue that so-called green policies such as the ESA place the needs of wildlife before those of humans. This issue will be explored at length in Chapter 2.

WHY SAVE ENDANGERED SPECIES?

Proponents of conservation believe that saving species from extinction is important for many reasons. Species have both aesthetic and recreational value, as shown by the tremendous popularity of zoos, wildlife safaris, and wildlife watching. Wildlife also have educational and scientific value. In addition, because all species depend on other species for resources, the impact of a single lost species could potentially be immense. Scientists have shown that habitats with greater biodiversity are more stable—that is, better able to adjust to and recover from disturbances. This is because different species may perform overlapping functions in a biologically diverse ecosystem. Habitats with less diversity are more vulnerable, because a disturbance affecting one species may cause the entire network

FIGURE 1.2

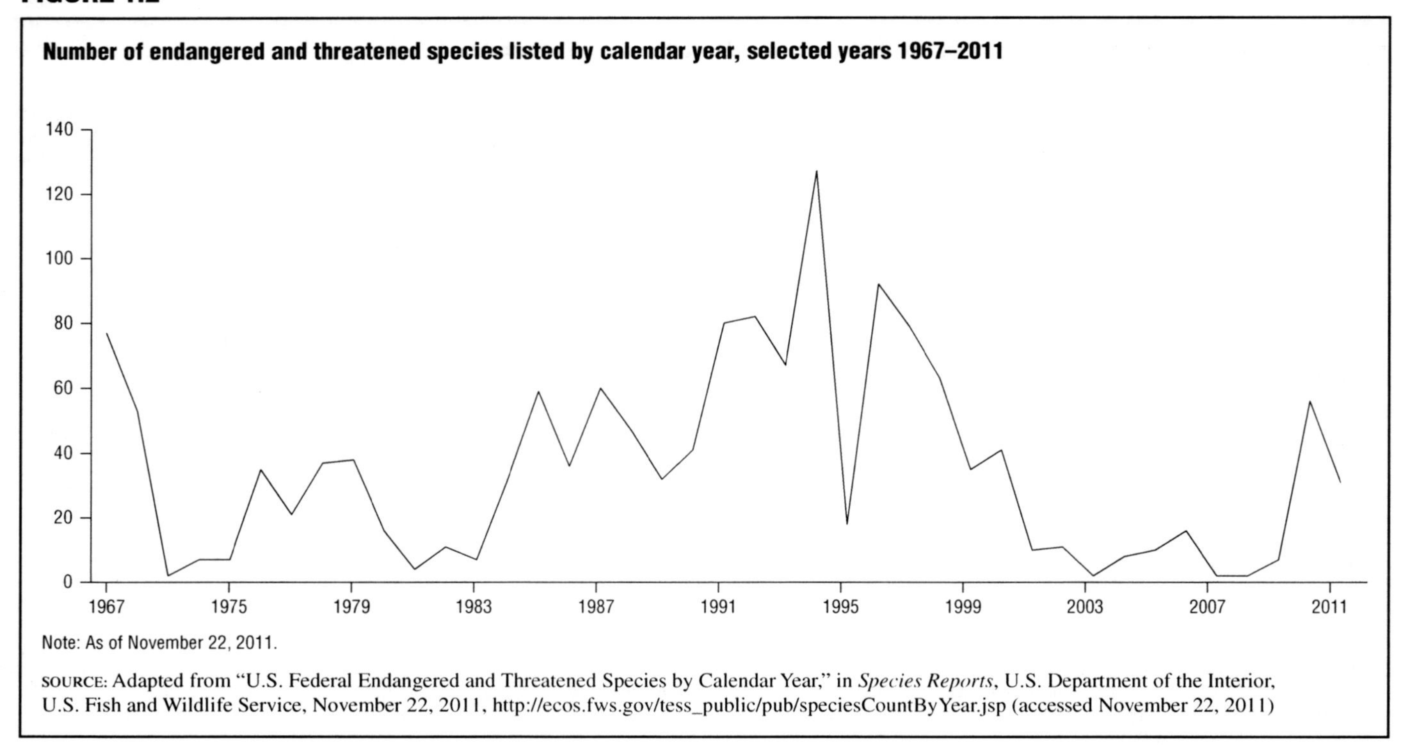

Number of endangered and threatened species listed by calendar year, selected years 1967–2011

Note: As of November 22, 2011.

SOURCE: Adapted from "U.S. Federal Endangered and Threatened Species by Calendar Year," in *Species Reports*, U.S. Department of the Interior, U.S. Fish and Wildlife Service, November 22, 2011, http://ecos.fws.gov/tess_public/pub/speciesCountByYear.jsp (accessed November 22, 2011)

of interactions to collapse. Furthermore, many species have great economic value to human beings. Plants provide the genetic diversity that is used to breed new strains of agricultural crops, and many have been used to develop pharmaceutical products. Aside from the economic or utilitarian reasons for preserving species, many people think that humankind has a moral responsibility to maintain the earth's biodiversity.

The Gallup Organization conducts an annual poll on environmental issues. In *Water Issues Worry Americans Most, Global Warming Least* (March 28, 2011, http://www.gallup.com/poll/146810/Water-Issues-Worry-Americans-Global-Warming-Least.aspx), Lydia Saad of the Gallup Organization indicates that in 2011 participants were asked to express their level of worry about various environmental problems. Sixty-four percent of those asked expressed a "great deal" or "fair amount" of concern regarding the extinction of plant and animal species. (See Table 1.4.) This placed extinction sixth in terms of the amount of worry. As shown in Table 1.5, concern over extinction has generally declined since 2000, when Gallup first polled people about the issue. In 2000, 45% of respondents expressed a "great deal" of concern. Between 2002 and 2010 the percentage fluctuated between a low of 31% and a high of 39%. By 2011, 34% of respondents had a "great deal" of concern.

Are Some Species More Important than Others?

In general, the public places high value on some endangered species and not on others. For example, whales and seals are popular animals for which protection measures receive widespread support. By contrast, several species listed under the ESA are considered pests or predators because they pose a threat to human livelihoods or safety. For example, Utah prairie dogs are burrowing animals that produce networks of underground tunnels. The resulting holes and dirt mounds can ruin cropland and trip and injure livestock. The protection of Utah prairie dogs and other imperiled rodents is a source of contention for people who believe that the ESA puts animal interests above human interests. The same debate rages over predators such as wolves and mountain lions that may prey on livestock, pets, and even people.

Biological Indicator Species

Some species are dying out due to pollution and environmental degradation, problems that also affect human health and well-being. Species that are particularly useful in reporting on the health of ecosystems are called biological indicator species. Environmental scientists rely on sensitive indicator species just as coal miners once relied on canaries to check air safety in underground tunnels, where dangerous gases frequently became concentrated enough to be poisonous. Miners carried a canary into the mineshaft, knowing that the air was safe to breathe as long as the canary lived. If the bird started to sicken, however, miners evacuated the tunnel. In the same way, the sudden deaths of large numbers of bald eagles and peregrine falcons during the 1960s warned people about the dangers of dichlorodiphenyltrichloroethane (DDT), a powerful pesticide in wide use at the time.

TABLE 1.3

Number of endangered and threatened species, by state or territory, November 1, 2011

State or territory	Listings
Alabama	123 listings
Alaska	19 listings
Arizona	62 listings
Arkansas	32 listings
California	313 listings
Colorado	35 listings
Connecticut	21 listings
Delaware	24 listings
District of Columbia	7 listings
Florida	118 listings
Georgia	77 listings
Hawaii	380 listings
Idaho	22 listings
Illinois	39 listings
Indiana	32 listings
Iowa	20 listings
Kansas	17 listings
Kentucky	46 listings
Louisiana	31 listings
Maine	16 listings
Maryland	31 listings
Massachusetts	27 listings
Michigan	27 listings
Minnesota	18 listings
Mississippi	44 listings
Missouri	33 listings
Montana	15 listings
Nebraska	18 listings
Nevada	40 listings
New Hampshire	15 listings
New Jersey	26 listings
New Mexico	48 listings
New York	33 listings
North Carolina	64 listings
North Dakota	10 listings
Ohio	29 listings
Oklahoma	22 listings
Oregon	61 listings
Pennsylvania	23 listings
Rhode Island	18 listings
South Carolina	43 listings
South Dakota	12 listings
Tennessee	96 listings
Texas	95 listings
Utah	43 listings
Vermont	12 listings
Virginia	66 listings
Washington	46 listings
West Virginia	23 listings
Wisconsin	19 listings
Wyoming	18 listings
American Samoa	4 listings
Guam	14 listings
Northern Mariana Islands	11 listings
Puerto Rico	77 listings
Virgin Islands	17 listings
Outlying Caribbean Islands	0 listings
Outlying Pacific Islands	0 listings

SOURCE: "Species Listed in Each State Based on Published Historic Range and Population Data," in *Species Reports*, U.S. Department of the Interior, U.S. Fish and Wildlife Service, November 1, 2011, http://ecos.fws.gov/tess_public/pub/stateListing.jsp (accessed November 1, 2011)

U.S. ENTITIES CONSERVING IMPERILED SPECIES

A variety of government and private entities in the United States conduct activities that help conserve imperiled species.

TABLE 1.4

Public opinion poll on greatest environmental worries, March 2011

DEGREE TO WHICH AMERICANS WORRY ABOUT ENVIRONMENTAL PROBLEMS

I'm going to read you a list of environmental problems. As I read each one, please tell me if you personally worry about this problem a great deal, a fair amount, only a little, or not at all.

	Great deal/ fair amount	Not much/ not at all
	%	%
Contamination of soil and water by toxic waste	79	20
Pollution of rivers, lakes, and reservoirs	79	22
Pollution of drinking water	77	23
Maintenance of the nation's supply of fresh water for household needs	75	24
Air pollution	72	28
Extinction of plant and animal species	64	36
The loss of tropical rain forests	63	35
Urban sprawl and loss of open spaces	57	42
Global warming	51	48

SOURCE: Lydia Saad, "Degree to Which Americans Worry about Environmental Problems," in *Water Issues Worry Americans Most, Global Warming Least*, The Gallup Organization, March 28, 2011, http://www.gallup.com/poll/146810/Water-Issues-Worry-Americans-Global-Warming-Least.aspx (accessed October 27, 2011).

Federal Government

As noted earlier, the USFWS and the NMFS conserve endangered and threatened species through the authority of the ESA, which allows them to issue rules and regulations that govern how imperiled species are managed. They also educate the public by maintaining websites and publishing fact sheets and other documents about the issue. Since 2006 the U.S. Congress has annually declared an Endangered Species Day on which the agencies work with private partners to raise awareness about species conservation. In 2011 the day was celebrated on May 20 and included events around the country.

The USFWS, like many other agencies within the DOI, is also responsible for land conservation. Protecting natural ecosystems and habitats helps ensure that imperiled species (and other native plants and animals) have places to flourish. The USFWS manages the National Wildlife Refuge System (NWRS), the only network of federal lands and waters that are managed principally for the protection of fish and wildlife. In 2011 the NWRS (2011, http://www.fws.gov/refuges/refugeLocatorMaps/index.html) included 548 refuges and thousands of small wetlands around the country. Approximately one-third of the total refuge acreage was wetland habitat, reflecting the importance of wetlands for wildlife survival.

Fifty-nine of the refuges were established specifically for endangered species. (See Table 1.6.) Protected species include a variety of plants and animals. Many other listed animal species use refuge lands on a temporary basis for breeding or migratory rest stops. Virtually

TABLE 1.5

Public opinion poll on worry about extinction of plant and animal species, various dates, selected years 2000–11

Extinction of plant and animal species

	Great deal	Fair amount	Only a little	Not at all	No opinion
2011 Mar 3–6	34	30	23	13	1
2010 Mar 4–7	31	30	24	15	*
2009 Mar 5–8	37	28	22	12	*
2008 Mar 6–9	37	31	20	11	*
2007 Mar 11–14	39	30	19	12	*
2006 Mar 13–16	34	29	23	14	1
2004 Mar 8–11	36	26	23	15	*
2003 Mar 3–5	34	32	21	12	1
2002 Mar 4–7	35	30	22	12	1
2001 Mar 5–7	43	30	19	7	1
2000 Apr 3–9	45	33	14	8	*

* = less than 1%

SOURCE: Jeff Jones and Lydia Saad, "I'm going to read you a list of environmental problems. As I read each one, please tell me if you personally worry about this problem a great deal, a fair amount, only a little, or not at all. First, how much do you personally worry about—Extinction of plant and animal species?" in *Gallup Poll Social Series: Environment, Final Topline*, The Gallup Organization, March 3–6, 2011, http://www.gallup.com/poll/File/146801/Worry_environmental_problems_110328.pdf (accessed October 27, 2011).

every species of bird in North America has been recorded in the refuge system.

The USFWS, along with the U.S. Forest Service, the National Park Service (NPS), and the Bureau of Land Management, oversees the National Wilderness Preservation System of undeveloped federal land. According to Wilderness.net, in "Creation and Growth of the National Wilderness Preservation System" (2011, http://www.wilderness.net/index.cfm?fuse=NWPS&sec=fastfacts), as of 2011, 757 so-called wilderness areas had been designated across the country covering approximately 109.5 million acres (44.3 million ha). Alaska, California, and other western states are home to most of the wilderness areas.

The NPS (November 8, 2011, http://www.nps.gov/faqs.htm) notes that in 2011 there were 397 units in the National Park System covering more than 84 million acres (34 million ha). The units include national parks, monuments, preserves, lakeshores, seashores, wild and scenic rivers, trails, historic sites, military parks, battlefields, historical parks, recreation areas, memorials, and parkways. Besides preserving habitats that range from Arctic tundra to tropical rain forest, the system protects many imperiled plant and animal species.

The national parks have played a significant role in the return of several species, including red wolves and peregrine falcons. The national parks also contain designated critical habitat for many listed species. However, not all these are publicly disclosed, to protect rare species from collectors, vandals, or curiosity seekers.

In 2011 the U.S. Forest Service (2011, http://www.fs.fed.us/aboutus/meetfs.shtml) managed nearly 193 million acres (78.1 million ha) of public lands in 155 national forests and 20 national grasslands. National Forest lands also include many lakes and ponds. National Forest lands are, in general, not conserved to the same degree as NPS lands. For example, much logging occurs within these forests. Endangered, threatened, and sensitive species on National Forest lands are subjected to biological evaluations to determine the effects of management activities on them. Conservation measures are also incorporated to preserve these species.

The Debate over Use of Federally Protected Lands

Ever since federal conservation lands were first set aside, a national debate has raged over how they should be used. Many of these lands contain natural resources of great value in commercial markets, including timber, oil, gas, and minerals. Political and business interests that wish to harvest these resources are pitted against environmentalists, who want to preserve the lands. During the 1990s such a battle raged over the issue of logging in old-growth forests of the Pacific Northwest—the same forests that provided habitat for endangered northern spotted owls. A similar controversy has been brewing for decades over the drilling of oil and gas in the Arctic National Wildlife Refuge.

OIL DRILLING IN THE ARCTIC NATIONAL WILDLIFE REFUGE. The Arctic National Wildlife Refuge (ANWR) is located in northern Alaska. (See Figure 1.3.) Covering 19 million acres (7.7 million ha), it is the largest national wildlife refuge in the United States. The ANWR was established in 1980 by passage of the Alaska National Interest Lands Conservation Act. In Section 1002 of the act, Congress deferred a decision on the future management of 1.5 million acres (607,000 ha) of the ANWR, because of conflicting interests between potential oil and gas resources thought to be located there and the area's importance as a wildlife habitat. This disputed area of coastal plain came to be known as the 1002 area.

TABLE 1.6

National wildlife refuges established for endangered species

State	Unit name	Species of concern	Unit acreage
Alabama	Sauta Cave NWR	Indiana bat, gray bat	264
	Fern Cave NWR	Indiana bat, gray bat	199
	Key Cave NWR	Alabama cavefish, gray bat	1,060
	Watercress darter NWR	Watercress darter	7
Arkansas	Logan Cave NWR	Cave crayfish, gray bat, Indiana bat, Ozark cavefish	124
Arizona	Buenos Aires NWR	Masked bobwhite quail	116,585
	Leslie Canyon	Gila topminnow, Yaqui chub, Peregrine falcon	2,765
	San Bernardino NWR	Gila topminnow, Yaqui chub, Yaqui catfish, Beautiful shiner, Huachuca water umbel	2,369
California	Antioch Dunes NWR	Lange's metalmark butterfly, Antioch Dunes evening-primrose, Contra Costa wallflower	55
	Bitter Creek NWR	California condor	14,054
	Blue Ridge NWR	California condor	897
	Castle Rock NWR	Aleutian Canada goose	14
	Coachella Valley NWR	Coachello Valley fringe-toed lizard	3,592
	Don Edwards San Francisco Bay NWR	California clapper rail, California least tern, salt marsh harvest mouse	21,524
	Ellicott Slough NWR	Santa Cruz long-toed salamander	139
	Hopper Mountain NWR	California condor	2,471
	Sacramento River NWR	Valley Elderberry longhorn beetle, bald eagle, least Bell's vireo	7,884
	San Diego NWR	San Diego fairy shrimp, San Diego mesa mint, Otay mesa mint, California orcutt grass, San Diego button-celery	1,840
	San Joaquin River NWR	Aleutian Canada goose	1,638
	Seal Beach NWR	Light-footed clapper rail, California least tern	911
	Sweetwater Marsh NWR	Light-footed clapper rail	316
	Tijuana Slough NWR	Light-footed clapper rail	1,023
Florida	Archie Carr NWR	Loggerhead sea turtle, green sea turtle	29
	Crocodile Lake NWR	American crocodile	6,686
	Crystal River NWR	West Indian manatee	80
	Florida panther NWR	Florida panther	23,379
	Hobe Sound NWR	Loggerhead sea turtle, green sea turtle	980
	Lake Wales Ridge NWR	Florida scrub jay, snakeroot, scrub blazing star, Carter's mustard, papery Whitlow-wort, Florida bonamia, scrub lupine, highlands scrub hypericum, Garett's mint, scrub mint, pygmy gringe-tree, wire weed, Florida ziziphus, scrub plum, Eastern Indigo snake, bluetail mole skink, sand skink	659
	National Key Deer Refuge	Key deer	8,542
	St. Johns NWR	Dusky seaside sparrow	6,255
Hawaii	Hakalau Forest NWR	Akepa, akiapolaau, 'o'u, Hawaiian hawk, Hawaiian creeper	32,730
	Hanalei NWR	Hawaiian stilt, Hawaiian coot, Hawaiian moorhen, Hawaiian duck	917
	Huleia NWR	Hawaiian stilt, Hawaiian coot, Hawaiian moorhen, Hawaiian duck	241
	James C. Campbell NWR	Hawaiian stilt, Hawaiian coot, Hawaiian moorhen, Hawaiian duck	164
	Kakahaia NWR	Hawaiian stilt, Hawaiian coot	45
	Kealia Pond NWR	Hawaiian stilt, Hawaiian coot	691
	Pearl Harbor NWR	Hawaiian stilt	61
Iowa	Driftless Area NWR	Iowa pleistocene snail	521
Massachusetts	Massasoit NWR	Plymouth red-bellied turtle	184
Michigan	Kirtland's warbler WMA	Kirtland's warbler	6,535
Mississippi	Mississippi sandhill crane NWR	Mississippi sandhill crane	19,713
Missouri	Ozark cavefish NWR	Ozark cavefish	42
	Pilot Knob NWR	Indiana bat	90
Nebraska	Karl E. Mundt NWR	Bald eagle	19
Nevada	Ash Meadows NWR	Devil's Hole pupfish, Warm Springs pupfish, Ash Meadows amargosa pupfish, Ash Meadows speckled dace, Ash Meadows naucorid, Ash Meadows blazing star, amargosa niterwort, Ash Meadows milk-vetch, Ash Meadows sunray, spring-loving centaury, Ash Meadows gumplant, Ash Meadows invesia	13,268
	Moapa Valley NWR	Moapa dace	32
Oklahoma	Ozark Plateau NWR	Ozark big-eared bat, Gray bat	2,208
Oregon	Bear Valley NWR	Bald eagle	4,200
	Julia Butler Hansen Refuge for Columbian white-tail deer	Columbian white-tailed deer	2,750
	Nestucca Bay NWR	Aleutian Canada goose	457
South Dakota	Karl E. Mundt NWR	Bald eagle	1,044
Texas	Attwater Prairie chicken NWR	Attwater's Greater Prairie chicken	8,007
	Balcones Canyonlands NWR	Black-capped vireo, golden-cheeked warbler	14,144
Virgin Islands	Green Cay NWR	St. Croix ground lizard	14
	Sandy Point NWR	Leatherback sea turtle	327

There has been interest in tapping the oil deposits in northern Alaska since the mid-1900s. During the 1950s production of oil and gas in the refuge area—the 5% of Alaska's North Slope not already open to drilling—was prohibited unless specifically authorized by Congress.

TABLE 1.6

National wildlife refuges established for endangered species [CONTINUED]

State	Unit name	Species of concern	Unit acreage
Virginia	James River NWR	Bald eagle	4,147
	Mason Neck NWR	Bald eagle	2,276
Washington	Julia Butler Hansen Refuge for Columbian white-tail deer	Columbian white-tailed deer	2,777
Wyoming	Mortenson Lake NWR	Wyoming toad	1,776

NWR = National Wildlife Refuge

SOURCE: "National Wildlife Refuges Established for Endangered Species," in *National Wildlife Refuge System*, U.S. Department of the Interior, U.S. Fish and Wildlife Service, August 19, 2009, http://www.fws.gov/Refuges/whm/EndSpRefuges.html (accessed October 28, 2011)

FIGURE 1.3

Map of northern Alaska showing the National Petroleum Reserve—Alaska (NPRA) and Arctic National Wildlife Refuge (ANWR)

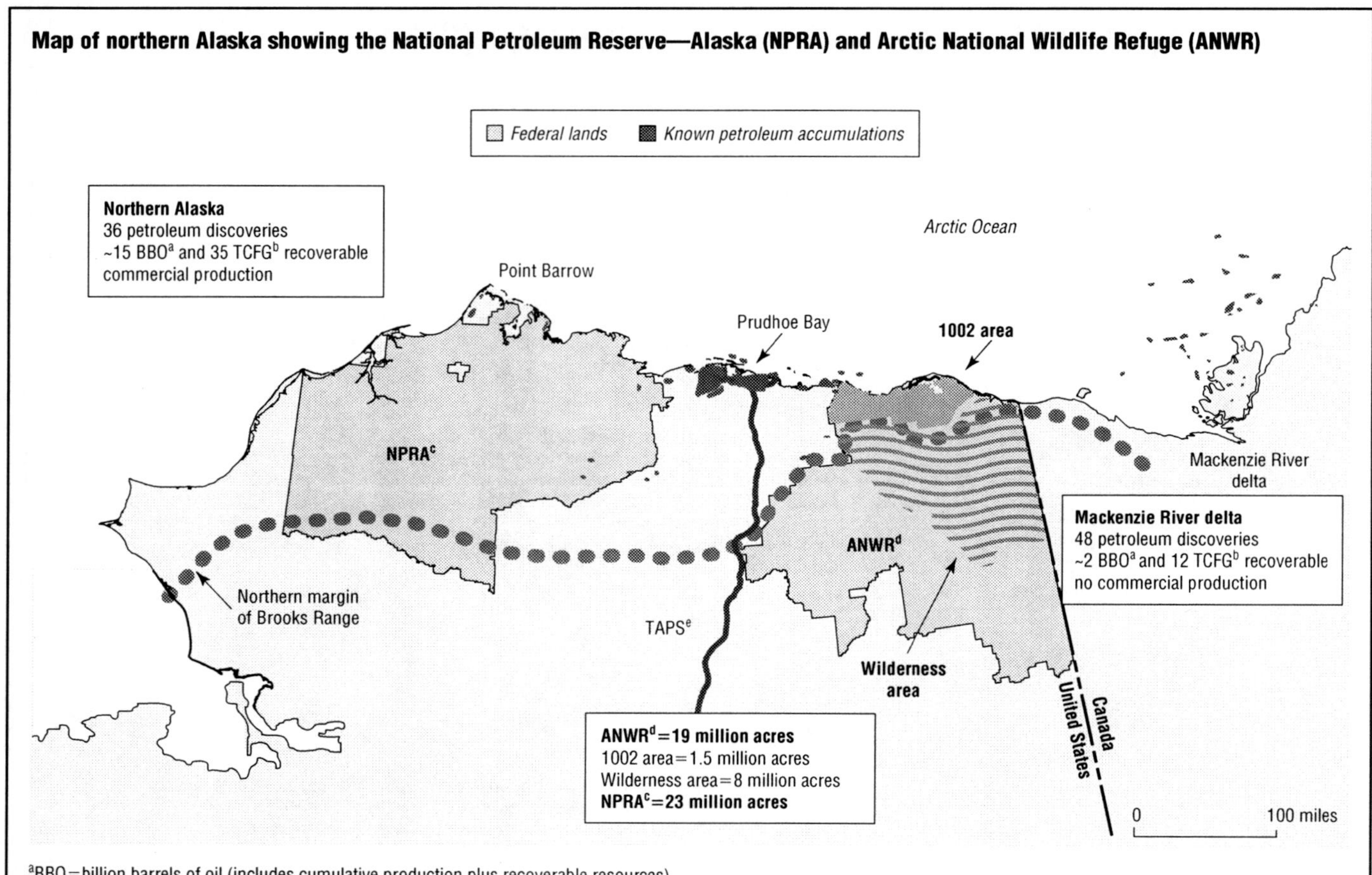

[a]BBO=billion barrels of oil (includes cumulative production plus recoverable resources).
[b]TCFG=trillion cubic feet of gas recoverable resources.
[c]NPRA=National Petroleum Reserve—Alaska.
[d]ANWR=Arctic National Wildlife Refuge.
[e]TAPS=Trans-Alaska Pipeline System.

SOURCE: "Figure 1. Map of Northern Alaska and Nearby Parts of Canada Showing Locations of the Arctic National Wildlife Refuge (ANWR), the 1002 Area, and the National Petroleum Reserve—Alaska (NPRA). Locations of Known Petroleum Accumulations and the Trans-Alaska Pipeline System (TAPS) Are Shown, as well as Summaries of Known Petroleum Volumes in Northern Alaska and the Mackenzie River Delta of Canada. BBO, Billion Barrels of Oil (Includes Cumulative Production Plus Recoverable Resources); TCFG, Trillion Cubic Feet of Gas Recoverable Resources," in *Arctic National Wildlife Refuge, 1002 Area, Petroleum Assessment, 1998, Including Economic Analysis*, U.S. Department of the Interior, U.S. Geological Survey, April 2001, http://pubs.usgs.gov/fs/fs-0028-01/fs-0028-01.pdf (accessed October 27, 2011)

In 1987 the DOI submitted a report to Congress on the resources of the 1002 area. At that time only a few oil accumulations had been found near the ANWR. Over the next decade, much larger oil fields were discovered nearby, particularly around Prudhoe Bay and in the National Petroleum Reserve–Alaska (NPRA). In 2002 and 2010 the U.S.

Geological Survey (USGS) estimated the volumes of oil and natural gas potentially available in the NPRA. In "2010 Updated Assessment of Undiscovered Oil and Gas Resources of the National Petroleum Reserve in Alaska (NPRA)" (October 2010, http://pubs.usgs.gov/fs/2010/3102/pdf/FS10-3102.pdf), the USGS estimates that 896 million barrels of oil and 53 trillion cubic feet (1.5 trillion cubic m) of natural gas are available in "conventional undiscovered accumulations."

Environmentalists argue that USFWS studies suggest that oil drilling in the refuge will harm many Arctic species by taking over habitat, damaging habitats through pollution, interfering with species' activities directly, or increasing opportunities for invasive species. The ANWR harbors the greatest number of plant and animal species of any park or refuge in the Arctic, including unique species such as arctic foxes, caribou, musk oxen, polar bears, and snow geese.

The protected status of the ANWR has been challenged by large oil companies and their political supporters. When Republicans took control of Congress in 1995, they passed legislation that allowed drilling in the ANWR, but President Bill Clinton (1946–) vetoed the bill. The succeeding administration under President George W. Bush (1946–) was much more supportive of drilling in the refuge.

In 2001 the U.S. House of Representatives again passed a bill allowing for drilling within the refuge. However, the U.S. Senate rejected this proposal in 2002. The September 11, 2001, terrorist attacks against the United States and the heightened tensions in the Middle East encouraged some politicians to emphasize the national security aspects of oil development in the ANWR. They argued that the United States cannot be truly secure until it reduces its dependence on foreign oil. The Bush administration continued to press for oil drilling in the ANWR; however, bills containing such measures ultimately failed to pass Congress. Likewise, similar bills have not been successful under the administration of Barack Obama (1961–), who took office in January 2009. Thus, as of January 2012 federal legislation had not been passed that allowed drilling in the ANWR.

State Governments

Every state has its own agency that is charged with managing and overseeing land and wildlife conservation. The USFWS (September 13, 2011, http://www.fws.gov/offices/statelinks.html) maintains a database of website links to agencies in all 50 states, the District of Columbia, and U.S. territories, such as Guam and the Virgin Islands. According to Jeffrey V. Wells et al., in "Global versus Local Conservation Focus of U.S. State Agency Endangered Bird Species List" (*PLoS ONE*, vol. 5, no. 1, January 6, 2010), 48 of the states also maintain their own lists of endangered, threatened, or "special concern" species. For example, as of January 2012 the New York Department of Environmental Conservation (http://www.dec.ny.gov/animals/7494.html) listed 53 animal and fish species as endangered, 35 as threatened, and 58 as "special concern." It should be noted that state lists of imperiled species often include species that are not federally listed.

Private Organizations and Individuals

Federal and state governments are not the only entities involved in land conservation and conserving imperiled species. Increasingly, environment-minded private organizations and citizens are purchasing land with the intent of preserving it for wildlife. National environmental groups such as the Nature Conservancy participate in these endeavors. The Nature Conservancy (2012, http://www.nature.org/aboutus/) notes that it helps protect more than 119 million acres (48.2 million ha) worldwide. Other major groups engaged in private land conservation include the Conservation Fund, the Land Trust Alliance, the Rocky Mountain Elk Foundation, the Society for the Protection of New Hampshire Forests, and the Trust for Public Land.

Every five years the Land Trust Alliance conducts a census on lands that are held for private conservation. The alliance states in *2010 National Land Trust Census Report: A Look at Voluntary Land Conservation in America* (November 16, 2011, http://www.landtrustalliance.org/land-trusts/land-trust-census/national-land-trust-census-2010/2010-final-report) that approximately 47 million acres (19 million ha) of land were held in local, regional, or national land trusts in 2010. Land trusts either purchase land outright or develop private, voluntary agreements called conservation easements or restrictions that limit future development of the land.

As will be explained in Chapter 2, the ESA allows private organizations and individuals to participate directly in the federal listing of endangered and threatened species by petitioning the USFWS or the NMFS on behalf of particular species. In addition, private parties can file lawsuits against the government for alleged failures to abide by the ESA.

INTERNATIONAL EFFORTS AT CONSERVATION

The UN Environment Programme (UNEP) was established to address diverse environmental issues on an international level. Many of its conventions have been extremely valuable in protecting global biodiversity and natural resources. The UNEP has also helped regulate pollution and the use of toxic chemicals.

Convention on International Trade in Endangered Species of Wild Fauna and Flora

The Convention on International Trade in Endangered Species of Wild Fauna and Flora (CITES) is an international agreement administered under the UNEP that regulates international trade in wildlife. CITES is perhaps the single most important international agreement relating to endangered species and has contributed critically to the protection of many threatened species. The international wildlife trade is estimated to involve hundreds of millions of specimens annually.

CITES was first drafted in 1963 at a meeting of the IUCN and went into effect in 1975. Protected plant and animals are listed in three separate CITES appendices, depending on the degree of endangerment. Appendix I includes species that are in immediate danger of extinction. CITES generally prohibits international trade of these species. Appendix II lists species that are likely to become in danger of extinction without strict protection from international trade. Permits may be obtained for the trade of Appendix II species only if trade will not harm the survival prospects of the species in the wild. Appendix III lists species whose trade is regulated in one or more nations. Any member nation can list a species in Appendix III to request international cooperation to prevent unsustainable levels of international trade. Nations agree to abide by CITES rules voluntarily. In 2012 there were 175 nations participating in the agreement (http://cites.org/eng/disc/parties/index.php).

Convention on Biological Diversity

The Convention on Biological Diversity was set up to conserve biodiversity and to promote the sustainable use of biodiversity. The convention supports national efforts in the documentation and monitoring of biodiversity, the establishment of refuges and other protected areas, and the restoration of degraded ecosystems. It also supports goals that are related to the maintenance of traditional knowledge of sustainable resource use, the prevention of invasive species introductions, and the control of invasive species that are already present. Finally, it funds education programs that promote public awareness of the value of natural resources.

Convention on the Conservation of Migratory Species of Wild Animals

The Convention on the Conservation of Migratory Species of Wild Animals (also known as the CMS or the Bonn Convention) recognizes that certain migratory species cross national boundaries and require protection throughout their range. The convention (2012, http://www.cms.int/about/intro.htm) aims to "conserve terrestrial, aquatic and avian migratory species throughout their range." It was originally signed in Bonn, Germany, in 1979 and went into force in November 1983. According to the CMS, in "Parties to the Convention on the Conservation of Migratory Species of Wild Animals and Its Agreements" (http://www.cms.int/about/part_lst.htm), as of July 1, 2011, 116 nations in Africa, Central and South America, Asia, Europe, and Oceania were involved in the agreement. The United States and several other nations were not official parties to the agreement but nonetheless abide by its rules.

The CMS provides two levels of protection to migratory species. Appendix I species are endangered and strictly protected. Appendix II species are less severely threatened but would nonetheless benefit from international cooperative agreements.

World Commission on Protected Areas

The UN's World Conservation Protection Centre is the leading international body dedicated to the selection, establishment, and management of national parks and protected areas. It helps establish natural areas around the world for the protection of plant and animal species and maintains a database of protected areas. Protected areas often consist of a core zone, in which wildlife cannot legally be disturbed by human beings, and a buffer zone, a transitional space that acts as a shield for the core zone. On the periphery are areas for managed human living. According to the World Conservation Protection Centre, in "About Protected Areas" (2012, http://www.unep-wcmc.org/about-protected-areas_163.html), a protected area is defined as "a clearly defined geographical space, recognised, dedicated and managed, through legal or other effective means, to achieve the long term conservation of nature with associated ecosystem services and cultural values."

Conservation biology theory suggests that protected areas should be as large as possible to increase biological diversity and to buffer refuges from outside pressures. The world's largest protected areas are the Ar-Rub'al-Khali Wildlife Management Area (Saudi Arabia), the Cape Churchill Wildlife Management Area (Canada), Great Barrier Reef Marine Park (Australia), Greenland National Park (Greenland), the Northern Wildlife Management Zone (Saudi Arabia), and the Qiangtang Nature Reserve (China).

FACTORS THAT CONTRIBUTE TO SPECIES ENDANGERMENT

Experts believe the increasing loss and decline of species can be attributed to the destructive effects of human activities, rather than to natural processes. People hunt and collect wildlife. They destroy natural habitats by clearing trees and filling swamps for development. Aquatic habitats are altered or destroyed by the building of dams. Humans also poison habitats with polluting chemicals and industrial waste. Indeed, there is widespread international scientific consensus that the air pollutants emitted during the burning of fossil fuels are causing changes in climate patterns on a global scale.

Habitat Destruction

Habitat destruction is probably the single most important factor leading to the endangerment of species. It plays a role in the decline of nearly all listed species and has had an impact on nearly every type of habitat and ecosystem.

Many types of human activity result in habitat destruction. Agriculture is a leading cause, with nearly half of the total land area in the United States being used for farming. Besides causing the direct replacement of natural habitat with fields, agricultural activity also results in soil erosion, pollution from pesticides and fertilizers, and runoff into aquatic habitats. Agriculture has compromised forest, prairie, and wetland habitats in particular.

Urban expansion has also destroyed wild habitat areas and is a primary factor in the endangerment of many plant species. As with agriculture, urbanization leads to the direct replacement of natural habitat. Furthermore, it results in the depletion of local resources, such as water, that are important to many species.

Logging, particularly the practice of clear-cutting, destroys important habitat for many species. Clear-cutting or extensive logging can also lead to significant erosion, harming both soils and aquatic habitats, which become blocked with soil.

Many other forms of human activity result in habitat destruction and degradation. Grazing by domestic livestock has a direct impact on many plant species, as well as on animal species that compete with livestock. Mining destroys vegetation and soil and degrades habitat through pollution. Dams destroy aquatic habitats in rivers and streams. Finally, human recreational activity, particularly the use of off-road vehicles, results in the destruction of natural habitat.

Habitat Fragmentation

Human land-use patterns often result in the fragmentation of natural habitat areas that are available to species. For example, building a road through a forest or converting part of a meadow into a farm or subdivision fragments the large area that used to be available to species into smaller separated pieces. Studies show that habitat fragmentation is occurring in most habitat types. Habitat fragmentation can have significant effects on species. Small populations can become isolated, so that dispersal from one habitat patch to another is impossible. Smaller populations are also more likely to become extinct. Finally, because there are more "edges" when habitats are fragmented, there can be increased exposure to predators and vulnerability to disturbances that are associated with human activity.

Climate Change

The earth's temperature is regulated by many factors, including energy inputs and outputs, chemical processes, and physical phenomena. Radiation from the sun passes through the earth's atmosphere and warms the planet. (See Figure 1.4.) In turn, the earth emits infrared radiation. Some of this outgoing infrared radiation does not escape into outer space but is trapped beneath the atmosphere to provide a warm "blanket" for the planet. The amount of trapped energy depends on several variables, including the composition of the atmosphere. Certain gases, such as carbon dioxide and methane, naturally trap heat beneath the atmosphere in the same way that glass panels keep heat from escaping from a greenhouse. This natural greenhouse effect keeps the earth warm and habitable for life. Scientists believe massive combustion (burning) of fossil fuels, such as oil and natural gas, introduced large amounts of carbon dioxide, methane, and other heat-trapping gases into the atmosphere over the last century. This buildup has been increasing the earth's temperature above that expected from the natural greenhouse effect, an effect known as global warming. Global warming is bringing about climate change, which has numerous consequences to the planet's environment, ecosystems, and inhabitants.

FIGURE 1.4

The greenhouse effect

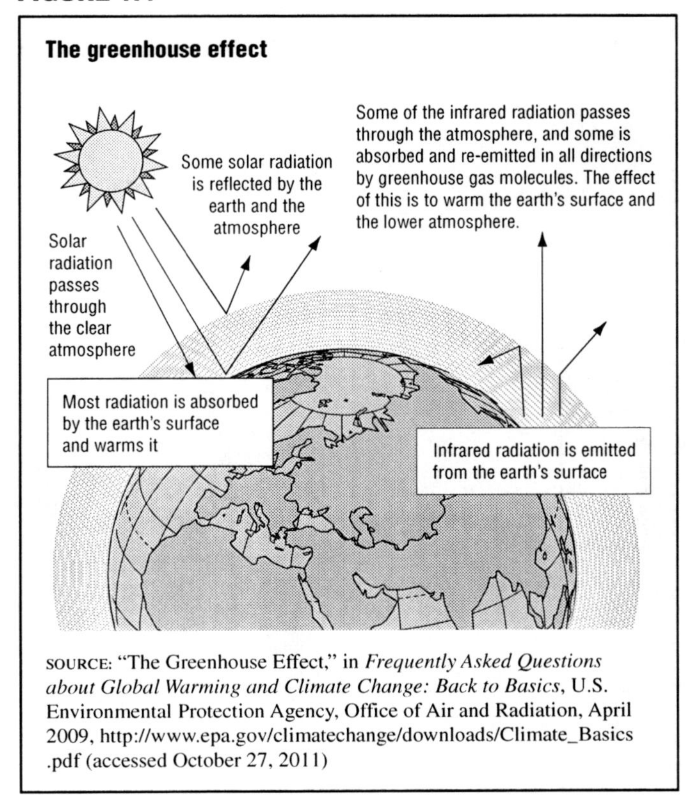

SOURCE: "The Greenhouse Effect," in *Frequently Asked Questions about Global Warming and Climate Change: Back to Basics*, U.S. Environmental Protection Agency, Office of Air and Radiation, April 2009, http://www.epa.gov/climatechange/downloads/Climate_Basics.pdf (accessed October 27, 2011)

Figure 1.5 shows rising global temperatures and atmospheric carbon dioxide concentrations between 1880 and 2009. The bars indicate temperatures that were above or below the average temperature during this period. An upward trend in both temperature and atmospheric carbon dioxide concentrations began around 1980 and continued to occur through 2009.

Chris D. Thomas et al. suggest in "Extinction Risk from Climate Change" (*Nature*, vol. 427, no. 6970, January 8, 2004), a study of habitats consisting of 20% of the earth's surface, that 15% to 35% of the world's species may be extinct by 2050 if recent warming trends continue.

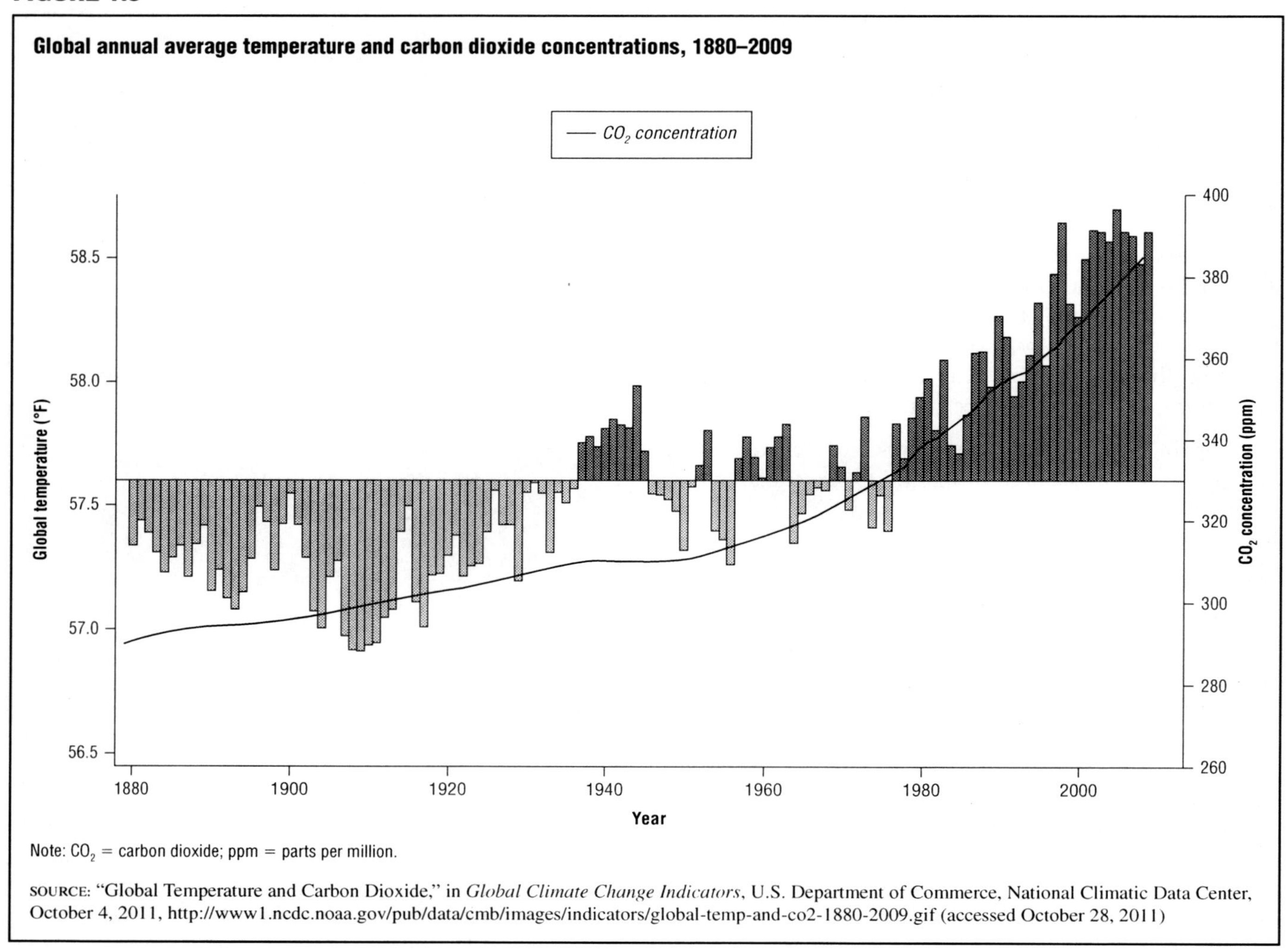

FIGURE 1.5

Global annual average temperature and carbon dioxide concentrations, 1880–2009

Note: CO_2 = carbon dioxide; ppm = parts per million.

SOURCE: "Global Temperature and Carbon Dioxide," in *Global Climate Change Indicators*, U.S. Department of Commerce, National Climatic Data Center, October 4, 2011, http://www1.ncdc.noaa.gov/pub/data/cmb/images/indicators/global-temp-and-co2-1880-2009.gif (accessed October 28, 2011)

Continued warming of the earth would alter habitats drastically, with serious consequences for many species. In Siberia and the northernmost regions of Canada habitats such as the tundra (permanently frozen land supporting only low-growing plant life such as mosses and lichens) and the taiga (expanses of evergreen forests located immediately south of the tundra) are shrinking. Deserts are expanding. Forests and grasslands are beginning to shift toward more appropriate climate regimes. Plant and animal species that cannot shift their ranges quickly enough or that have no habitat to shift into are dying out. Some plants and animals that are found in precise, narrow bands of temperature and humidity, such as edelweiss or monarch butterflies, are likely to find their habitats wiped out entirely. Global warming is already endangering some of the most diverse ecosystems on the earth, such as coral reefs and tropical cloud forests. The impact on endangered species, which are already in a fragile state, may be particularly great.

The Intergovernmental Panel on Climate Change (IPCC) is an international body of scientists and policy-makers that publishes reports on climate change. In *Climate Change 2007: Impacts, Adaptation, and Vulnerability* (2007, http://www.ipcc.ch/), the IPCC reports that "observational evidence from all continents and most oceans shows that many natural systems are being affected by regional climate changes, particularly temperature increases." The consequences include an earlier occurrence of springtime events and a poleward movement in the ranges of plant and animal species. Warming temperatures in natural water bodies are causing range changes for algae, plankton, and fish species. IPCC scientists estimate that approximately 20% to 30% of animal and plant species have an increased risk of extinction if the global average temperature increases by more than 2.7 to 4.5 degrees Fahrenheit (1.5 to 2.5 degrees Celsius).

Pollution

Pollution is caused by the release of industrial and chemical wastes into the land, air, and water. It can damage habitats and kill or sicken animals and plants. Pollution comes from a wide variety of sources, including industrial operations, mining, farming, and automobiles. Even animals that are not directly exposed to pollution can be affected, if other species they rely on die out.

Hunting and Trade

Humans have hunted many animal species to extinction, and hunting continues to be a major threat to some species. In the United States gray wolves were nearly wiped out because they were considered a threat to livestock. The Caribbean monk seal was exterminated because it was viewed as a competitor for fish. Other animals are hunted for the value of their hides, tusks, or horns, including elephants and rhinoceroses. Many exotic species, such as parrots and other tropical birds, are taken from their natural habitats for the pet trade.

Invasive Species

Invasive species are those that have been introduced from their native habitat into a new, nonnative habitat and cause environmental harm. Some introductions of invasive species are accidental, resulting from "stowaways" on ships and planes. Other introductions are done purposely, for example, to stock lakes or rivers with popular food fish. Invasive species harm native life forms by competing with them for food and other resources, or by preying on them or parasitizing them. Even though there are sometimes beneficial effects from introducing nonnative species, the effects are usually harmful.

The introduction of invasive species can lead to genetic swamping. This is a condition that arises when large numbers of one species breed with a much smaller population of another related species. The genetic material of the invasive species becomes overwhelming, causing the resulting generations to lose many of the characteristics that made the smaller population a unique species in the first place.

CHAPTER 2
THE ENDANGERED SPECIES ACT

The Endangered Species Act (ESA) of 1973 is generally considered to be one of the most far-reaching laws ever enacted by any nation for the preservation of wildlife. The passage of the ESA resulted from alarm at the decline of many species worldwide, as well as from recognition of the importance of preserving species diversity. The purpose of the ESA is to identify species that are either endangered (at risk of extinction throughout all or a significant portion of their range) or threatened (likely to become endangered in the future). Except for recognized insect pests, all animals and plants are eligible for listing under the ESA. Listed species are protected without regard to either commercial or sport value.

The ESA is also one of the most controversial and contentious laws ever passed. It affects the rights of private landowners and how they manage their property if endangered species are found there. It also allows private individuals and groups to sue federal agencies for alleged failures in carrying out the law. The result has been a flood of litigation since the 1990s by conservation organizations. They believe that full and effective implementation of the ESA will help ensure the survival of imperiled species. However, critics charge that the ESA has saved virtually no species, puts too many restrictions on land and water development projects, and is too expensive for the results that it achieves.

THE ESA: A LANDMARK PROTECTION

The roots of the ESA lay in the Endangered Species Preservation Act, which was passed in 1966. This act established a process for listing species as endangered and provided some measure of protection. The first species to be listed are shown in Table 2.1. The Endangered Species Conservation Act of 1969 provided protection to species facing worldwide extinction, prohibiting their import and sale within the United States.

Passed by Congress in 1973, the ESA was substantially amended in 1978, 1982, and 1987. Table 2.2 describes the key sections of the amended law. The ESA is administered by the U.S. Department of the Interior (DOI) through the U.S. Fish and Wildlife Service (USFWS). The U.S. Department of Commerce, through the National Marine Fisheries Service (NMFS), is responsible for most marine (ocean-based) species and those that are anadromous (migrate between freshwaters and marine waters). The Biological Resources Division of the U.S. Geological Survey conducts research on species for which the USFWS has management authority.

It should be noted that the original ESA defined the word *species* to include species, subspecies, or "smaller taxa." Taxa is the plural of taxon, which is a grouping on the taxonomic table. In 1978 the ESA was amended to define a smaller taxon for vertebrates (animals with a backbone) as a distinct population segment (DPS). A DPS is a distinct population of vertebrates capable of interbreeding with each other that live in a specific geographical area. A DPS is usually described using geographical terms, such as northern or southern, or by a given latitude or longitude. In 1991 the NMFS developed a policy defining the DPS for Pacific salmon populations. Salmon are anadromous, and most salmon migrate in groups at particular times of the year. Each of these groups is called a stock. The NMFS developed a new term, the evolutionarily significant unit (ESU), to refer to a distinct stock of Pacific salmon.

In summary, the word *species* as used in the ESA can mean a species, a subspecies, a DPS (vertebrates only), or an ESU (Pacific salmon only).

LISTING UNDER THE ESA

According to the USFWS, in "Listing a Species as Threatened or Endangered" (June 2011, http://www.fws.gov/endangered/esa-library/pdf/listing.pdf), the ESA

TABLE 2.1

First list of endangered species, 1967

In accordance with section 1(c) of the Endangered Species Preservation Act of October 15, 1966 (80 Stat. 926; 16 U.S.C. 668aa(c) I [the Secretary of the Interior] find after consulting the states, interested organizations, and individual scientists, that the following listed native fish and wildlife are threatened with extinction.

Mammals

- Indiana bat—*Myotis sodalis*
- Delmarva Peninsula fox squirrel—*Sciurus niger cinereus*
- Timber wolf—*Canis lupus lycaon*
- Red wolf—*Canis niger*
- San Joaquin kit fox—*Vulpes macrotis mutica*
- Grizzly bear—*Ursus horribilis*
- Black-footed ferret—*Mustela nigripes*
- Florida panther —*Felis concolor coryi*
- Caribbean monk seal—*Monachus tropicalis*
- Guadalupe fur seal—*Arctocephalus philippi townsendi*
- Florida manatee or Florida sea cow—*Trichechus manatus latirostris*
- Key deer—*Odocoileus virginianus clavium*
- Sonoran pronghorn—*Antilocapra americana sonoriensis*

Birds

- Hawaiian dark-rumped petrel—*Pterodroma phaeopygia sandwichensis*
- Hawaiian goose (nene)—*Branta sandvicensis*
- Aleutian Canada goose—*Branta canadensis leucopareia*
- Tule white-fronted goose—*Anser albifrons gambelli*
- Laysan duck—*Anas laysanensis*
- Hawaiian duck (or koloa)—*Anas wyvilliana*
- Mexican duck—*Anas diazi*
- California condor—*Gymnogyps californianus*
- Florida Everglade kite (Florida Snail Kite)—*Rostrhamus sociabilis plumbeus*
- Hawaiian hawk (or ii)—*Buteo solitarius*
- Southern bald eagle—*Haliaeetus t. leucocephalus*
- Attwater's greater prairie chicken—*Tympanuchus cupido attwateri*
- Masked bobwhite—*Colinus virginianus ridgwayi*
- Whooping crane —*Grus americana*
- Yuma clapper rail—*Rallus longirostris yumanensis*
- Hawaiian common gallinule—*Gallinula chloropus sandvicensis*
- Eskimo curlew —*Numenius borealis*
- Puerto Rican parrot—*Amazona vittata*
- American ivory-billed woodpecker—*Campephilus p. principalis*
- Hawaiian crow (or alala)—*Corvus hawaiiensis*
- Small Kauai thrush (puaiohi)—*Phaeornia pulmeri*
- Nihoa millerbird—*Acrocephalus kingi*
- Kauai oo (or oo aa)—*Moho braccatus*
- Crested honeycreeper (or akohekohe)—*Palmeria dolei*
- Akiapolaau—*Hemignathus wilsoni*
- Kauai akialoa—*Hemignathus procerus*
- Kauai nukupuu —*Hemignathus lucidus hanapepe*
- Laysan finchbill (Laysan finch)— *Psittirostra c. cantans*
- Nihoa finchbill (Nihoa finch)— *Psittirostra cantans ultima*
- Ou—*Psittirostra psittacea*
- Palila—*Psittirostra bailleui*
- Maui parrotbill—*Pseudonestor xanthophyrys*
- Bachman's warbler—*Vermivora bachmanii*
- Kirtland's warbler—*Dendroica kirtlandii*
- Dusky seaside sparrow—*Ammospiza nigrescens*
- Cape Sable sparrow—*Ammospiza mirabilis*

Reptiles and Amphibians

- American alligator—*Alligator mississippiensis*
- Blunt-nosed leopard lizard—*Crotaphytus wislizenii silus*
- San Francisco garter snake—*Thamnophis sirtalis tetrataenia*
- Santa Cruz long-toed salamander—*Ambystoma macrodactylum croceum*
- Texas blind salamander—*Typhlomolge rathbuni*
- Black toad, Inyo County toad—*Bufo exsul*

Fishes

- Shortnose sturgeon—*Acipenser brevirostrum*
- Longjaw Cisco—*Coregonus alpenae*
- Paiute cutthroat trout—*Salmo clarki seleniris*
- Greenback cuttthroat trout—*Salmo clarki stomias*
- Montana Westslope cutthroat trout—*Salmo clarki*
- Gila trout—*Salmo gilae*
- Arizona (*Apache*) trout—*Salmo sp.*
- Desert dace—*Eremichthys acros*
- Humpback chub—*Gila cypha*
- Little Colorado spinedace—*Lepidomeda vittata*
- Moapa dace—*Moapa coriacea*
- Colorado River squawfish—*Ptychocheilus lucius*
- Cui-ui—*Chasmistes cujus*
- Devils Hole pupfish—*Cyprinodon diabolis*
- Commanche Springs pupfish—*Cyprinodon elegans*
- Owens River pupfish —*Cyprinodon radiosus*
- Pahrump killifish—*Empetrichythys latos*
- Big Bend gambusia—*Gambusia gaigei*
- Clear Creek gambusia—*Gambusia heterochir*
- Gila topminnow—*Poeciliopsis occidentalis*
- Maryland darter—*Etheostoma sellare*
- Blue pike—*Stizostedion vitreum glaucum*

SOURCE: Stewart L. Udall, "Native Fish and Wildlife Endangered Species," in *Federal Register*, vol. 32, no. 48, March 11, 1967

TABLE 2.2

Key sections of the Endangered Species Act

Section	Description
4	Addresses the listing and recovery of species and designation of critical habitat.
6	Focuses on cooperation with the states and authorizes FWS and NOAA fisheries to provide financial assistance to states that have entered into cooperative agreements supporting the conservation of endangered and threatened species.
7	Requires all federal agencies, in consultation with FWS or NOAA fisheries, to use their authorities to further the purpose of the ESA and to ensure that their actions are not likely to jeopardize the continued existence of listed species or result in destruction or adverse modification of critical habitat.
8	Outlines the procedures for international cooperation.
9	Defines prohibited actions, including the import and export, take, illegally taken possession of illegally taken species, transport, or sale of endangered or threatened species.
10	Lays out the guidelines under which a permit may be issued to authorize prohibited activities, such as take of endangered or threatened species.
10(a)(1)(A)	Allows for permits for the taking of threatened or endangered species for scientific purposes or for purposes of enhancement of propagation or survival.
10(a)(1)(B)	Allows for permits for incidental taking of threatened or endangered species.

FWS = Fish and Wildlife Services.
NOAA = National Oceanic and Atmospheric Administration.
ESA = Endangered Species Act.

SOURCE: Adapted from "Endangered Species Act of 1973, As Amended," in *Endangered Species Glossary*, U.S. Department of the Interior, U.S. Fish and Wildlife Service, April 2005, http://www.fws.gov/endangered/esa-library/pdf/glossary.pdf (accessed November 30, 2011)

stipulates that there are five criteria that must be evaluated before a decision is made to list a species:

- The present or threatened destruction, modification, or curtailment of its habitat or range
- Overutilization for commercial, recreational, scientific, or educational purposes
- Disease or predation
- The inadequacy of existing regulatory mechanisms
- Other natural or manmade factors affecting its survival

The primary status codes assigned to listed species are E for endangered and T for threatened. However, there are many other status codes for specific types of listings. (See Table 2.3.)

As of November 2011, there were 1,066 U.S. species (422 animals and 644 plants) and 553 foreign species (552 animals and 1 plant) listed as endangered, and 317 U.S. species (167 animals and 150 plants) and 54 foreign species (52 animals and 2 plants) listed as threatened under the ESA. (See Table 1.2 in Chapter 1.)

The Listing Process

The process by which a species becomes listed under the ESA is a legal process with specifically defined steps.

TABLE 2.3

Endangered Species Act status codes

E	Endangered
T	Threatened
EmE	Emergency listing, endangered
EmT	Emergency listing, threatened
EXPE, XE	Experimental population, essential
EXPN, XN	Experimental population, non-essential
SAE, E(S/A)	Similarity of appearance to an endangered taxon
SAT, T(S/A)	Similarity of appearance to a threatened taxon
PE	Proposed endangered
PT	Proposed threatened
PEXPE, PXE	Proposed experimental population, essential
PEXPN, PXN	Proposed experimental population, non-essential
PSAE, PE(S/A)	Proposed similarity of appearance to an endangered taxon
PSAT, PT(S/A)	Proposed similarity of appearance to a threatened taxon
C	Candidate taxon, ready for proposal
D3A	Delisted taxon, evidently extinct
D3B	Delisted taxon, invalid name in current scientific opinion
D3C	Delisted taxon, recovered
DA	Delisted taxon, amendment of the act
DM	Delisted taxon, recovered, being monitored first five years
DO	Delisted taxon, original commercial data erroneous
DP	Delisted taxon, discovered previously unknown additional populations and/or habitat
DR	Delisted taxon, taxonomic revision (improved understanding)
AD	Proposed delisting
AE	Proposed reclassification to endangered
AT	Proposed reclassification to threatened

SOURCE: "Endangered Species Act Status Codes," in *USFWS Threatened and Endangered Species System (TESS)*, U.S. Department of the Interior, U.S. Fish and Wildlife Service, undated, http://ecos.fws.gov/tess_public/html/db-status.html (accessed October 28, 2011)

Successful listing results in regulations that are legally enforceable within all U.S. jurisdictions. At various stages of the listing process, the USFWS or the NMFS publishes its actions in the *Federal Register* (http://www.gpo.gov/fdsys/browse/collection.action?collectionCode=FR), an official document that is compiled daily by the National Archives and Records Administration in Washington, D.C., and printed/published digitally by the U.S. Government Printing Office. The *Federal Register* details specific legal actions of the federal government, such as rules, proposed rules, notices from federal agencies, executive orders, and miscellaneous presidential documents.

There are three ways for the listing process to be initiated:

- Submittal of a petition to the USFWS or the NMFS
- Initiative of the USFWS or the NMFS
- Emergency designation by the USFWS or the NMFS

Figure 2.1 diagrams the most common listing process under the ESA, one that begins with a petition submittal. The listing process is described in detail by Joy Nicholopoulos in "Endangered Species Listing Program" (*Endangered Species Bulletin*, vol. 24, no. 6, November–December 1999) and by the USFWS in "Listing a Species as Threatened or Endangered: Section 4 of the Endangered Species Act" (February 2001, http://library.fws.gov/Pubs9/listing.pdf).

PETITION SUBMITTAL. The process for listing a new species as endangered or threatened begins with a formal petition from a person, organization, or government agency. This petition is submitted to the USFWS for terrestrial and freshwater species or to the NMFS for marine and anadromous species. All petitions must be backed by published scientific data supporting the need for listing. Within 90 days the USFWS or the NMFS is supposed to determine whether there is "substantial information" to suggest that a species might require listing under the ESA.

STATUS REVIEW. A status review is triggered when a petition is found to suggest that listing "may be warranted" or on the initiative of the USFWS or the NMFS. The purpose of a status review is to determine whether a listing is warranted and what that listing should be.

In *Endangered Species Petition Management Guidance* (July 1996, http://www.nmfs.noaa.gov/pr/pdfs/laws/petition_management.pdf), the USFWS and the NMFS note that status reviews are required under section 4(b)(1)(A) of the ESA. The agencies define a status review as "the act of reviewing all the available information on a species to determine if it should be provided protection under the ESA. A status review should also use the knowledge of experts; the greater the extent to

FIGURE 2.1

The petition process under the Endangered Species Act

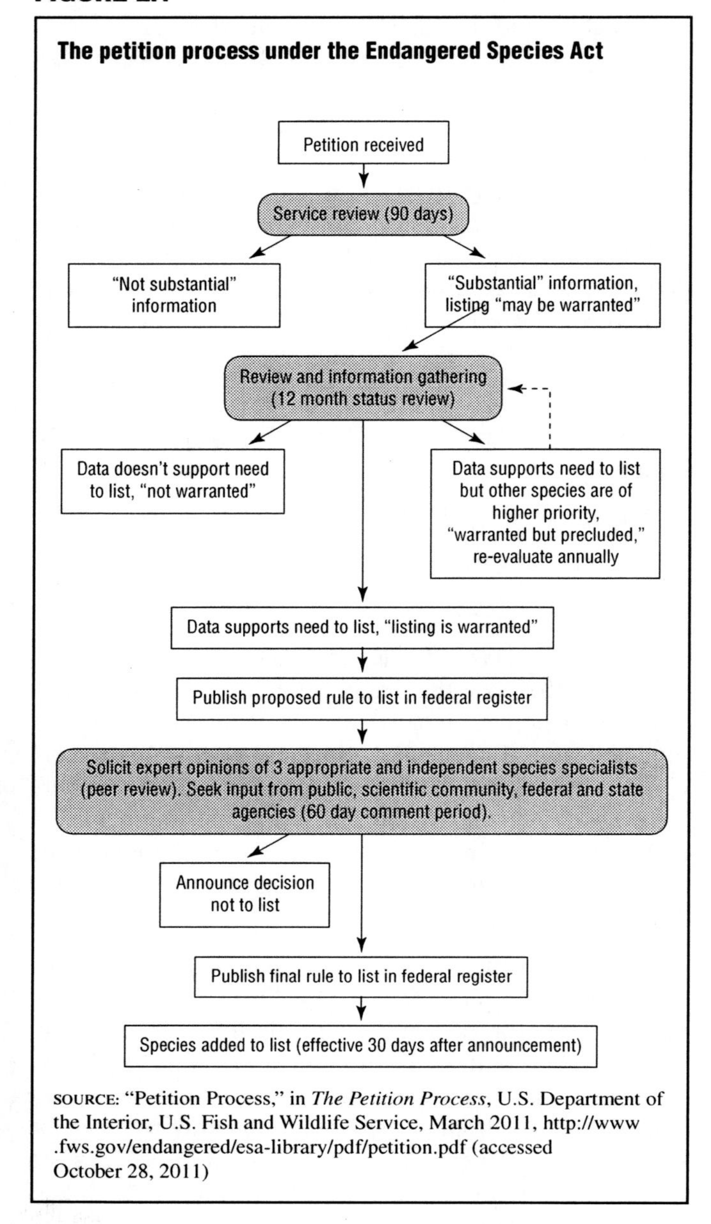

SOURCE: "Petition Process," in *The Petition Process*, U.S. Department of the Interior, U.S. Fish and Wildlife Service, March 2011, http://www.fws.gov/endangered/esa-library/pdf/petition.pdf (accessed October 28, 2011)

which Service biologists can build an external consensus using the expertise of various parties (e.g., Federal, State, Tribal, University, Heritage programs), the better." Comments and information are also requested from the general public through publication of a notice in the *Federal Register*.

In general, a status review covers the following elements:

- Biology, life history, and reproductive factors of the species
- Genetic information
- Habitat characteristics
- Abundance and population trends
- Threats to species survival and well-being
- Conservation measures
- Existing regulatory mechanisms protecting the species

A status review is supposed to be completed within 12 months. There are three possible determinations from a status review:

- Listing is not warranted
- Listing is warranted but precluded
- Listing is warranted

LISTING IS NOT WARRANTED. A finding that listing is not warranted must be accompanied by information explaining why the data presented do not support the petitioned action or why there are not enough data to make an appropriate determination.

LISTING IS WARRANTED BUT PRECLUDED—CANDIDATE SPECIES. In some cases it may be decided that a species should be proposed for listing, but development of the listing regulation is precluded by other listing activities with higher priorities. In other words, the USFWS or the NMFS acknowledges that a species deserves protection under the ESA, but the USFWS or the NMFS has other priorities that it believes must come first. The species may be designated as a "candidate species." This process is described by the USFWS in "Candidate Species: Section 4 of the Endangered Species Act" (March 2011, http://www.fws.gov/endangered/esa-library/pdf/candidate_species.pdf).

Candidate species are assigned a listing priority number (LPN) ranging from 1 to 12, with lower numbers (1 to 3) indicating greater priority compared with other candidates. (See Table 2.4.) Priority is determined based on three considerations:

- The magnitude of the threats facing the species
- The immediacy of the threats facing the species
- The taxonomic uniqueness of the species

The NMFS has a different definition for candidate species. It calls them "species of concern" and does not propose them for listing on the basis that available information is inadequate to justify doing so.

Candidate species are reevaluated annually to confirm that their listing status and LPN continue to be appropriate. The USFWS describes the reevaluation of each candidate species in a publication called a Candidate Notice of Review (CNOR). CNORs are published (typically annually) in the *Federal Register*. CNORs dating back to 1994 are available at http://www.fws.gov/endangered/what-we-do/cnor.html.

The candidate reevaluations continue until the species is proposed for listing or until its status improves sufficiently to remove it from consideration for listing.

TABLE 2.4

Listing priority numbers for candidate species

Threat			
Magnitude	**Immediacy**	**Taxonomy**	**Priority**
High	Imminent	Monotypic genus	1
		Species	2
		Subspecies/population	3
	Non-imminent	Monotypic genus	4
		Species	5
		Subspecies/population	6
Moderate to low	Imminent	Monotypic genus	7
		Species	8
		Subspecies/population	9
	Non-imminent	Monotypic genus	10
		Species	11
		Subspecies/population	12

SOURCE: John J. Fay and W. L. Thomas, "Table 1. Priorities for Listing or Reclassification from Threatened to Endangered," in "Fish and Wildlife Service: Endangered and Threatened Species Listing and Recovery Priority Guidelines," *Federal Register*, vol. 48, no. 184, September 21, 1983, http://www.fws.gov/endangered/esa-library/pdf/1983_LPN_Policy_FR_pub.pdf (accessed October 28, 2011)

The USFWS or the NMFS works with state wildlife agencies and other groups to help preserve and improve the status of candidate species, hoping that populations may recover enough that species will not require listing.

As of January 2012, the most recent CNOR was published in October 2011 (http://www.fws.gov/endangered/esa-library/pdf/CNOR2011.pdf) and included a total of 244 candidate species. (See Table 2.5.) Plant species dominated the list. As shown in Table 2.6, Hawaii had the most candidate species of any state (55), followed by Texas (26), Arizona (24), California (23), and Florida (22). It should be noted that some species are found in more than one state.

LISTING IS WARRANTED. A determination that listing is warranted means that a species is officially proposed for listing through the publication of this action in the *Federal Register*. At this point the USFWS or the NMFS asks at least three independent biological experts to verify that the petitioned species requires listing under either threatened or endangered status. After that, input from the public, from other federal and state agencies, and from the scientific community is welcomed. This period of public comment typically lasts 60 days, but may be extended in some cases. Within 45 days of proposal issuance interested parties can request that public hearings be held on the issues involved with listing. Such hearings are also held in cases where public interest is high in the listing outcome.

As shown in Table 2.7, there were 60 species proposed for listing under the ESA as of November 2011. Most were birds and flowering plants, many of which are found only in Hawaii.

TABLE 2.5

Candidate species as of October 2011

Priority	Scientific name	Common name
	Mammals	
2	Eumops floridanus	Bat, Florida bonneted
3	Emballonura semicaudata rotensis.	Bat, Pacific sheath-tailed (Mariana Islands subspecies).
3	Emballonura semicaudata semicaudata.	Bat, Pacific sheath-tailed (American Samoa DPS).
2	Sylvilagus transitionalis	Cottontail, New England
6	Martes pennanti	Fisher (west coast DPS)
3	Zapus hudsonius luteus	Mouse, New Mexico meadow jumping.
3	Thomomys mazama couchi.	Pocket gopher, Shelton
3	Thomomys mazama douglasii.	Pocket gopher, Brush Prairie.
3	Thomomys mazama glacialis.	Pocket gopher, Roy Prairie.
3	Thomomys mazama louiei.	Pocket gopher, Cathlamet.
3	Thomomys mazama melanops.	Pocket gopher, Olympic
3	Thomomys mazama pugetensis.	Pocket gopher, Olympia
3	Thomomys mazama tacomensis.	Pocket gopher, Tacoma
3	Thomomys mazama tumuli.	Pocket gopher, Tenino
3	Thomomys mazama yelmensis.	Pocket gopher, Yelm
3	Cynomys gunnisoni	Prairie dog, Gunnison's (populations in central and south-central Colorado, north-central New Mexico).
9	Spermophilus brunneus endemicus.	Squirrel, Southern Idaho ground.
5	Spermophilus washingtoni.	Squirrel, Washington ground.
9	Odobenus rosmarus divergens.	Walrus, Pacific
6	Gulo gulo luscus	Wolverine, North American (contiguous U.S. DPS).
	Birds	
3	Porzana tabuensis	Crake, spotless (American Samoa DPS).
3	Coccyzus americanus	Cuckoo, yellow-billed (western U.S. DPS).
9	Gallicolumba stairi	Ground-dove, friendly (American Samoa DPS).
3	Eremophila alpestris strigata.	Horned lark, streaked
3	Calidris canutus rufa	Knot, red
8	Gavia adamsii	Loon, yellow-billed
8	Brachyramphus brevirostris.	Murrelet, Kittlitz's
5	Synthliboramphus hypoleucus.	Murrelet, Xantus's
8	Anthus spragueii	Pipit, Sprauge's
2	Tympanuchus pallidicinctus.	Prairie-chicken, lesser
8	Centrocercus urophasianus.	Sage-grouse, greater
3	Centrocercus urophasianus.	Sage-grouse, greater (bi-state DPS).
6	Centrocercus urophasianus.	Sage-grouse, greater (Columbia Basin DPS).
2	Centrocercus minimus	Sage-grouse, Gunnison

TABLE 2.5

Candidate species as of October 2011 [CONTINUED]

Priority	Scientific name	Common name
3	Oceanodroma castro	Storm-petrel, band-rumped (Hawaii DPS).
11	Dendroica angelae	Warbler, elfin-woods
	Reptiles	
3	Thamnophis eques megalops.	Gartersnake, northern Mexican.
8	Sistrurus catenatus	Massasauga (=rattle-snake), eastern.
3	Pituophis melanoleucus lodingi.	Snake, black pine
5	Pituophis ruthveni	Snake, Louisiana pine
3	Chionactis occipitalis klauberi.	Snake, Tucson shovel-nosed.
6	Gopherus agassizii	Tortoise, desert (Sonoran DPS).
8	Gopherus polyphemus	Tortoise, gopher (eastern population).
3	Kinosternon sonoriense longifemorale.	Turtle, Sonoyta mud
	Amphibians	
9	Rana luteiventris	Frog, Columbia spotted (Great Basin DPS).
3	Rana muscosa	Frog, mountain yellow-legged (Sierra Nevada DPS).
2	Rana pretiosa	Frog, Oregon spotted
8	Lithobates onca	Frog, relict leopard
8	Notophthalmus perstriatus.	Newt, striped
2	Eurycea waterlooensis	Salamander, Austin blind
8	Gyrinophilus gulolineatus	Salamander, Berry Cave
8	Eurycea naufragia	Salamander, George-town.
2	Plethodon neomexicanus	Salamander, Jemez Mountains.
8	Eurycea tonkawae	Salamander, Jollyville Plateau.
2	Eurycea chisholmensis	Salamander, Salado
11	Anaxyrus canorus	Toad, Yosemite
3	Hyla wrightorum	Treefrog, Arizona (Huachuca/Canelo DPS).
8	Necturus alabamensis	Waterdog, black warrior (=Sipsey Fork).
	Fishes	
8	Gila nigra	Chub, headwater
7	Iotichthys phlegethontis	Chub, least
9	Gila robusta	Chub, roundtail (lower Colorado River Basin DPS).
11	Etheostoma cragini	Darter, Arkansas
2	Crystallaria cincotta	Darter, diamond
3	Etheostoma sagitta spilotum.	Darter, Kentucky arrow
8	Percina aurora	Darter, Pearl
3	Thymallus arcticus	Grayling, Arctic (upper Missouri River DPS).
5	Moxostoma sp.	Redhorse, sicklefin
2	Cottus sp.	Sculpin, grotto
5	Notropis oxyrhynchus	Shiner, sharpnose
5	Notropis buccula	Shiner, smalleye
3	Catostomus discobolus yarrowi.	Sucker, Zuni bluehead
9	Oncorhynchus clarki virginalis.	Trout, Rio Grande cutthroat.

TABLE 2.5

Candidate species as of October 2011 [CONTINUED]

Priority	Scientific name	Common name
	Clams	
8	Popenaias popei	Hornshell, Texas
2	Ptychobranchus subtentum	Kidneyshell, fluted
2	Lampsilis rafinesqueana	Mucket, Neosho
2	Lexingtonia dolabelloides	Pearlymussel, slabside
9	Quadrula cylindrica cylindrica.	Rabbitsfoot
	Snails	
8	Elimia melanoides	Mudalia, black
2	Planorbella magnifica	Ramshorn, magnificent
2	Ostodes strigatus	Sisi snail
2	Pseudotryonia adamantina.	Snail, Diamond Y Spring
2	Samoana fragilis	Snail, fragile tree
2	Partula radiolata	Snail, Guam tree
2	Partula gibba	Snail, Humped tree
2	Partulina semicarinata	Snail, Lanai tree
2	Partulina variabilis	Snail, Lanai tree
2	Partula langfordi	Snail, Langford's tree
2	Cochliopa texana	Snail, Phantom cave
2	Newcombia cumingi	Snail, Newcomb's tree
2	Eua zebrina	Snail, Tutuila tree
11	Pyrgulopsis notidicola	Springsnail, elongate mud meadows.
2	Tryonia circumstriata (=stocktonensis).	Springsnail, Gonzales
11	Pyrgulopsis thompsoni	Springsnail, Huachuca
8	Pyrgulopsis morrisoni	Springsnail, Page
2	Tryonia cheatumi	Springsnail (=Tryonia), Phantom.
5	Sonorella rosemontensis	Talussnail, Rosemont
	Insects	
2	Hylaeus anthracinus	Bee, Hawaiian yellow-faced.
2	Hylaeus assimulans	Bee, Hawaiian yellow-faced.
2	Hylaeus facilis	Bee, Hawaiian yellow-faced.
2	Hylaeus hilaris	Bee, Hawaiian yellow-faced.
2	Hylaeus kuakea	Bee, Hawaiian yellow-faced.
2	Hylaeus longiceps	Bee, Hawaiian yellow-faced.
2	Hylaeus mana	Bee, Hawaiian yellow-faced.
3	Plebejus shasta charlestonensis.	Blue, Mt. Charleston
3	Strymon acis bartrami	Butterfly, Bartram's hairstreak.
3	Anaea troglodyta floridalis.	Butterfly, Florida leafwing.
3	Hypolimnas octucula mariannensis.	Butterfly, Mariana eight-spot.
2	Vagrans egistina	Butterfly, Mariana wandering.
2	Atlantea tulita	Butterfly, Puerto Rican harlequin.
5	Glyphopsyche sequatchie.	Caddisfly, Sequatchie
5	Pseudanophthalmus insularis.	Cave beetle, Baker Station (=insular).
5	Pseudanophthalmus caecus.	Cave beetle, Clifton
11	Pseudanophthalmus colemanensis.	Cave beetle, Coleman
5	Pseudanophthalmus fowlerae.	Cave beetle, Fowler's
5	Pseudanophthalmus frigidus.	Cave beetle, icebox

TABLE 2.5
Candidate species as of October 2011 [CONTINUED]

Priority	Scientific name	Common name
5	Pseudanophthalmus tiresias.	Cave beetle, Indian Grave Point (=Soothsayer).
5	Pseudanophthalmus inquisitor.	Cave beetle, inquirer
5	Pseudanophthalmus troglodytes.	Cave beetle, Louisville
5	Pseudanophthalmus paulus.	Cave beetle, Noblett's
5	Pseudanophthalmus parvus.	Cave beetle, Tatum
3	Euphydryas editha taylori.	Checkerspot butterfly, Taylor's (=Whulge).
5	Hermelycaena [Lycaena] hermes.	Copper, Hermes
8	Megalagrion xanthomelas.	Damselfly, orangeblack Hawaiian.
5	Ambrysus funebris	Naucorid bug (=Furnace Creek), Nevares Spring.
2	Drosophila digressa	Fly, Hawaiian Picture-wing.
8	Heterelmis stephani	Riffle beetle, Stephan's
8	Hesperia dacotae	Skipper, Dakota
8	Polites mardon	Skipper, Mardon
2	Oarisma poweshiek	Skipperling, Poweshiek
5	Lednia tumana	Stonefly, melwater lednian.
2	Cicindela albissima	Tiger beetle, Coral Pink Sand Dunes.
5	Cicindela highlandensis	Tiger beetle, highlands
	Arachnids	
8	Cicurina wartoni	Meshweaver, Warton's cave.
	Crustaceans	
2	Gammarus hyalleloides	Amphipod, diminutive
8	Stygobromus kenki	Amphipod, Kenk's
5	Metabetaeus lohena	Shrimp, anchialine pool
5	Palaemonella burnsi	Shrimp, anchialine pool
5	Procaris hawaiana	Shrimp, anchialine pool
4	Vetericaris chaceorum	Shrimp, anchialine pool
	Flowering plants	
11	Abronia alpina	Sand-verbena, Ramshaw Meadows.
8	Agave eggersiana	No common name
8	Arabis georgiana	Rockcress, Georgia
11	Argythamnia blodgettii	Silverbush, Blodgett's
3	Artemisia borealis var. wormskioldii.	Wormwood, northern
5	Astragalus anserinus	Milkvetch, Goose Creek
3	Astragalus cusickii var. packardiae.	Milkvetch, Packard's
8	Astragalus microcymbus	Milkvetch, skiff
8	Astragalus schmolliae	Milkvetch, Schmoll
11	Astragalus tortipes	Milkvetch, Sleeping Ute
3	Bidens campylotheca pentamera.	Ko`oko`olau
3	Bidens campylotheca waihoiensis.	Ko`oko`olau
8	Bidens conjuncta	Ko`oko`olau
3	Bidens micrantha ctenophylla.	Ko`oko`olau
8	Boechera (Arabis) pusilla	Rockcress, Fremont County or small.
8	Brickellia mosieri	Brickell-bush, Florida
2	Calamagrostis expansa	Reedgrass, Maui
2	Calamagrostis hillebrandii.	Reedgrass, Hillebrand's
5	Calochortus persistens	Mariposa lily, Siskiyou
2	Canavalia pubescens	`Awikiwiki

TABLE 2.5
Candidate species as of October 2011 [CONTINUED]

Priority	Scientific name	Common name
8	Castilleja christii	Paintbrush, Christ's
9	Chamaecrista lineata var. keyensis.	Pea, Big Pine partridge
12	Chamaesyce deltoidea pinetorum.	Sandmat, pineland
9	Chamaesyce deltoidea serpyllum.	Spurge, wedge
6	Chorizanthe parryi var. fernandina.	Spineflower, San Fernando Valley.
2	Chromolaena frustrata	Thoroughwort, Cape Sable.
8	Cirsium wrightii	Thistle, Wright's
2	Consolea corallicola	Cactus, Florida semaphore.
5	Cordia rupicola	No common name
2	Cyanea asplenifolia	Haha
2	Cyanea kunthiana	Haha
2	Cyanea obtusa	Haha
2	Cyanea tritomantha	`Aku
2	Cyrtandra filipes	Ha`iwale
2	Cyrtandra oxybapha	Ha`iwale
3	Dalea carthagenensis var. floridana.	Prairie-clover, Florida
5	Dichanthelium hirstii	Panic grass, Hirst Brothers'.
5	Digitaria pauciflora	Crabgrass, Florida pine-land.
3	Echinomastus erectocentrus var. acunensis.	Cactus, Acuna
8	Erigeron lemmonii	Fleabane, Lemmon
2	Eriogonum codium	Buckwheat, Umtanum Desert.
6	Eriogonum corymbosum var. nilesii.	Buckwheat, Las Vegas
5	Eriogonum diatomaceum	Buckwheat, Churchill Narrows.
5	Eriogonum kelloggii	Buckwheat, Red Mountain.
8	Eriogonum soredium	Buckwheat, Frisco
2	Festuca hawaiiensis	No common name
11	Festuca ligulata	Fescue, Guadalupe
2	Gardenia remyi	Nanu
8	Geranium hanaense	Nohoanu
8	Geranium hillebrandii	Nohoanu
5	Gonocalyx concolor	No common name
2	Harrisia aboriginum	Pricklyapple, aboriginal (shellmound applecactus).
5	Hazardia orcuttii	Orcutt's hazardia
2	Hedyotis fluviatilis	Kampua`a
8	Helianthus verticillatus	Sunflower, whorled
2	Hibiscus dasycalyx	Rose-mallow, Neches River.
5	Ivesia webberi	Ivesia, Webber
3	Joinvillea ascendens ascendens.	`Ohe
5	Leavenworthia crassa	Gladecress, unnamed
3	Leavenworthia exigua var. laciniata.	Gladecress, Kentucky
2	Leavenworthia texana	Gladecress, Texas golden.
8	Lepidium ostleri	Peppergrass, Ostler's
5	Linum arenicola	Flax, sand
3	Linum carteri var. carteri	Flax, Carter's small-flowered.
3	Mimulus fremontii var. vandenbergensis.	Monkeyflower, Vandenberg.
2	Myrsine fosbergii	Kolea
2	Myrsine vaccinioides	Kolea
8	Narthecium americanum	Asphodel, bog
2	Nothocestrum latifolium	`Aiea
2	Ochrosia haleakalae	Holei
3	Pediocactus peeblesianus var. fickeiseniae	Cactus, Fickeisen plains
9	Penstemon scariosus var. albifluvis.	Beardtongue, White River.
2	Peperomia subpetiolata	`Ala`ala wai nui
5	Phacelia stellaris	Phacelia, Brand's
2	Phyllostegia bracteata	No common name
8	Phyllostegia floribunda	No common name

TABLE 2.5

Candidate species as of October 2011 [CONTINUED]

Priority	Scientific name	Common name
9	Physaria douglasii tuplashensis.	Bladderpod, White Bluffs
8	Physaria globosa	Bladderpod, Short's
2	Pinus albicaulis	Pine, whitebark
8	Platanthera integrilabia	Orchid, white fringeless
2	Platydesma remyi	No common name
2	Pleomele fernaldii	Hala pepe
11	Potentilla basaltica	Cinquefoil, Soldier Meadow.
3	Pseudognaphalium (=Gnaphalium) sandwicensium var. molokaiense.	`Ena`ena
2	Ranunculus hawaiensis	Makou
2	Ranunculus mauiensis	Makou
8	Rorippa subumbellata	Cress, Tahoe yellow
2	Schiedea pubescens	Ma`oli`oli
2	Schiedea salicaria	No common name
5	Sedum eastwoodiae	Stonecrop, Red Mountain.
2	Sicyos macrophyllus	`Anunu
12	Sideroxylon reclinatum austrofloridense.	Bully, Everglades
2	Solanum conocarpum	Bacora, marron
8	Solanum nelsonii	Popolo
8	Solidago plumosa	Goldenrod, Yadkin River
2	Sphaeralcea gierischii	Mallow, Gierisch
2	Stenogyne cranwelliae	No common name
8	Streptanthus bracteatus	Twistflower, bracted
8	Symphyotrichum georgianum.	Aster, Georgia
8	Trifolium friscanum	Clover, Frisco
	Ferns and allies	
8	Cyclosorus boydiae	No common name
2	Huperzia (=Phlegmariurus) stemmermanniae.	Wawae`iol e
3	Microlepia strigosa var. mauiensis (=Microlepia mauiensis).	Palapalai
3	Trichomanes punctatum floridanum.	Florida bristle fern

DPS = distinct population segments.

SOURCE: Adapted from "Table 1. Candidate Notice of Review (Animals and Plants)," in "Endangered and Threatened Wildlife and Plants; Review of Native Species That Are Candidates for Listing As Endangered or Threatened; Annual Notice of Findings on Resubmitted Petitions; Annual Description of Progresson Listing Actions," *Federal Register*, vol. 76, no. 207, October 26, 2011, http://www.fws.gov/endangered/esa-library/pdf/CNOR2011.pdf (accessed November 30, 2011)

FINAL DECISION ON LISTING. After a listing has been proposed, the USFWS or the NMFS must take one of three possible actions:

- Withdraw the proposal—the biological information is found not to support listing the species
- Extend the proposal period—there is substantial disagreement within the scientific community regarding the listing; only one six-month extension is allowed, and then a final decision must be made
- Publish a final listing rule in the *Federal Register*—the listing becomes effective 30 days after publication, unless otherwise indicated

TABLE 2.6

Number of candidate species, by state or territory, November 1, 2011

State or territory	Candidates
Alabama	12 candidates
Alaska	3 candidates
Arizona	24 candidates
Arkansas	3 candidates
California	23 candidates
Colorado	13 candidates
Connecticut	2 candidates
Delaware	3 candidates
District of Columbia	1 candidates
Florida	22 candidates
Georgia	9 candidates
Hawaii	55 candidates
Idaho	8 candidates
Illinois	4 candidates
Indiana	4 candidates
Iowa	3 candidates
Kansas	4 candidates
Kentucky	12 candidates
Louisiana	6 candidates
Maine	2 candidates
Maryland	1 candidates
Massachusetts	2 candidates
Michigan	2 candidates
Minnesota	5 candidates
Mississippi	4 candidates
Missouri	5 candidates
Montana	6 candidates
Nebraska	2 candidates
Nevada	16 candidates
New Hampshire	2 candidates
New Jersey	3 candidates
New Mexico	13 candidates
New York	3 candidates
North Carolina	8 candidates
North Dakota	5 candidates
Ohio	3 candidates
Oklahoma	5 candidates
Oregon	12 candidates
Pennsylvania	2 candidates
Rhode Island	2 candidates
South Carolina	5 candidates
South Dakota	4 candidates
Tennessee	16 candidates
Texas	26 candidates
Utah	14 candidates
Vermont	1 candidates
Virginia	5 candidates
Washington	22 candidates
West Virginia	2 candidates
Wisconsin	2 candidates
Wyoming	5 candidates
American Samoa	5 candidates
Guam	6 candidates
Northern Mariana Islands	6 candidates
Puerto Rico	4 candidates
Virgin Islands	2 candidates
Outlying Caribbean Islands	0 candidates
Outlying Pacific Islands	0 candidates

SOURCE: "Candidate Species Are in Each State Based on Published Population Data," in *Species Reports*, U.S. Department of the Interior, U.S. Fish and Wildlife Service, November 1, 2011, http://ecos.fws.gov/tess_public/pub/stateListing.jsp?status=candidate (accessed November 1, 2011)

Section 4(c)(2)(A) of the ESA requires a review of the condition and situation of a listed species at least every five years to decide whether it still requires government protection. However, USFWS records indicate that these so-called five-year reviews are seldom performed on time. In fact, they are frequently late by several years or even decades.

TABLE 2.7

Species proposed for listing under the Endangered Species Act, November 2011

Common name	Scientific name	Species group	Listing status*
A`e	Zanthoxylum oahuense	Flowering plants	PE
Alabama pearlshell	Margaritifera marrianae	Clams	PE
Alani	Melicope christophersenii	Flowering plants	PE
Alani	Melicope makahae	Flowering plants	PE
Alani	Melicope hiiakae	Flowering plants	PE
Ash-breasted tit-tyrant	Anairetes alpinus	Birds	PE
Blackline Hawaiian damselfly	Megalagrion nigrohamatum nigrolineatum	Insects	PE
Blue-billed curassow	Crax alberti	Birds	PE
Brown-banded antpitta	Grallaria milleri	Birds	PE
Cauca guan	Penelope perspicax	Birds	PE
Choctaw bean	Villosa choctawensis	Clams	PE
Chupadera springsnail	Pyrgulopsis chupaderae	Snails	PE
Cook's petrel	Pterodroma cookii	Birds	PT
coqui, Llanero	Eleutherodactylus juanariveroi	Amphibians	PE
Crimson Hawaiian damselfly	Megalagrion leptodemus	Insects	PE
Dunes sagebrush lizard	Sceloporus arenicolus	Reptiles	PE
Esmeraldas woodstar	Chaetocercus berlepschi	Birds	PE
Fuzzy pigtoe	Pleurobema strodeanum	Clams	PT
Gorgeted wood-quail	Odontophorus strophium	Birds	PE
Graham beardtongue	Penstemon grahamii	Flowering plants	PT
Haha	Cyanea lanceolata	Flowering plants	PE
Haha	Cyanea calycina	Flowering plants	PE
Ha`iwale	Cyrtandra kaulantha	Flowering plants	PE
Ha`iwale	Cyrtandra sessilis	Flowering plants	PE
Hala pepe	Pleomele forbesii	Flowering plants	PE
Hulumoa	Korthalsella degeneri	Flowering plants	PE
Junin grebe	Podiceps taczanowskii	Birds	PE
Junin rail	Laterallus tuerosi	Birds	PE
Kaulu	Pteralyxia macrocarpa	Flowering plants	PE
Ko`oko`olau	Bidens amplectens	Flowering plants	PE
Miami Blue butterfly	Cyclargus (5 Hemiargus) thomasi	Insects	PE
Narrow pigtoe	Fusconaia escambia	Clams	PT
No common name	Cyrtandra waiolani	Flowering plants	PE
No common name	Cyanea purpurellifolia	Flowering plants	PE
No common name	Tetraplasandra lydgatei	Flowering plants	PE
No common name	Cyrtandra gracilis	Flowering plants	PE
No common name	Platydesma cornuta cornuta	Flowering plants	PE
No common name	Platydesma cornuta decurrens	Flowering plants	PE
No common name	Doryopteris takeuchii	Ferns and allies	PE
Oahu wild coffee (5 kopiko)	Psychotria hexandra ssp. oahuensis var.	Flowering plants	PE
Oceanic Hawaiian damselfly	Megalagrion oceanicum	Insects	PE
Peruvian plantcutter	Phytotoma raimondii	Birds	PE
Philippine cockatoo	Cacatua haematuropygia	Birds	PE
Queen Charlotte goshawk	Accipiter gentilis laingi	Birds	PE
Rayed bean	Villosa fabalis	Clams	PE
Round ebonyshell	Fusconaia rotulata	Clams	PE
Royal cinclodes	Cinclodes aricomae	Birds	PE
San Bernardino springsnail	Pyrgulopsis bernardina	Snails	PE
San Francisco manzanita	Arctostaphylos franciscana	Flowering plants	PE
Sheepnose mussel	Plethobasus cyphyus	Clams	PE
Snuffbox mussel	Epioblasma triquetra	Clams	PE
Southern kidneyshell	Ptychobranchus jonesi	Clams	PE
Southern sandshell	Hamiota (5 Lampsilis) australis	Clams	PE
Spectaclecase (mussel)	Cumberlandia monodonta	Clams	PE
Tapered pigtoe	Fusconaia burkei	Clams	PT
Three Forks springsnail	Pyrgulopsis trivialis	Snails	PE
White-browed tit-spinetail	Leptasthenura xenothorax	Birds	PE
White cockatoo	Cacatua alba	Birds	PT
Yellow-billed parrot	Amazona collaria	Birds	PT
Yellow-crested cockatoo	Cacatua sulphurea	Birds	PE

*PE = proposed endangered; PT = proposed threatened.

SOURCE: "Generate Species List," in *Species Reports*, U.S. Department of the Interior, U.S. Fish & Wildlife Service, November 2011, http://ecos.fws.gov/tess_public/pub/adHocSpeciesForm.jsp (accessed November 30, 2011)

EMERGENCY LISTING. The ESA authorizes the USFWS or the NMFS to issue temporary emergency listings for species when evidence indicates an immediate and significant risk to the well-being of a species (e.g., following a natural disaster). Two designations are possible: endangered emergency listing and threatened emergency listing. The listing must be published in the *Federal Register* and is effective for only 240 days. During this time the normal status review procedure continues.

Petitioners also have the right under the ESA to ask the USFWS or the NMFS for an emergency listing for a species.

The USFWS (http://ecos.fws.gov/tess_public/pub/adHocSpeciesCountForm.jsp) indicates that as of November 2011 there were no emergency listings in effect for endangered or threatened species listed under the ESA.

PROTECTIONS AND ACTIONS TRIGGERED BY LISTING

In "Listing and Critical Habitat: Overview" (December 28, 2011, http://www.fws.gov/endangered/what-we-do/listing-overview.html), the USFWS describes the protections or actions that are triggered when a species is listed under the ESA.

Taking, Possessing, or Trading Listed Species

The ESA makes illegal the taking of listed species. For animal species, taking is defined as killing, harming, harassing, pursuing, or removing the species from the wild. For plants, taking means collecting or "maliciously" damaging endangered plants on federal lands. The ESA also outlaws removing or damaging listed plants on state and private lands "in knowing violation of State law or in the course of violating a State criminal trespass law." The USFWS notes that some state laws specifically prohibit the taking of federally listed plants and animals.

As the USFWS explains in "Permits for Native Species under the Endangered Species Act" (October 2011, http://www.fws.gov/endangered/esa-library/pdf/permits.pdf), it is also illegal under the ESA to possess, ship, deliver, carry, transport, sell, or receive any listed species that was taken in violation of the law. Likewise, the import, export, and interstate or foreign sales of listed species are prohibited except for permitted conservation purposes.

Civil and criminal penalties can be levied for violations of these provisions; however, exemptions are allowed under certain sections of the ESA.

SECTION 4(D) EXEMPTIONS. Section 4(d) of the ESA (http://www.fws.gov/endangered/laws-policies/section-4.html) allows the USFWS to issue regulations governing the conservation of threatened species. As a result, "special" rules have been issued that grant exemptions from the taking, possession, and import/export/sales provisions for some threatened species. (See Table 2.8.) For example, grizzly bears can be harmed "in self-defense or in defense of others." However, such taking has to be reported to authorities within a specified time period.

INCIDENTAL TAKE PERMITS AND HABITAT CONSERVATION PLANS. When the original ESA was passed, it included exceptions that allowed the taking of listed species only for scientific research or other conservation activities authorized by the act. In 1982 Congress added a provision in section 10 of the ESA that allows "incidental take" of listed species of wildlife by nonfederal entities. Incidental take is defined as take that is incidental to, but not the purpose of, an otherwise lawful activity. Incidental taking cannot appreciably reduce the likelihood of the survival and recovery of listed species in the wild. The incidental take provision was added to allow private landowners some freedom to develop their land even if it provides habitat to listed species.

To obtain an incidental take permit, an applicant has to prepare a Habitat Conservation Plan (HCP). The HCP process is described by the USFWS in "Habitat Conservation Plans under the Endangered Species Act" (April 2011, http://www.fws.gov/endangered/esa-library/pdf/hcp.pdf). An HCP describes the impacts that are likely to result from the taking of the species and the measures the applicant will take to minimize and mitigate the impacts. HCPs are generally partnerships drawn up by people at the local level who are working with officials from the USFWS or the NMFS. The plans frequently represent compromises between developers and environmentalists.

Included in the agreement is a "no surprise" provision that assures landowners that the overall cost of species protection measures will be limited to what has been agreed to under the HCP. In return, landowners make a long-term commitment to conservation as negotiated in the HCP. Many HCPs include the preservation of significant areas of habitat for endangered species.

ENHANCEMENT OF SURVIVAL PERMITS. Another type of permit issued under the ESA is the enhancement of survival permit. This permit applies to species that are candidate species or are likely to become candidate species. It authorizes future incidental take (should the species become listed) by nonfederal landowners in exchange for proactive management of the species on their property. Permit applicants must agree to participate in a Candidate Conservation Agreement with Assurances (CCAA) or a Safe Harbor Agreement (SHA). These agreements provide assurances to landowners that no additional future regulatory restrictions will be imposed.

SUBSISTENCE TAKING. Besides the previously mentioned taking exemptions offered by the ESA, the native peoples of Alaska who rely on certain endangered or threatened animals for food or other products needed for subsistence are exempted from the taking rule under the Marine Mammal Protection Act. This act is described in detail in Chapter 3.

Critical Habitat

Under the ESA, the USFWS or the NMFS must decide whether critical habitat should be designated for a listed species. Critical habitat is specific geographical

TABLE 2.8

Threatened species covered by "Special Rules"

Common name	Scientific name	Species group	Special rules. See title 50 Code of Federal Regulations part(s):
African elephant	Loxodonta africana	Mammals	17.40(e)
Apache trout	Oncorhynchus apache	Fishes	17.44(a)
Argali	Ovis ammon	Mammals	17.40(j)
Bayou darter	Etheostoma rubrum	Fishes	17.44(b)
Beautiful shiner	Cyprinella formosa	Fishes	17.44(h)
Big Spring spinedace	Lepidomeda mollispinis pratensis	Fishes	17.44(i)
Black howler monkey	Alouatta pigra	Mammals	17.40(c)
Bluetail mole skink	Eumeces egregius lividus	Reptiles	17.42(d)
Bog (=Muhlenberg) turtle	Clemmys muhlenbergii	Reptiles	17.42(f)
Bull Trout	Salvelinus confluentus	Fishes	17.44(w), 17.44(x)
California tiger salamander	Ambystoma californiense	Amphibians	17.43
Canada lynx	Lynx canadensis	Mammals	17.40(k), 23.54
Chihuahua chub	Gila nigrescens	Fishes	17.44(g)
Chimpanzee	Pan troglodytes	Mammals	17.40(c)
Chinook salmon	Oncorhynchus (=Salmo) tshawytscha	Fishes	223.203
Chiricahua leopard frog	Rana chiricahuensis	Amphibians	17.43
Chum salmon	Oncorhynchus (=Salmo) keta	Fishes	223.203
Coastal California gnatcatcher	Polioptila californica californica	Birds	17.41(b)
Desert dace	Eremichthys acros	Fishes	17.44(m)
Desert tortoise	Gopherus agassizii	Reptiles	17.42(e)
Formosan rock macaque	Macaca cyclopis	Mammals	17.40(c)
Foskett speckled dace	Rhinichthys osculus ssp.	Fishes	17.44(j)
Gelada baboon	Theropithecus gelada	Mammals	17.40(c)
Gray wolf	Canis lupus	Mammals	17.40(d), 17.84(i), 17.84(k), 17.84(n)
Greenback cutthroat trout	Oncorhynchus clarki stomias	Fishes	17.44(f)
Green sea turtle	Chelonia mydas	Reptiles	17.42(b), 223.205, 223.206, 223.207, 224.104
Grizzly bear	Ursus arctos horribilis	Mammals	17.40(b), 17.84(l)
Guadalupe fur seal	Arctocephalus townsendi	Mammals	223.201
Gulf sturgeon	Acipenser oxyrinchus desotoi	Fishes	17.44(v)
Hutton tui chub	Gila bicolor ssp.	Fishes	17.44(j)
Japanese macaque	Macaca fuscata	Mammals	17.40(c)
Lahontan cutthroat trout	Oncorhynchus clarki henshawi	Fishes	17.44(a)
Leopard	Panthera pardus	Mammals	17.40(f)
Leopard darter	Percina pantherina	Fishes	17.44(d)
Lesser slow loris	Nycticebus pygmaeus	Mammals	17.40(c)
Little Colorado spinedace	Lepidomeda vittata	Fishes	17.44(t)
Little Kern golden trout	Oncorhynchus aguabonita whitei	Fishes	17.44(e)
Loach minnow	Tiaroga cobitis	Fishes	17.44(q)
Loggerhead sea turtle	Caretta caretta	Reptiles	17.42(b), 223.205, 223.206, 223.207
Long-tailed langur	Presbytis potenziani	Mammals	17.40(c)
Louisiana black bear	Ursus americanus luteolus	Mammals	17.40(i)
Madison Cave isopod	Antrolana lira	Crustaceans	17.46(a)
Niangua darter	Etheostoma nianguae	Fishes	17.44(k)
Nile crocodile	Crocodylus niloticus	Reptiles	17.42(c)
Olive ridley sea turtle	Lepidochelys olivacea	Reptiles	17.42(b), 223.205, 223.206, 223.207, 224.104
Paiute cutthroat trout	Oncorhynchus clarki seleniris	Fishes	17.44(a)
Pecos bluntnose shiner	Notropis simus pecosensis	Fishes	17.44(r)
Philippine tarsier	Tarsius syrichta	Mammals	17.40(c)
Preble's meadow jumping mouse	Zapus hudsonius preblei	Mammals	17.40(l)
Purple-faced langur	Presbytis senex	Mammals	17.40(c)
Pygmy sculpin	Cottus paulus (=pygmaeus)	Fishes	17.44(u)
Railroad Valley springfish	Crenichthys nevadae	Fishes	17.44(n)
Saltwater crocodile	Crocodylus porosus	Reptiles	17.42(c)
Sand skink	Neoseps reynoldsi	Reptiles	17.42(d)
San Marcos salamander	Eurycea nana	Amphibians	17.43(a)
Slackwater darter	Etheostoma boschungi	Fishes	17.44(c)
Slender chub	Erimystax cahni	Fishes	17.44(c)
Sockeye salmon	Oncorhynchus (=Salmo) nerka	Fishes	223.203
Sonora chub	Gila ditaenia	Fishes	17.44(o)
Southern sea otter	Enhydra lutris nereis	Mammals	17.84(d)
Spikedace	Meda fulgida	Fishes	17.44(p)
Spotfin chub	Erimonax monachus	Fishes	17.44(c), 17.84(m), 17.84(o)
Steelhead	Oncorhynchus (=Salmo) mykiss	Fishes	223.203

TABLE 2.8

Threatened species covered by "Special Rules" [CONTINUED]

Common name	Scientific name	Species group	Special rules. See title 50 Code of Federal Regulations part(s):
Steller sea-lion	Eumetopias jubatus	Mammals	223.102, 223.202
Stump-tailed macaque	Macaca arctoides	Mammals	17.40(c)
Toque macaque	Macaca sinica	Mammals	17.40(c)
Utah prairie dog	Cynomys parvidens	Mammals	17.40(g)
Waccamaw silverside	Menidia extensa	Fishes	17.44(s)
Warner sucker	Catostomus warnerensis	Fishes	17.44(l)
White-footed tamarin	Saguinus leucopus	Mammals	17.40(c)
Yacare caiman	Caiman yacare	Reptiles	17.42(g)
Yaqui catfish	Ictalurus pricei	Fishes	17.44(h)
Yellowfin madtom	Noturus flavipinnis	Fishes	17.44(c), 17.84(e)

SOURCE: Adapted from "Generate Species List," in *Species Reports*, U.S. Department of the Interior, U.S. Fish & Wildlife Service, November 2011, http://ecos.fws.gov/tess_public/pub/adHocSpeciesForm.jsp (accessed November 30, 2011)

areas of land, water, and/or air space that contain features essential for the conservation of a listed species and that may require special management and protection. For example, these could be areas that are used for breeding, resting, and feeding. If the agency decides that critical habitat should be designated, a proposal notice is published in the *Federal Register* for public comment. If it is decided that critical habitat is needed, then the final boundaries are published in the *Federal Register*.

The role of critical habitat is often misunderstood by the public. Critical habitat designation does not set up a refuge or sanctuary for a species in which no development can take place. It can provide additional protection for a specific geographical area that might not occur without the designation. For example, if the USFWS determines that an area not currently occupied by a species is needed for species recovery and designates that area as critical habitat, any federal actions involving that area have to avoid adverse modifications. Critical habitat designation has no regulatory impact on private landowners unless they wish to take actions on their land that involve federal funding or permits.

The original ESA did not provide a time limit for the setting of critical habitat. In 1978 the law was amended to require that critical habitat be designated at the same time a species is listed. However, the designation is only required "when prudent." For example, the USFWS or the NMFS can refuse to designate critical habitat for a species if doing so would publicize the specific locations of organisms known to be targets for illegal hunting or collection. Historically, both agencies have broadly used the "when prudent" clause to justify not setting critical habitat for many listed species. This has been a contentious issue between the government and conservation groups.

As of November 2011, critical habitat had been designated for 610 species. (See Table 2.9.) This represented 44% of the 1,383 U.S. species listed under the ESA at that time. (See Table 1.2 in Chapter 1.) Plants make up more than half of the listed species for which critical habitat has been designated.

TABLE 2.9

Number of U.S. endangered and threatened species with critical habitat specified, November 1, 2011

Mammals	31
Birds	27
Reptiles	14
Amphibians	10
Fish	67
Clams	27
Snails	8
Insects	32
Arachnids	6
Crustaceans	11
Corals	2
Flowering plants	360
Ferns and allies	15

SOURCE: Adapted from "Listed Species with Critical Habitat," in *Species Reports*, U.S. Department of the Interior, U.S. Fish and Wildlife Service, November 1, 2011, http://ecos.fws.gov/tess_public/pub/criticalHabitat.jsp?nmfs=1 (accessed November 1, 2011)

Experimental Populations

For some species, primarily mammals, birds, fish, and aquatic invertebrates, recovery efforts include the introduction of individuals into new areas. Typically, this is accomplished by moving a small group of imperiled animals from an established area to one or more other locations within the species' historical range of distribution.

Experimental populations of a species are not subject to the same rigorous protections under the ESA as other members of the species. Experimental populations can be considered threatened, even if the rest of the species is listed as endangered. In addition, the USFWS can designate an experimental population as essential or nonessential. A nonessential designation indicates that the survival of this population is not believed essential to the survival of the species as a whole. A nonessential experimental population is treated under the law as if it is proposed for listing, not already listed. This results in less protection under the ESA.

As of November 2011, there were 63 experimental populations listed under the ESA, all nonessential. (See Table 2.10.) Note that some species have multiple experimental populations.

Actions by Federal Agencies

Section 7 of the ESA includes restrictions on federal agencies (and their nonfederal agency permit applicants) regarding endangered and threatened species. These restrictions are described by the USFWS in "Consultations with Federal Agencies: Section 7 of the Endangered Species Act" (April 2011, http://www.fws.gov/endangered/esa-library/pdf/consultations.pdf). Specifically, the federal agencies must "aid in the conservation of listed species" and "ensure that their activities are not likely to jeopardize the continued existence of listed species or adversely modify designated critical habitats." This applies to any activities that are funded, authorized, or permitted by these federal agencies or carried out on lands or waters managed by them.

According to the USFWS, there are two processes through which federal agencies (known as action agencies) comply with section 7 of the ESA as they plan projects or activities that might impact listed species: informal consultation or formal consultation. In either case the first step is for the action agency to coordinate with the USFWS or the NMFS early in the process to determine which, if any, listed species are within the project area and to determine whether the project might affect the listed species or its critical habitat. If the action agency and the USFWS or the NMFS (whichever has jurisdiction) agree that the project will not jeopardize the listed species, then the project may proceed. The same is true if the project can be easily modified to prevent adverse effects. In either case an informal consultation is said to have taken place.

If the action agency makes a preliminary determination that the project is likely to adversely affect listed species, then it may initiate a formal consultation. The action agency provides detailed information about the project and the listed species to the USFWS or the NMFS. The governing agency then issues a report called a "biological opinion" that determines whether or not the proposed action is likely to have an adverse effect, and if so, recommends "reasonable and prudent alternatives that could allow the project to move forward."

The USFWS notes that it participated in more than 30,000 section 7 consultations during fiscal year 2010 (i.e., October 1, 2009, through September 30, 2010). The vast majority of these consultations were of the informal type in which action agencies modified their projects to avoid adverse impacts to listed species.

Recovery Plans

The ESA requires that a recovery plan be developed and implemented for every listed species unless "such a plan will not promote the conservation of the species." The USFWS and the NMFS are directed to give priority to those species that are most likely to benefit from having a plan in place. The recovery potential of species is ranked from 1 to 18 by the USFWS. (See Table 2.11.) Low rankings indicate a greater likelihood that the species can be recovered. Priority is based on the degree of threat, the potential for recovery, and taxonomy (genetic distinctiveness). In addition, rankings can be appended with the letter C when species recovery is in conflict with economic activities. Species with a C designation have higher priority than other species within the same numerical ranking. The NMFS uses a different rating system that ranges from 1 (highest recovery potential) to 12 (lowest recovery potential). (See Table 2.12.)

Each recovery plan must include the following three elements:

- Site-specific management actions to achieve the plan's goals
- Objective and measurable criteria for determining when a species is recovered
- Estimates of the amount of time and money that will be required to achieve recovery

In "Endangered Species Recovery Program" (June 2011, http://www.fws.gov/endangered/esa-library/pdf/recovery.pdf), the USFWS describes a recovery plan as "a road map with detailed site-specific management actions for private, Federal, and State cooperation in conserving listed species and their ecosystems." Recovery plans include precisely defined milestones for recovery achievement. For example, recovery may be considered accomplished when a certain number of individuals is reached and specifically named threats are eliminated. However, the USFWS notes that a recovery plan is not a regulatory document.

Notices regarding proposed new or revised recovery plans must be placed in the *Federal Register* so that public comment can be obtained and considered before a plan is finalized.

DELISTING UNDER THE ESA

Delisting occurs when a species is removed from the candidate list, the proposed list, or the final list of endangered and threatened species. Delisting takes place for a variety of reasons, as indicated by the D codes in

TABLE 2.10

Experimental populations, November 2011

Inverted common name	Scientific name	Species group	Where listed
Bean, Cumberland (pearlymussel)	Villosa trabalis	Clams	U.S.A. (AL; The free-flowing reach of the Tennessee R. from the base of Wilson Dam downstream to the backwaters of Pickwick Reservoir [about 12 RM (19 km)] and the lower 5 RM [8 km] of all tributaries to this reach in Colbert and Lauderdale Cos.)
Bean, Cumberland (pearlymussel)	Villosa trabalis	Clams	U.S.A. (TN—specified portions of the French Broad and Holston Rivers)
Bear, grizzly	Ursus arctos horribilis	Mammals	U.S.A. experimental non-essential (portions of ID and MT)
Blossom, tubercled (pearlymussel)	Epioblasma torulosa torulosa	Clams	U.S.A. (AL; The free-flowing reach of the Tennessee R. from the base of Wilson Dam downstream to the backwaters of Pickwick Reservoir [about 12 RM (19 km)] and the lower 5 RM [8 km] of all tributaries to this reach in Colbert and Lauderdale Cos.)
Blossom, turgid (pearlymussel)	Epioblasma turgidula	Clams	U.S.A. (AL; The free-flowing reach of the Tennessee R. from the base of Wilson Dam downstream to the backwaters of Pickwick Reservoir [about 12 RM (19 km)] and the lower 5 RM [8 km] of all tributaries to this reach in Colbert and Lauderdale Cos.)
Blossom, yellow (pearlymussel)	Epioblasma florentina florentina	Clams	U.S.A. (AL; The free-flowing reach of the Tennessee R. from the base of Wilson Dam downstream to the backwaters of Pickwick Reservoir [about 12 RM (19 km)] and the lower 5 RM [8 km] of all tributaries to this reach in Colbert and Lauderdale Cos.)
Chub, slender	Erimystax cahni	Fishes	U.S.A. (TN—specified portions of the French Broad and Holston Rivers)
Chub, spotfin	Erimonax monachus	Fishes	U.S.A. (TN—specified portions of the French Broad and Holston Rivers)
Chub, spotfin	Erimonax monachus	Fishes	Shoal Creek
Chub, spotfin	Erimonax monachus	Fishes	Tellico River, between the backwaters of the Tellico Reservoir and the Tellico Ranger Station, in Monroe County, Tennessee
Clubshell	Pleurobema clava	Clams	U.S.A. (AL; The free-flowing reach of the Tennessee R. from the base of Wilson Dam downstream to the backwaters of Pickwick Reservoir [about 12 RM (19 km)] and the lower 5 RM [8 km] of all tributaries to this reach in Colbert and Lauderdale Cos.)
Combshell, Cumberlandian	Epioblasma brevidens	Clams	U.S.A. (AL; The free-flowing reach of the Tennessee R. from the base of Wilson Dam downstream to the backwaters of Pickwick Reservoir [about 12 RM (19 km)] and the lower 5 RM [8 km] of all tributaries to this reach in Colbert and Lauderdale Cos.)
Combshell, Cumberlandian	Epioblasma brevidens	Clams	U.S.A. (TN—specified portions of the French Broad and Holston Rivers)
Condor, California	Gymnogyps californianus	Birds	U.S.A. (specific portions of Arizona, Nevada, and Utah)
Crane, whooping	Grus americana	Birds	U.S.A. (AL, AR, GA, IL, IN, IA, KY, LA, MI, MN, MS, MO, NC, OH, SC, TN, VA, WI, WV)
Crane, whooping	Grus americana	Birds	U.S.A (Southwestern Louisiana)
Crane, whooping	Grus americana	Birds	U.S.A. (CO, ID, FL, NM, UT, and the western half of Wyoming)
Darter, boulder	Etheostoma wapiti	Fishes	Shoal Creek
Darter, duskytail	Etheostoma percnurum	Fishes	U.S.A. (TN—specified portions of the French Broad and Holston Rivers)
Darter, duskytail	Etheostoma percnurum	Fishes	Tellico River, between the backwaters of the Tellico Reservoir and the Tellico Ranger Station, in Monroe County, Tennessee
Falcon, northern aplomado	Falco femoralis septentrionalis	Birds	Southwestern population
Fanshell	Cyprogenia stegaria	Clams	U.S.A. (TN—specified portions of the French Broad and Holston Rivers)
Ferret, black-footed	Mustela nigripes	Mammals	U.S.A. (specific portions of AZ, CO, MT, SD, UT, and WY)
Lampmussel, Alabama	Lampsilis virescens	Clams	U.S.A. (AL; The free-flowing reach of the Tennessee R. from the base of Wilson Dam downstream to the backwaters of Pickwick Reservoir [about 12 RM (19 km)] and the lower 5 RM [8 km] of all tributaries to this reach in Colbert and Lauderdale Cos.)
Madtom, pygmy	Noturus stanauli	Fishes	U.S.A. (TN—specified portions of the French Broad and Holston Rivers)
Madtom, smoky	Noturus baileyi	Fishes	Tellico River, between the backwaters of the Tellico Reservoir and the Tellico Ranger Station, in Monroe County, Tennessee
Madtom, yellowfin	Noturus flavipinnis	Fishes	Tellico River between the backwaters of the Tellico Reservoir and the Tellico Ranger Station, in Monroe County, Tennessee
Madtom, yellowfin	Noturus flavipinnis	Fishes	N. Fork Holston R., VA, TN; S. Fork Holston R., upstream to Ft. Patrick Henry Dam, TN; Holston R., downstream to John Sevier Detention Lake Dam, TN; and all tributaries thereto
Madtom, yellowfin	Noturus flavipinnis	Fishes	U.S.A. (TN—specified portions of the French Broad and Holston Rivers)
Mapleleaf, winged	Quadrula fragosa	Clams	U.S.A. (AL; The free-flowing reach of the Tennessee R. from the base of Wilson Dam downstream to the backwaters of Reservoir [about 12 RM (19 km)] and the lower 5 RM [8 km] of all tributaries to this reach in Colbert and Lauderdale.)
Minnow, Rio Grande silvery	Hybognathus amarus	Fishes	Rio Grande, from Little Box Canyon (approximately 10.4 river miles downstream of Fort Quitman, TX) to Amistad Dam; and on the Pecos River, from its confluence with Independence Creek to its confluence with the Rio Grande.

Table 2.3. In general, delisting occurs when the USFWS or the NMFS finds that a species has recovered or become extinct, or on various procedural grounds, including discovery of additional habitats or populations. The USFWS notes in "Delisting a Species: Section 4 of the Endangered Species Act" (April 2011, http://www.fws.gov/endangered/esa-library/pdf/delisting.pdf) that the delisting process for a species believed recovered is similar to the

TABLE 2.10
Experimental populations, November 2011 [CONTINUED]

Inverted common name	Scientific name	Species group	Where listed
Monkeyface, Appalachian (pearlymussel)	Quadrula sparsa	Clams	USA (TN—specified portions of the French Broad and Holston Rivers)
Monkeyface, Cumberland (pearlymussel)	Quadrula intermedia	Clams	U.S.A. (AL; The free-flowing reach of the Tennessee R. from the base of Wilson Dam downstream to the backwaters of Pickwick Reservoir [about 12 RM (19 km)] and the lower 5 RM [8 km] of all tributaries to this reach in Colbert and Lauderdale Cos.)
Monkeyface, Cumberland (pearlymussel)	Quadrula intermedia	Clams	U.S.A. (TN—specified portions of the French Broad and Holston Rivers)
Mussel, oyster	Epioblasma capsaeformis	Clams	U.S.A. (TN—specified portions of the French Broad and Holston Rivers)
Mussel, oyster	Epioblasma capsaeformis	Clams	U.S.A. (AL; The free-flowing reach of the Tennessee R. from the base of Wilson Dam downstream to the backwaters of Pickwick Reservoir [about 12 RM (19 km)] and the lower 5 RM [8 km] of all tributaries to this reach in Colbert and Lauderdale Cos.)
Otter, southern sea	Enhydra lutris nereis	Mammals	All areas subject to U.S. jurisdiction south of Pt. Conception, CA (34026.9' N. Lat.)
Pearlymussel, birdwing	Conradilla caelata	Clams	U.S.A. (AL; The free-flowing reach of the Tennessee R. from the base of Wilson Dam downstream to the backwaters of Pickwick Reservoir [about 12 RM (19 km)] and the lower 5 RM [8 km] of all tributaries to this reach in Colbert and Lauderdale Cos.)
Pearlymussel, birdwing	Conradilla caelata	Clams	U.S.A. (TN—specified portions of the French Broad and Holston Rivers)
Pearlymussel, cracking	Hemistena lata	Clams	U.S.A. (AL; The free-flowing reach of the Tennessee R. from the base of Wilson Dam downstream to the backwaters of Pickwick Reservoir [about 12 RM (19 km)] and the lower 5 RM [8 km] of all tributaries to this reach in Colbert and Lauderdale Cos.)
Pearlymussel, cracking	Hemistena lata	Clams	U.S.A. (TN—specified portions of the French Broad and Holston Rivers)
Pearlymussel, dromedary	Dromus dromas	Clams	U.S.A. (TN—specified portions of the French Broad and Holston Rivers)
Pearlymussel, dromedary	Dromus dromas	Clams	U.S.A. (AL; The free-flowing reach of the Tennessee R. from the base of Wilson Dam downstream to the backwaters of Pickwick Reservoir [about 12 RM (19 km)] and the lower 5 RM [8 km] of all tributaries to this reach in Colbert and Lauderdale Cos.)
Pigtoe, finerayed	Fusconaia cuneolus	Clams	U.S.A. (TN—specified portions of the French Broad and Holston Rivers)
Pigtoe, finerayed	Fusconaia cuneolus	Clams	U.S.A. (AL; The free-flowing reach of the Tennessee R. from the base of Wilson Dam downstream to the backwaters of Pickwick Reservoir [about 12 RM (19 km)] and the lower 5 RM [8 km] of all tributaries to this reach in Colbert and Lauderdale Cos.)
Pigtoe, rough	Pleurobema plenum	Clams	U.S.A. (TN—specified portions of the French Broad and Holston Rivers)
Pigtoe, shiny	Fusconaia cor	Clams	U.S.A. (AL; The free-flowing reach of the Tennessee R. from the base of Wilson Dam downstream to the backwaters of Pickwick Reservoir [about 12 RM (19 km)] and the lower 5 RM [8 km] of all tributaries to this reach in Colbert and Lauderdale Cos.)
Pigtoe, shiny	Fusconaia cor	Clams	U.S.A. (TN—specified portions of the French Broad and Holston Rivers)
Pikeminnow (=squawfish), Colorado	Ptychocheilus lucius	Fishes	Salt and Verde R. drainages, AZ
Pimpleback, orangefoot (pearlymussel)	Plethobasus cooperianus	Clams	U.S.A. (TN—specified portions of the French Broad and Holston Rivers)
Pronghorn, Sonoran	Antilocapra americana sonoriensis	Mammals	U.S.A. (AZ), Mexico In Arizona, an area north of Interstate 8 and south of Interstate 10, bounded by the Colorado River on the west and Interstate 10 on the east; and an area south of Interstate 8, bounded by Highway 85 on the west, Interstates 10 and 19 on the east, and the U.S.—Mexico border on the south.
Purple cat's paw (=purple cat's paw pearlymussel)	Epioblasma obliquata obliquata	Clams	U.S.A. (AL;The free-flowing reach of the Tennessee R. from the base of Wilson Dam downstream to the backwaters of Pickwick Reservoir [about 12 RM (19 km)] and the lower 5 RM [8 km] of all tributaries to this reach in Colbert and Lauderdale Cos.)
Rail, Guam	Rallus owstoni	Birds	Rota
Ring pink (mussel)	Obovaria retusa	Clams	U.S.A. (TN—specified portions of the French Broad and Holston Rivers)
Riversnail, Anthony's	Athearnia anthonyi	Snails	U.S.A. (TN—specified portions of the French Broad and Holston Rivers)

listing process. The agency assesses population data, recovery achievements, and threats to the species. A proposal for delisting is published in the *Federal Register* for review and comment by scientists and the public. Expert opinions are obtained from three independent species specialists. All the collected information is analyzed and a final decision on delisting is published in the *Federal Register*. Under the ESA, the USFWS or the NMFS (in cooperation with state agencies) is required to monitor for at least five years any species that has been delisted due to recovery. This is accomplished through a postdelisting monitoring strategy that goes through peer review (review by qualified scientists) and public comment before being finalized.

TABLE 2.10
Experimental populations, November 2011 [CONTINUED]

Inverted common name	Scientific name	Species group	Where listed
Riversnail, Anthony's	Athearnia anthonyi	Snails	U.S.A. (AL; The free-flowing reach of the Tennessee R. from the base of Wilson Dam downstream to the backwaters of Pickwick Reservoir [about 12 RM (19 km)] and the lower 5 RM [8 km] of all tributaries to this reach in Colbert and Lauderdale Cos.)
Squirrel, Delmarva Peninsula fox	Sciurus niger cinereus	Mammals	U.S.A. (DE, Sussex Co.)
Trout, bull	Salvelinus confluentus	Fishes	Clackamas River subbasin and the mainstem Willamette River, from Willamette Falls to its points of confluence with the Columbia River, including Multnomah Channel
Wartyback, white (pearlymussel)	Plethobasus cicatricosus	Clams	U.S.A. (TN—specified portions of the French Broad and Holston Rivers)
Wolf, gray	Canis lupus	Mammals	U.S.A. (WY)
Wolf, gray	Canis lupus	Mammals	U.S.A. (portions of AZ, NM and TX)
Wolf, red	Canis rufus	Mammals	U.S.A. (portions of NC and TN)
Woundfin	Plagopterus argentissimus	Fishes	Gila R. drainage, AZ, NM

R. = river; RM = river mile; km = kilometer; Co. = county.

SOURCE: Adapted from "Experimental Populations," in *Species Reports*, U.S. Department of the Interior, U.S. Fish and Wildlife Service, November 1, 2011, http://ecos.fws.gov/tess_public/pub/experimentalPopulations.jsp (accessed November 1, 2011)

Entities (species or distinct populations) that have been delisted as of November 2011 are shown in Table 2.13. Of the 51 delisted entities, 23 had recovered, 10 became extinct, and the remainder had procedural issues.

PRIVATE-PARTY PETITIONS

As noted earlier, the ESA allows private parties (e.g., individuals and nongovernmental organizations [NGOs]) to submit listing petitions to the federal government. The NGOs are typically environmental or conservation groups that have the funding and expertise to develop petitions containing the required scientific data supporting the need for listing. Since the passage of the law in 1973, some NGOs have flooded the USFWS and the NMFS with listing petitions. According to D. Noah Greenwald, Kieran Suckling, and Martin Taylor, in "The Listing Record" (Dale D. Goble, J. Michael Scott, and Frank W. Davis, eds., *The Endangered Species Act at Thirty: Renewing the Conservation Promise*, 2006), the vast majority of the listings that occurred between 1996 and 2004 were driven by private-party petitions. Since 2004 private parties have submitted petitions covering hundreds of additional species.

The USFWS or the NMFS must determine within 90 days whether a petition contains "substantial information" suggesting that a species may require listing under the ESA. If listing is found to be warranted, the responsible agency is supposed to complete a status review for the species within 12 months. These time constraints have proven infeasible given the large number of petitions coming into the agencies and the large number of species involved. As a result, NGOs have filed numerous lawsuits against the agencies for failing to comply with ESA-mandated deadlines.

Since the 1990s the USFWS has repeatedly complained that many of its decisions and activities are driven by court orders, rather than by scientific priorities. The agency has also suggested that NGOs have taken advantage of ESA provisions that allow citizen involvement in lawsuits. Critics claim the groups flood the USFWS with petitions so that lawsuits can be brought when the agency is unable to respond in a timely manner. Deborah Zabarenko notes in "Deal Aims to Cut Endangered Species Red Tape" (*Reuters*, May 10, 2011) that between 1994 and 2006 the USFWS received an average of approximately 17 petitions annually; however, between 2007 and May 2011 the agency received 1,230 petitions. As of May 2011, the agency reportedly had a backlog of over 600 petitions still waiting to be reviewed.

Environmentalists argue that the lawsuits are necessary because the USFWS fails to do the job that has been assigned to it under the ESA.

Candidate Species

The USFWS has also been repeatedly sued for designating species as candidate species. As noted earlier, this designation means that the USFWS acknowledges that the species deserves protection under the ESA, but believes that more pressing priorities must come first. The agency's September 1997 CNOR (http://frwebgate.access.gpo.gov/cgi-bin/getdoc.cgi?dbname=1997_register&docid=fr19se97-41) included 207 candidate species. Subsequent annual CNORs (http://www.fws.gov/endangered/what-we-do/earlier-notices.html) have continued to include 200 or more candidate species, much to the displeasure of conservation and wildlife NGOs.

TABLE 2.11

U.S. Fish and Wildlife Service recovery potential priority ranking system

Degree of threat	Recovery potential	Taxonomy	Priority	Conflict
	High	Monotypic genus	1	1C 1
	High	Species	2	2C 2
	High	Subspecies	3	3C 3
High	Low	Monotypic genus	4	4C 4
	Low	Species	5	5C 5
	Low	Subspecies	6	6C 6
	High	Monotypic genus	7	7C 7
	High	Species	8	8C 8
	High	Subspecies	9	9C 9
Moderate	Low	Monotypic genus	10	10C 10
	Low	Species	11	11C 11
	Low	Subspecies	12	12C 12
	High	Monotypic genus	13	13C 13
	High	Species	14	14C 14
	High	Subspecies	15	15C 15
Low	Low	Monotypic genus	16	16C 16
	Low	Species	17	17C 17
	Low	Subspecies	18	18C 18

SOURCE: John J. Fay and W. L. Thomas, "Table 3. Recovery Priority," in "Fish and Wildlife Service: Endangered and Threatened Species Listing and Recovery Priority Guidelines," *Federal Register*, vol. 48, no. 184, September 21, 1983, and *Federal Register*, vol. 48, no. 221, November 15, 1983 (correction), http://www.fws.gov/endangered/esa-library/pdf/48fr43098-43105.pdf (accessed October 28, 2011)

In "Improving ESA Implementation" (December 8, 2011, http://www.fws.gov/endangered/improving_ESA/listing_workplan.html), the USFWS notes that multiple lawsuits dealing with candidate species were bundled together and resolved in late 2011 through legal settlements that were reached separately with the suing NGOs. Overall, the agency agreed to make listing decisions by 2017 for all 251 species included in the November 2010 CNOR (http://www.fws.gov/endangered/what-we-do/cnor-2010.html) and to make critical habitat determinations for any of these species that become listed. In addition, the agency set a timetable for reviewing previously submitted petitions for hundreds of species. In exchange the NGOs involved in the settlements agreed to restrictions on their ESA petitioning and litigation activities through 2017.

TABLE 2.12

National Marine Fisheries Service recovery potential priority ranking system

Magnitude of threat	Recovery potential	Conflict	Priority
High	High	Conflict	1
		No conflict	2
	Low to moderate	Conflict	3
		No conflict	4
Moderate	High	Conflict	5
		No conflict	6
	Low to moderate	Conflict	7
		No conflict	8
Low	High	Conflict	9
		No conflict	10
	Low to moderate	Conflict	11
		No Conflict	12

SOURCE: William J. Fox, Jr., "Table 3. Species Recovery Priority," in "National Oceanic and Atmospheric Administration: Endangered and Threatened Species Listing and Recovery Priority Guidelines," *Federal Register*, vol. 55, no. 116, June 15, 1990, http://www.nmfs.noaa.gov/pr/pdfs/fr/fr55-24296.pdf (accessed October 28, 2011)

The work called for under the legal settlements represents an increase in workload for the USFWS. It remains to be seen whether Congress will fund the agency sufficiently to allow the work to be completed as planned. The timetable can also be derailed if other NGOs submit numerous petitions or file numerous ESA-related lawsuits over the coming years.

Equal Access to Judgment Act

Critics assert that NGOs benefit financially from ESA-related lawsuits at the taxpayers' expense because of the Equal Access to Judgment Act (EAJA). The EAJA, passed by Congress in 1980 and signed into law by President Ronald Reagan (1911–2004), requires the federal government to pay the attorneys' fees of eligible individuals, nonprofit organizations, and other entities when these parties sue the federal government and prevail in court. NGOs have filed and won so many ESA-related lawsuits that they are believed to have collected many millions of dollars in legal fees from the government. The exact figures are unknown because the government does not compile or disclose the information. The U.S. Government Accountability Office (GAO) is an independent investigatory agency that reports to Congress. In *Equal Access to Justice Act: Its Use in Selected Agencies* (January 14, 1998, http://www.gao.gov/assets/90/87317.pdf), the GAO indicates that the government quit tracking EAJA payments in 1994.

Critics claim that litigious NGOs abuse the EAJA by submitting numerous ESA listing petitions and then suing when the government is unable to meet the mandated deadlines. The groups then collect attorneys' fees that fund additional listing petitions and lawsuits.

TABLE 2.13

Delisted U.S. and foreign species, November 1, 2011

Date species first listed	Date delisted	Species name	Reason delisted
8/27/1984	2/23/2004	Broadbill, Guam (Myiagra freycineti)	Extinct
3/11/1967	9/2/1983	Cisco, longjaw (Coregonus alpenae)	Extinct
4/30/1980	12/4/1987	Gambusia, Amistad (Gambusia amistadensis)	Extinct
12/8/1977	2/23/2004	Mallard, Mariana (Anas oustaleti)	Extinct
6/14/1976	1/9/1984	Pearlymussel, Sampson's (Epioblasma sampsoni)	Extinct
3/11/1967	9/2/1983	Pike, blue (Stizostedion vitreum glaucum)	Extinct
10/13/1970	1/15/1982	Pupfish, Tecopa (Cyprinodon nevadensis calidae)	Extinct
4/10/1979	10/28/2008	Seal, Caribbean monk (Monachus tropicalis)	Extinct
3/11/1967	12/12/1990	Sparrow, dusky seaside (Ammodramus maritimus nigrescens)	Extinct
6/4/1973	10/12/1983	Sparrow, Santa Barbara song (Melospiza melodia graminea)	Extinct
4/28/1976	8/31/1984	Butterfly, Bahama swallowtail (Heraclides andraemon bonhotei)	Original data in error—act amendment
6/14/1976	2/29/1984	Turtle, Indian flap-shelled (Lissemys punctata punctata)	Original data in error—erroneous data
4/29/1986	6/18/1993	Globeberry, Tumamoc (Tumamoca macdougalii)	Original data in error—new information discovered
5/27/1978	9/14/1989	Milk-vetch, Rydberg (Astragalus perianus)	Original data in error—new information discovered
7/13/1982	9/22/1993	Pennyroyal, Mckittrick (Hedeoma apiculatum)	Original data in error—new information discovered
9/26/1986	2/28/2000	Shrew, Dismal Swamp southeastern (Sorex longirostris fisheri)	Original data in error—new information discovered
12/14/1992	9/24/2010	Snail, Utah valvata (Valvata utahensis)	Original data in error—new information discovered
12/18/1977	11/22/1983	Treefrog, pine barrens FL pop. (Hyla andersonii)	Original data in error—new information discovered
5/18/1984	6/19/2006	Agave, Arizona (Agave arizonica)	Original data in error—not a listable entity
12/7/1979	9/22/1993	Cactus, spineless hedgehog (Echinocereus triglochidiatus var. inermis)	Original data in error—not a listable entity
3/10/1997	4/14/2006	Pygmy-owl, cactus ferruginous AZ pop. (Glaucidium brasilianum cactorum)	Original data in error—not a listable entity
12/6/1979	10/1/2003	Barberry, Truckee (Berberis (=Mahonia) sonnei)	Original data in error—taxonomic revision
2/17/1984	2/6/1996	Bidens, cuneate (Bidens cuneata)	Original data in error—taxonomic revision
11/28/1979	6/24/1999	Cactus, Lloyd's hedgehog (Echinocereus lloydii)	Original data in error—taxonomic revision
3/11/1967	7/25/1978	Duck, Mexican U.S.A. only (Anas diazi)	Original data in error—taxonomic revision
10/11/1979	11/27/1989	Hedgehog cactus, purple-spined (Echinocereus engelmannii var. purpureus)	Original data in error—taxonomic revision
12/14/1992	9/5/2007	Springsnail, Idaho (Pyrgulopsis idahoensis)	Original data in error—taxonomic revision
9/13/1996	4/26/2000	Trout, coastal cutthroat Umpqua R. (Oncorhynchus clarki clarki)	Original data in error—taxonomic revision
7/27/1979	6/4/1987	Alligator, American (Alligator mississippiensis)	Recovered
9/17/1980	8/27/2002	Cinquefoil, Robbins' (Potentilla robbinsiana)	Recovered
7/5/1979	9/2/2011	Coneflower, Tennessee purple (Echinacea tennesseensis)	Recovered
9/5/1985	2/18/2011	Daisy, Maguire (Erigeron maguirei)	Recovered
7/24/2003	7/24/2003	Deer, Columbian white-tailed Douglas County DPS (Odocoileus virginianus leucurus)	Recovered
6/2/1970	9/12/1985	Dove, Palau ground (Gallicolumba canifrons)	Recovered
3/11/1967	8/8/2007	Eagle, bald lower 48 States (Haliaeetus leucocephalus)	Recovered
6/2/1970	8/25/1999	Falcon, American peregrine (Falco peregrinus anatum)	Recovered
6/2/1970	10/5/1994	Falcon, Arctic peregrine (Falco peregrinus tundrius)	Recovered
6/2/1970	9/12/1985	Flycatcher, Palau fantail (Rhipidura lepida)	Recovered
3/11/1967	3/20/2001	Goose, Aleutian Canada (Branta canadensis leucopareia)	Recovered
12/30/1974	3/9/1995	Kangaroo, eastern gray (Macropus giganteus)	Recovered
12/30/1974	3/9/1995	Kangaroo, red (Macropus rufus)	Recovered
12/30/1974	3/9/1995	Kangaroo, western gray (Macropus fuliginosus)	Recovered
6/2/1970	9/21/2004	Monarch, Tinian (old world flycatcher) (Monarcha takatsukasae)	Recovered
6/2/1970	9/12/1985	Owl, Palau (Pyrroglaux podargina)	Recovered
6/2/1970	12/17/2009	Pelican, brown except U.S. Atlantic coast, FL, AL (Pelecanus occidentalis)	Recovered
6/2/1970	2/4/1985	Pelican, brown U.S. Atlantic coast, FL, AL (Pelecanus occidentalis)	Recovered
8/30/1999	9/15/2011	Snake, Lake Erie water subspecies range clarified (Nerodia sipedon insularum)	Recovered
5/22/1997	8/18/2005	Sunflower, Eggert's (Helianthus eggertii)	Recovered
6/16/1994	6/16/1994	Whale, gray except where listed (Eschrichtius robustus)	Recovered
5/5/2011	5/5/2011	Wolf, gray Northern Rocky Mountain DPS (delisted, except WY) (Canis lupus)	Recovered
7/19/1990	10/7/2003	Woolly-star, Hoover's (Eriastrum hooveri)	Recovered

DPS = distinct population segments.

SOURCE: Delisting Report, in *Species Reports*, U.S. Department of the Interior, U.S. Fish and Wildlife Service, November 1, 2011, http://ecos.fws.gov/tess_public/pub/delistingReport.jsp (accessed November 1, 2011)

Overprotection Lawsuits

Overprotection lawsuits are typically filed by private parties whose economic interests are threatened by ESA measures. Examples include farmers, ranchers, landowners and developers, industry groups and associations, and the operators or users of natural resources, such as water and irrigation districts. The parties may be opposed to listings or designation of critical habitat for particular species or they may attempt to compel the federal government to delist species.

One of the most famous overprotection cases in ESA history concerns the polar bear, which was proposed in 2007 for listing as threatened. As will be explained in Chapter 3, the listing is unique because the most significant threat to the species' survival is believed to be climate change, specifically the loss of habitat due to continuing global warming. In "PLF Challenges Unwarranted Polar Bear Listing" (2011, http://www.pacificlegal.org/page.aspx?pid=667), the Pacific Legal Foundation (PLF) notes that it sued the government in 2008 on behalf of the California Cattlemen's Association on various legal grounds. In June 2011 a federal court ruled against the PLF; however, the organization filed an appeal. As of January 2012, the appeal had not been heard.

As noted earlier, the USFWS has historically been late in completing five-year reviews that are required by the ESA. Groups that allege the ESA is overprotective use legal action to force the agency to conduct the reviews for species the groups believe have recovered or improved in population. In "Time to Delist California Species That Aren't Endangered or Threatened" (2012, http://www.pacificlegal.org/page.aspx?pid=1329), the PLF notes that in 2005 it entered into a settlement agreement with the agency over the latter's failure to conduct the reviews for nearly 100 listed species in California. By 2008 the reviews for more than half of the species had been completed and resulted in recommendations for the delisting or downlisting of six of the species. The PLF complains that under USFWS policy the agency "will not act on these recommendations until it is petitioned to do so." As a result, the PLF "petitioned the agency to make these changes" and states that in January 2011 it succeeded in obtaining status reviews, which had to be completed in 12 months. As of January 2012, the results of the reviews had not been published.

Environment and Natural Resources Division

The U.S. Department of Justice's Environment and Natural Resources Division (ENRD) handles litigation cases that are associated with the nation's environmental and conservation laws, including the ESA. The ENRD provides in *Overview of the Endangered Species Act and Highlights of Recent Litigation* (January 2004, http://www.abanet.org/environ/committees/endangered/OverviewoftheESA.pdf) a brief summary of the several hundred ESA-related lawsuits that had reached the courts as of early 2004. The ENRD notes that some ESA litigation involved broad issues, such as the function of the law and the scope of the duties of the USFWS and the NMFS under the law. However, the vast majority of lawsuits focused on particular species. ENRD-litigation summaries for individual years since 2004 are available at http://www.justice.gov/enrd/Current_topics.html.

U.S. Supreme Court Cases

One of the foremost legal cases in ESA history involved a tiny fish called the snail darter in the Little Tennessee River. The ENRD reports in "*Tennessee Valley Authority v. Hill*: The Snail Darter Case" (November 2010, http://www.justice.gov/enrd/4724.htm) that a biologist discovered the fish in the river in 1973, the same year that the ESA became law. At that time the Tennessee Valley Authority (TVA; a federally owned corporation) was building the Tellico Dam near Knoxville, Tennessee. The dam was to provide hydroelectric power and flood control by backing up the river into a 30-mile (48.3-km) reservoir. In 1975 the USFWS declared the snail darter an endangered species and designated the Little Tennessee River as its critical habitat. Completion of the nearly finished dam—on which $78 million of tax money had already been spent—was put on hold while a massive legal battle was fought.

In 1978 the U.S. Supreme Court ruled in *Tennessee Valley Authority v. Hill* (437 U.S. 153) that protection of the snail darter outweighed the economic loss of abandoning the dam. The ENRD notes that the Supreme Court acknowledged "the perceived absurdity in forfeiting tens of millions of dollars of public funds for a small fish," but points out that the ESA did not include any provisions for cost-benefit analysis. In other words, the apparent intent of the ESA was to protect imperiled species regardless of the cost.

The decision was hailed as a great victory by environmental and conservation groups, but was extremely unpopular otherwise. Congress added new language to section 7 of the ESA allowing costs to be considered when federally funded or managed projects conflict with the needs of imperiled species. In addition, a separate bill was passed exempting the Tellico Dam from the ESA requirements for the snail darter. In 1979 the dam was completed, and the tiny fish was presumed exterminated. However, during the early 1980s more populations of the snail darter were discovered in Tennessee rivers and streams. In 1984 the USFWS downlisted the species from endangered to threatened and rescinded its critical habitat designation. As of January 2012, the snail darter in Tennessee remained on the list of threatened species.

In "ESA in the Supreme Court" (http://www.justice.gov/enrd/4714.htm), the ENRD notes that four other notable ESA-related cases had reached the U.S. Supreme Court as of November 2010:

- *Babbitt v. Sweet Home Chapter of Communities for a Better Oregon* (515 U.S. 687 [1985])—the court affirmed the DOI's interpretation of taking under the ESA to include "significant modification or degradation where it actually kills or injures wildlife."
- *Bennett v. Spear* (520 U.S. 154 [1997])—the court broadened the right of individuals (and groups) to sue the federal government over ESA issues.

- *National Association of Home Builders v. Defenders of Wildlife* (No. 06-340 [2007])—the court affirmed that only specific types of federal actions fall under section 7 of the ESA.
- *Winter v. Natural Resources Defense Council* (No. 07-1239 [2008])—the court overruled lower court decisions that had placed restrictions on the U.S. Navy's use of sonar during submarine training exercises off the coast of California. The lower courts had found in favor of plaintiffs, which argued that the sonar harmed marine mammals protected by federal laws, including the ESA.

ESA SPENDING

Various federal agencies spend money to uphold the ESA. The primary agencies are the USFWS and the NMFS. For accounting purposes, the federal government operates on a fiscal year (FY) that runs from October through September. Thus, FY 2012 covers October 1, 2011, through September 30, 2012. Each year by the first Monday in February the U.S. president must present a proposed budget to the U.S. House of Representatives. This is the amount of money that the president estimates will be required to operate the federal government during the next fiscal year. It should be noted that Congress can take many months, even more than a year, to finalize and enact a federal budget. This was the case for the FY 2011 budget, which was proposed in February 2010, but not enacted until April 2011 due to fierce disagreements within Congress about government spending.

The Wolf Rider

The FY 2011 budget bill was particularly noteworthy because it included an unprecedented rider. A rider is an amendment (typically tacked onto a bill at the last moment) that is not directly relevant to the intent of the bill, but affects a law or program covered by the bill. The 2011 rider essentially delisted a gray wolf DPS in the northern Rocky Mountains by reinstating an April 2009 USFWS (http://edocket.access.gpo.gov/2009/pdf/E9-5991.pdf) decision to delist the DPS. The 2009 decision had set off a complex legal battle that remained unresolved at the time the budget bill was passed. Details about the delisting and corresponding litigation are provided in Chapter 7.

According to the article "White House Set to Pass Budget Bill That Sets Precedent on Endangered Wolves" (Associated Press, April 14, 2011), the rider was notable because it marked the first time since passage of the ESA in 1973 that Congress has specifically targeted an ESA-listed species. The article indicates that Congress has intervened in ESA-related controversies before, including the TVA dam project that was held up by the snail darter during the 1980s and the battle in the Northwest over logging versus the northern spotted owl during the 1990s (discussed in Chapter 8). However, these interventions targeted the economic interests involved, rather than the listing statuses of the species themselves.

The USFWS Budget

Figure 2.2 shows USFWS funding enacted for FYs 2009 and 2010 and requested for FY 2011. The president proposed a $1.6 billion budget for the USFWS for FY 2011. However, by the time the FY 2011 budget was enacted, the agency's funding had been reduced to just over $1.5 billion through across-the-board rescission (elimination of spending that had been previously approved). (See Table 2.14.)

The funding requested each year by the president is called current, temporary, or discretionary funding. Agencies also receive additional appropriations that are known as permanent or mandatory funding. This is money that is allocated to the agency on a continuing basis and is not requested each year. As shown in Figure 2.2, an additional $1.2 billion was available to the USFWS in FY 2011 under permanent appropriations. According to the DOI, in *Budget Justifications and Performance Information Fiscal Year 2012* (August 25, 2011, http://www.doi.gov/budget/2012/data/pdf/summary_2012_house_Interior_Accttble20110825.pdf), most USFWS permanent appropriations are turned over to the states for restoration and conservation of fish and wildlife resources.

FIGURE 2.2

U.S. Fish and Wildlife Service funding, fiscal years 2009–11

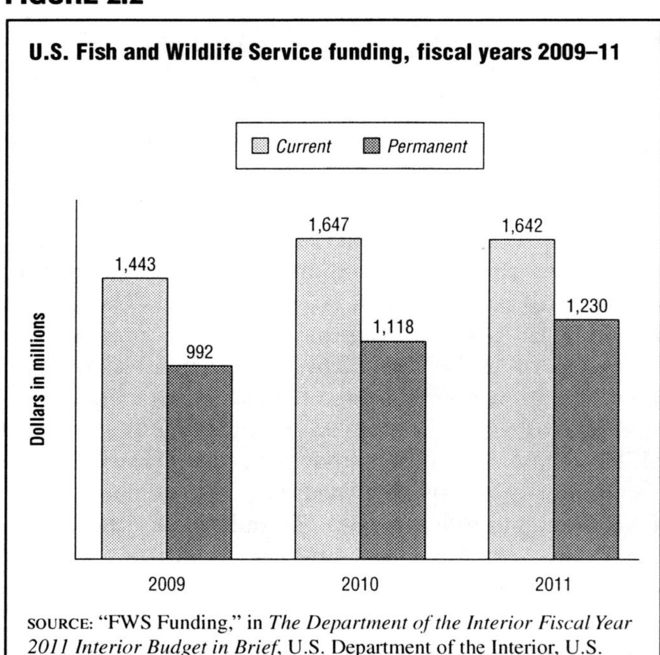

SOURCE: "FWS Funding," in *The Department of the Interior Fiscal Year 2011 Interior Budget in Brief*, U.S. Department of the Interior, U.S. Fish and Wildlife Service, February 2010, http://www.doi.gov/budget/2011/11Hilites/BH055.pdf (accessed October 28, 2011)

TABLE 2.14

U.S. Fish and Wildlife Service budgets enacted in fiscal years 2010 and 2011 and requested for fiscal year 2012

Bureaus/offices Accounts	2010 Enacted	2011 Enacted (PL 112–10) with ATB rescission	2012 President's budget
Fish and wildlife service			
Resource management	1,269,406	1,244,861	1,271,867
Construction	37,439	20,804	23,088
Land acquisition	86,340	54,890	140,000
Coop endangered species conservation fund	85,000	59,880	100,000
National wildlife refuge fund	14,500	14,471	0
North American wetlands conservation fund	47,647	37,425	50,000
Neotropical migratory birds conservation fund	5,000	3,992	5,000
Multinational species conservation fund	11,500	9,980	9,750
State and tribal wildlife grants	90,000	61,876	95,000
Landowner incentive program (LIP)* (rescind avail. balances)		(3,049)	
Private stewardship grants			
Total, current appropriations	**1,646,832**	**1,505,130**	**1,694,705**

*Actual LIP balances as of 4/18/2011 for rescission are $3,097,530.27.
Note: Dollars in thousands.
P.L. = Public Law, ATB = Across-the-board.

SOURCE: Adapted from "FY 2012 DOI Formulation President's Budget," in *What's New at Budget?*, U.S. Department of the Interior, Office of Budget, August 25, 2011, http://www.doi.gov/budget/2012/data/pdf/summary_2012_house_Interior_Accttble20110825.pdf (accessed December 5, 2011)

TABLE 2.15

U.S. Fish and Wildlife Service appropriations for Endangered Species program, fiscal years 2009–11

	2009 Actual 2010	Enacted 2011	Request from	Change 2010
Ecological services				
Endangered species				
Conservation candidate	10,670	12,580	11,471	−1,109
Listing	19,266	22,103	20,945	−1,158
Consultation	53,462	59,307	63,299	+3,992
Recovery	74,575	85,319	85,611	+292
Subtotal, endangered species	**157,973**	**179,309**	**181,326**	**+2,017**

Note: Dollars in thousands.

SOURCE: Adapted from "Highlights of Budget Changes by Appropriation Activity/Subactivity, Appropriation: Resource Management," in *The Department of the Interior Fiscal Year 2011 Interior Budget in Brief*, U.S. Department of the Interior, U.S. Fish and Wildlife Service, February 2010, http://www.doi.gov/budget/2011/11Hilites/BH055.pdf (accessed October 28, 2011)

Table 2.15 shows that $181.3 million was requested for USFWS endangered species programs for FY 2011. Nearly half of the money was allocated to recovery programs ($85.6 million), followed by consultations with other agencies and groups ($63.3 million), listing activities ($20.9 million), and candidate species conservation ($11.5 million).

The FY 2012 budget was proposed in February 2011 and passed in December 2011. As shown in Table 2.14, the budget request for the USFWS was nearly $1.7 billion. In "Department of the Interior Budget Snapshot—Proposed 2012 Appropriation" (December 2011, http://www.doi.gov/budget/2012/data/pdf/summary_hr2055_consolidated.pdf), the DOI reports that by the time the budget was enacted in December 2011 that figure had been reduced to $1.5 billion, which included $176 million devoted to endangered species.

FY 2012 SPENDING CAPS. In March 2011 USFWS officials (http://www.gpo.gov/fdsys/pkg/CHRG-112hhrg66897/html/CHRG-112hhrg66897.htm) testified before a congressional appropriations subcommittee regarding the agency's budget request for FY 2012. The officials requested that Congress cap the amount of money that can be spent on processing listing petitions. The USFWS hoped that a congressionally imposed spending cap would provide the agency with a legally defensible reason for missing ESA-mandated listing deadlines.

The spending cap plan drew strong criticism from both sides of the ESA debate. According to Lawrence Hurley, in "Obama Plan to Cap Funding for Endangered Species Act Petitions Angers Litigants" (*New York Times*, March 24, 2011), groups on both sides of the spectrum complained that the caps do not provide a sensible solution for the underlying problems.

The final enacted FY 2012 budget bill (December 15, 2011, http://www.doi.gov/budget/2012/data/pdf/summary_hr2055_Bill_Report.pdf) appropriated $1.2 billion for USFWS resource management. It also set spending caps for various activities that are required under section 4 of the ESA, including a cap of $7.5 million for critical habitat activities (excluding litigation support) for species that were listed before October 1, 2010, and $1.5 million for activities related to listing petitions for U.S. species.

NMFS Budget

The NMFS handles ESA management of marine mammals, such as whales, seals, and anadromous fish. According to the NMFS, in *FY 2012 Budget Highlights* (February 15, 2011, http://www.corporateservices.noaa.gov/~nbo/fy12_budget_highlights/NMFS_FY12_One_pager.pdf), the FY 2012 budget request for the NMFS was just over $1 billion. The agency does not break down ESA activities within its budget summary; however, the NMFS reports that funding for environmental consultations and conservation and recovery actions for marine mammals was expected to increase by $5.5 million in FY 2012.

ESA Expenditures

Section 18 of the ESA requires the USFWS to file an annual report detailing certain expenditures that were made for the conservation of threatened and endangered species under the act. In *Federal and State Endangered and Threatened Species Expenditures: Fiscal Year 2010* (September 2011, http://www.fws.gov/endangered/esa-library/pdf/2010.EXP.FINAL.pdf), the USFWS indicates that nearly $1.2 billion was spent by federal agencies in FY 2010 to protect more than 1,400 specific entities (species, subspecies, DPS, or ESU) under the ESA.

ESA EXPENDITURES BY SPECIES. Table 2.16 shows the 20 species with the highest reported expenditures under the ESA in FY 2010. Together, these species accounted for $763.8 million in spending. The list is dominated by fish species. Approximately $230.3 million was spent on the Chinook salmon, followed by $195.7 million for the steelhead. Both species are anadromous, as are coho salmon, sockeye salmon, and chum salmon, and are found in the waters of the Pacific Northwest. The other fish on the list (excluding the Atlantic salmon) are freshwater species. There are three marine (ocean-based) mammals on the list: the Steller sea lion, the right whale, and the West Indian manatee. The list also contains several bird species and reptiles.

TABLE 2.16

The twenty listed species with the highest expenditures under the Endangered Species Act, fiscal year 2010

Ranking	Species	Expenditure
1	Salmon, chinook	$230,268,061
2	Steelhead	$195,729,612
3	Trout, bull	$43,809,838
4	Salmon, coho	$37,596,542
5	Sea-lion, Steller	$33,247,899
6	Woodpecker, red-cockaded	$25,194,067
7	Sturgeon, pallid	$22,696,252
8	Salmon, sockeye	$22,535,318
9	Salmon, chum	$18,562,958
10	Tortoise, Desert	$16,747,485
11	Owl, northern spotted	$14,086,571
12	Whale, right	$13,596,253
13	Salmon, Atlantic	$13,145,296
14	Plover, Piping	$12,118,527
15	Minnow, Rio Grande silvery	$11,976,201
16	Sea turtle, loggerhead	$11,285,625
17	Bear, grizzly	$10,680,384
18	Flycatcher, southwestern willow	$10,579,072
19	Manatee, West Indian	$10,353,294
20	Tern, Least	$9,614,921

SOURCE: Adapted from "Table 2. Species Ranked in Descending Order of Total FY 2010 Reported Expenditures, Not Including Land Acquisition Costs," in *Federal and State Endangered and Threatened Species Expenditures: Fiscal Year 2010*, U.S. Department of the Interior, U.S. Fish and Wildlife Service, 2010, http://www.fws.gov/endangered/esa-library/pdf/2010.EXP.FINAL.pdf (accessed October 25, 2011)

How best to use the funds that are allocated to endangered species has been a contentious issue for years. The majority (nearly two-thirds) of the money spent in FY 2010 was devoted to only a few species.

The number of species being added to the federal threatened and endangered species list is expected to grow. Even though vertebrate species dominated the list during the early years of the act, by the 21st century plants and invertebrate animals made up a much greater proportion of the listed species. (See Table 1.2 in Chapter 1.) These species are politically more difficult to defend than either mammals or birds, which are more inherently appealing to most Americans because of the "warm and fuzzy" factor. These circumstances raise questions about the continued feasibility of a species-by-species preservation strategy, and the USFWS struggles under intense legal and political pressures to decide which species to protect first.

THE ECOSYSTEM APPROACH

During the 1990s there was a growing concern that traditional methods of species protection, which took a species-by-species approach, were ineffective. Many alternatives were proposed. One of the most popular was a method variously called the ecosystem, habitat, or community approach. In "What Is an Ecosystem?" (August 8, 2008, http://www.fws.gov/midwest/EcosystemConservation/ecosystem.html), the USFWS defines an ecosystem as a "a geographic area and all its living components (e.g., people, plants, animals, and microorganisms), their physical surroundings (e.g., soil, water, and air), and the natural cycles that sustain them." Central to this approach is a focus on conservation of large intact areas of habitat. It is hoped that by focusing on entire habitats, rather than on individual species recovery, many species will be protected before they reach critically low population sizes.

Ecosystem conservation considers entire communities of species as well as their interactions with the physical environment and aims to develop integrated plans that involve wildlife, physical resources, and sustainable use. Such an approach sometimes requires compromise between environmentalists and developers.

ASSESSING THE ESA

The ESA has become one of the most controversial, litigious, and politically polarizing laws in U.S. history. Oddly enough, it did not start out that way. According to J. Michael Scott, Frank W. Davis, and Dale D. Goble, in "Introduction" to the *Endangered Species Act at Thirty*, the original act enjoyed broad bipartisan support (support from both major political parties). The bills establishing the ESA were passed unanimously by the U.S. Senate and almost unanimously (390 to 12 in the first vote and 355 to 4 in the second vote) in the U.S. House of Representatives. Even though Congress had a Democratic majority at the time, the Republican president Richard M. Nixon (1913–1994) supported the ESA and signed the final bill into law in 1973. Environmental issues were a high priority during the early 1970s, and numerous sweeping environmental laws were passed around the same time, including the Clean Water Act and legislation that would become the Clean Air Act. Scott, Davis, and Goble describe passage of the ESA as "an idealistic and perhaps naive attempt to preserve humanity by preserving other species in the ecological support system that makes life possible." Certainly, its original supporters did not foresee the bitter contention that this law would later arouse.

The original law was very restrictive, in that it prohibited taking under almost all circumstances. The controversial snail darter court case and other litigation prompted Congress to make the ESA more permissive and flexible via amendments in 1978 and 1982. The 1982 amendment was particularly striking because it allowed for incidental take under certain circumstances, a concession that bothered environmentalists. The amendment was also notable for ending the possibility of using economic considerations in listing species. In 1981 President Reagan signed Executive Order 12291 (http://www.archives.gov/federal-register/codification/executive-order/12291.html) to "reduce the burdens of existing and future regulations." One requirement of this broad executive order was that federal agencies had to weigh the economic costs against the benefits of regulations. Reagan was a conservative Republican who opposed what he saw as the overregulation of business and private interests. Congress responded by adding language in the 1982 ESA amendment that listing determinations would be made "solely on the basis of the best scientific and commercial data available." Thus, economic costs are not a consideration during the ESA listing process. This decision remains highly controversial and is bitterly opposed by many industrial and commercial interests.

Property Rights and Economic Issues

By the early 1990s conservative Republicans enjoyed much greater political power in the United States. The November 1994 elections signaled the so-called Republican Revolution in which Republicans (primarily conservative Republicans) gained control of both the House and the Senate for the first time in 40 years and won large victories in state legislatures and governor's offices around the country. In general, Republicans believe the ESA violates private property rights and stifles economic growth by curbing development. They charge that environmental protection often results in the loss of jobs and business profits. This was a popular refrain during the early 1990s as the United States witnessed the birth of two long and bitter ESA disputes over high-profile species:

- Northern spotted owl—declared threatened in 1990. Subsequent designation of millions of acres of forestland in the Pacific Northwest as critical habitat halted logging on federal lands, causing lost jobs and negative economic effects. (See Chapter 8.)
- Pacific salmon—the first of several stocks in the Pacific Northwest was declared threatened in 1990, setting off protracted battles over fishing and water rights and economic effects on the fishing industry and farmers, particularly in the Klamath River basin. (See Chapter 4.)

Overall, hundreds of ESA listings were made during the early 1990s. As shown in Figure 1.2 in Chapter 1, 397 species were listed between 1990 and 1994. This five-year period accounted for 31% of all listings made since the law was passed in 1973. In 1994, 127 species were listed under the ESA—the most listings for any year previous or since that time.

MORATORIUM AND REFORM. During the spring of 1995 the Republican-controlled Congress imposed a moratorium on new ESA listings, and the law itself was in jeopardy. The authorization for funding the ESA had expired at the end of FY 1992. In 1993 and 1994 the Democratic-controlled Congress appropriated annual funds to keep the ESA going; however, the Republican-controlled Congress that took power in 1995 refused to do so. As shown in Figure 1.2 in Chapter 1, only 18 listings were made under the ESA in 1995. According to Scott, Davis, and Goble, this open congressional "hostility" toward the ESA spurred Bruce Babbitt (1938–), the U.S. secretary of the interior, to dramatically reform the law. He added so-called incentive-based strategies "to try and reconcile endangered species conservation with economic development." These reforms included the HCPs, CCAAs, and SHAs described earlier in this chapter for private landowners. During the spring of 1996 funding was restored for the ESA, and the listing moratorium ended.

FOCUS ON PRIVATE PROPERTY RIGHTS. The issue of private property rights is a prime concern among ESA critics, particularly industry and development groups. One measure they strongly support is compensation for the loss of use of land. For example, the National Stone, Sand, and Gravel Association states in *Endangered Species Act Reform* (November 2008, http://www.nssga.org/government/Position_Paper/esar.html), "ESA mandates have severely restricted the use and value of privately owned property. When severe restrictions occur without compensation by the federal government, the Act shifts to individual citizens the costs and burdens that should be shared by all citizens. The ESA must be modified to compensate land owners justly in a timely fashion when private property is preserved in a habitat conservation plan."

EVADING THE ESA. ESA critics also complain that the law provides no incentive for private landowners to participate in the conservation process. In fact, there is evidence that some people actively evade the law to prevent government restrictions on their land use. This claim is supported by Dean Lueck and Jeffrey A. Michael in "Preemptive Habitat Destruction under the Endangered Species Act" (*Journal of Law and Economics*, vol. 46, no. 1, 2003). The researchers examined land-use data between 1984 and 1990 for approximately 1,000 forest plots in North Carolina. As will be described in Chapter 8, the red-cockaded woodpecker is an endangered bird found in scattered forested habitats in the Southeast. It has been listed under the ESA since 1970. Lueck and Michael find that landowners with forest plots near areas known to contain the imperiled birds were much more likely to harvest and sell their timber than landowners with plots located farther from the birds. In addition, the plots closest to the birds were harvested when the trees were younger (and subsequently smaller and worth less). The researchers surmise that landowners with plots close to the birds rushed to harvest their trees before the trees became home to the imperiled species. Lueck and Michael call this "preemptive habitat destruction." The landowners evaded potential future ESA restrictions on their land use by destroying the very habitat that might attract and harbor the birds.

A more dire evasion technique allegedly used by some landowners is described by the colorful phrase "shoot, shovel, and shut up." The phrase was supposedly coined by Ralph R. Reiland, an economics professor known for his conservative views, in a column for the *Pittsburgh Tribune-Review*. In "Shoot, Shovel & Shut Up" (April 5, 2005, http://www.pittsburghlive.com/x/pittsburghtrib/opinion/columnists/reiland/s_187542.html), Reiland discusses the Lueck and Michael study and provides anecdotal stories about private landowners he claims have been economically harmed by ESA land-use restrictions. Reiland, like many ESA opponents, asserts that preemptive habitat destruction and "shoot, shovel and shut up" strategies are practiced by private landowners to avoid government interference with how they manage their land. As a result, these strategies thwart the preservation of the very species that the ESA is supposed to protect. This is presented as evidence that the ESA is a failed law.

Recovery Rate Controversy

One of the most frequent criticisms leveled against the ESA is that it has achieved recovery for few species since its passage in 1973. As shown in Table 2.13 and discussed earlier, only 23 U.S. entities (species or distinct populations) had been delisted due to recovery under the ESA as of November 2011. (Some of the recovered mammal entities are not entire species, but subpopulations of the wider population.) Considering that 1,990 species were listed under the ESA as of November 2011, the recovery rate was around 1%. (See Table 1.2 in Chapter 1.) This rate is often touted by ESA opponents as proof that the act has failed.

ESA proponents take a different stance on this issue. Verlyn Klinkenborg asks in "Last One" (*National Geographic*, January 2009): "How many species might have vanished without it?" Conservation groups maintain that rates of full recovery and delisting are not adequate measures of ESA success because it can take many decades for a species to recover after being listed under the ESA.

ESA Politics in the 21st Century

As noted earlier, the ESA survived a political assault by congressional conservatives during the mid-1990s. In 2000 the Republican George W. Bush (1946–) was elected president. Bush favored less government regulation of business and industry. He espoused a so-called new environmentalism that promoted conservation and cooperation at the local level over federal government prohibitions. During his eight years in office only 61 new listings were made under the ESA. (See Figure 1.2 in Chapter 1.) This historically low number has been roundly criticized and hailed as proof that the administration tried to undermine the intent of the law. There were also many accusations of political bullying of DOI scientists by administration officials. In 2007 Julie A. McDonald (1955–) resigned from her post as DOI deputy assistant secretary of fish, wildlife, and parks amid allegations that she pressured DOI scientists regarding ESA decisions. In "Interior Dept. Official Facing Scrutiny Resigns" (*Washington Post*, May 2, 2007), Elizabeth Williamson reports that an internal investigation at the DOI found that McDonald "repeatedly instructed Fish and Wildlife scientists to change their recommendations on identifying 'critical habitats.'" McDonald was also accused of arguing with staff scientists and mocking and even revising reports they produced that recommended the listing of species under the ESA.

The validity (or soundness) of scientific decisions within the ESA process became a source of fierce debate during the Bush administration. Conservative critics of the law claimed that USFWS and NMFS scientists were not using "sound science" to make listing and critical habitat decisions. By contrast, administration critics accused the agencies of bowing to political pressure and ignoring scientific data that supported protective measures for imperiled species.

In January 2007 Eugene H. Buck, M. Lynne Corn, and Pamela Baldwin of the Congressional Research Service (the investigative arm of Congress) addressed this conflict in *The Endangered Species Act and "Sound Science"* (http://www.fas.org/sgp/crs/misc/RL32992.pdf). The researchers examined claims from both sides of the debate and noted "by law, ESA decisions must have as sound a basis in science as is available, but this requirement can mean different things to different people." They added that "incomplete data, different interpretations among scientists, and evolving disciplines in science can make the consideration of relevant science challenging for the regulatory agencies." In addition, there were political considerations involved in setting ESA policies and making decisions. Buck, Corn, and Baldwin pointed out that "the complexity, uncertainty, and risk associated with many ESA issues, and the predictive nature of science with its emphasis on the probability of various outcomes rather than on absolute certainty, can make the interaction of scientists and decision-makers frustrating for both." The researchers discussed this conflict between science and politics within the Bush administration, but noted that the agencies administering the ESA "have procedures and policies in place to ensure the objectivity and integrity of the science that underpins agency decisions." In addition, Buck, Corn, and Baldwin indicated that scientists within and outside of the agencies were opposed to fundamental changes in the ESA scientific process. These changes were being promoted through legislation proposed by conservative members of Congress.

The Bush administration supported such legislative efforts to amend the ESA to change its scientific process and to be less restrictive on private property owners, businesses, and industry. However, these legislative efforts failed to pass. The administration also attempted to reform the ESA through procedural policy changes and rule interpretations. Two of these measures—limiting section 7 consultations and reinterpreting the definition of the word *range*—attracted widespread media attention and were subsequently addressed by Bush's successor, President Barack Obama (1961–). Obama, a Democrat, took office in January 2009. Because his campaign platform included several important environmental proposals, environmentalists were hopeful that his administration would be more friendly to the ESA than the Bush administration had been.

SECTION 7 CONSULTATIONS. In December 2008, the last full month of the Bush administration, the DOI announced in the press release "Interior Publishes Final Narrow Changes to Regulations, Clarifies Role of Global Processes in Consultation" (http://www.doi.gov/archive/news/08_News_Releases/121108a.html) the issuance of a rule "to clarify the consultation process under the Endangered Species Act." The rule relaxed the mandatory requirement that federal agencies consult with the USFWS or the NMFS regarding actions or projects that might affect ESA-listed species. These consultations became optional under certain circumstances. Dirk Kempthorne (1951–), the secretary of the interior, said the change would help the government "focus on protecting endangered species as it strives to rebuild the American economy." The rule was greeted positively by business interests and negatively by environmentalists, who considered it to be a last-ditch effort by the Bush administration to weaken the ESA.

In March 2009 Obama issued "Memorandum for the Heads of Executive Departments and Agencies" (http://www.whitehouse.gov/the_press_office/Memorandum-for-the-Heads-of-Executive-Departments-and-Agencies/), which effectively reversed the Bush administration's rule regarding section 7 consultations under the ESA. Obama ordered a review of the rule by the secretaries of the interior and commerce. During the review process he asked federal agencies "to follow the prior longstanding consultation and concurrence practices" that had been established under the ESA. As of January 2012, the rule review had not been completed.

THE DEFINITION OF *RANGE*. In 2007 the DOI solicitor David Longly Bernhardt (1969–) issued an internal memorandum that contained legal guidance on how the word *range* should be interpreted in the ESA requirement to provide protection for any species in danger of extinction or threatened with extinction in "a significant portion of its range." The article "2007 Legal Opinion of Endangered Species Act Is a Threat to Imperiled Species, Experts Say" (*ScienceDaily*, August 3, 2009) reports that this guidance was prompted by a 2005 court ruling that dismissed the long-standing interpretation that the word *range* refers to habitat the species has historically inhabited. The new interpretation limits the word to the present range, rather than to the historical range. Thus, in Bernhardt's legal opinion, a species could not be deemed imperiled "in an area where it no longer exists." The article notes that scientists complained the interpretation would result in the setting of "the smallest possible geographical unit" for an imperiled species.

Environmental and conservation groups were disappointed when the Obama administration continued to follow the Bush administration's legal guidance on what constitutes a "range" for an imperiled species. As a result, numerous lawsuits were filed over the new "range" legal guidance, which came to be known as the "M-Opinion." In December 2011 the Obama administration (http://www.gpo.gov/fdsys/pkg/FR-2011-12-09/pdf/2011-31782.pdf) published a draft policy that provided a new interpretation of the phrase "significant portion of its range" (SPR). The new policy contained the following three key elements:

- A species' "range" is the "general geographical area" within which the species is found at the time the USFWS or the NMFS conducts its status determinations.
- A portion of a species' range is "significant" if the portion is so important to the species' viability that the species would be in danger of extinction without it.
- If a species is endangered or threatened within an SPR, but not across its entire range, the ESA's protections would apply across the species' entire range. The entire species would be listed unless the population within the SPR is a valid DPS. In the latter case, the DPS, rather than the entire species, would be listed.

The new proposed policy elicited harsh criticism from both sides of the political debate. For example, in "Proposed Obama Policy on Endangered Species Act Is Recipe for Extinction" (December 8, 2011, http://www.biologicaldiversity.org/news/press_releases/2011/endangered-species-act-12-08-2011.html), the Center for Biological Diversity (CBD) complains that, similar to Bush administration policy, "range" is confined to where a species is currently found, rather than to where it has historically been found. The CBD notes that "a species could be absolutely gone or close to vanishing almost everywhere it's always lived—but not qualify for protection because it can still be called secure on one tiny patch of land." Industry groups and other organizations that advocate for less restrictive ESA regulations were also unhappy with the proposed policy, but for different reasons. For example, Robert Horton of the law firm Nossaman LLP notes in "Services Issue Notice of Controversial New Interpretation of Threatened and Endangered Species" (December 9, 2011, http://www.endangeredspecieslawandpolicy.com/2011/12/articles/regulatory-reform/services-issue-notice-of-controversial-new-interpretation-of-threatened-and-endangered-species/index.html) that landowners may be concerned because a species would be listed throughout its entire range even if it is only endangered or threatened within an SPR. Horton states that "a species may be listed in areas where it is currently thriving, resulting in unnecessary and costly over regulation in some areas."

As of January 2012, the proposed policy had not been finalized.

CHAPTER 3
MARINE MAMMALS

Marine mammals live in and around the ocean. They are warm-blooded, breathe air, have hair at some point during their life, give birth to live young (as opposed to laying eggs), and nourish their young by secreting milk. Whales, dolphins, porpoises, seals, sea lions, sea otters, manatees, dugongs (manatee relatives), and polar bears fall into this category.

Historically, marine mammals have garnered a high level of public support and legal protection. During the 1960s the television show *Flipper* entertained American audiences with stories about a highly intelligent and loveable dolphin that befriended and helped a family. Tourist attractions such as Marineland in Florida and SeaWorld in California began featuring acrobatic dolphins and whales in popular shows. The growing environmental movement seized on the public interest in marine mammals and lobbied for measures to protect animals that many people believed to be extremely smart and sociable.

At the time, purse-seine fishing was widely practiced by commercial tuna fishers in the eastern tropical Pacific Ocean. This fishing method involved the use of enormous nets, often hundreds of miles long, that were circled around schools of tuna. Many dolphins were inadvertently captured because they tend to mingle with fleets of tuna in that part of the ocean. Nontargeted animals that are captured during commercial fishing activities are called bycatch. Dolphin bycatch became a major public issue. Hauling in the enormous tuna-filled nets was a long process. As a result, the air-breathing dolphins were trapped for long periods underwater and often drowned. Public outcry over these killings and general concern for the welfare of marine mammals led Congress to pass the Marine Mammal Protection Act of 1972.

THE MARINE MAMMAL PROTECTION ACT

The Marine Mammal Protection Act (MMPA) was passed in 1972 and was substantially amended in 1994. The original act noted that "certain species and population stocks of marine mammals are, or may be, in danger of extinction or depletion as a result of man's activities." However, it was acknowledged that "inadequate" information was available concerning the population dynamics of the animals being protected.

The MMPA prohibits the taking (hunting, killing, capturing, and harassing) of marine mammals. The act also bars the importation of most marine mammals or their products. Exceptions are occasionally granted for scientific research, public display in aquariums, traditional subsistence hunting by Alaskan Natives, and some incidental capture during commercial fishing operations. The goal of the MMPA is to maintain marine populations at or above "optimum sustainable" levels.

Whales, dolphins, seals, and sea lions were put under the jurisdiction of the National Marine Fisheries Service (NMFS), an agency of the National Oceanic and Atmospheric Administration (NOAA) in the U.S. Department of Commerce. Polar bears, sea otters, manatees, and dugongs were placed under the jurisdiction of the U.S. Fish and Wildlife Service (USFWS), an agency of the U.S. Department of the Interior.

The MMPA requires the NMFS and the USFWS to conduct periodic surveys to estimate populations and to predict population trends for marine mammals in three regions of U.S. waters: the Pacific Ocean coast (excluding Alaska), the Atlantic Ocean coast (including the Gulf of Mexico), and the Alaskan coast. The survey results are published by the NMFS in the annual *Stock Assessment Report* (http://www.nmfs.noaa.gov/pr/sars/).

The MMPA was passed a year before the Endangered Species Act (ESA). The MMPA was driven largely by public affection for marine mammals, rather than by specific knowledge about impending species extinction. Eugene H. Buck and Harold F. Upton of the Congressional Research Service (the investigative arm of Congress) explain in *Fishery, Aquaculture, and Marine Mammal Legislation in the 112th Congress* (November 4, 2011, http://www.nationalaglawcenter.org/assets/crs/R41613.pdf) that "while some critics assert that the MMPA is scientifically irrational because it identifies one group of organisms for special protection unrelated to their abundance or ecological role, supporters note that the MMPA has accomplished much by way of promoting research and increased understanding of marine life as well as encouraging attention to incidental bycatch mortalities of marine life by commercial fishing and other maritime industries."

THE ENDANGERED SPECIES ACT

As shown in Table 2.1 in Chapter 2, the first list of native endangered species issued in 1967 included only three marine mammal species: the Caribbean monk seal, the Guadalupe fur seal, and the Florida manatee (or Florida sea cow). Over the following decades additional marine mammals were added as information became available on their population status. As of November 2011, there were 16 species of marine mammals listed as endangered or threatened in the United States. (See Table 3.1.) In addition, there were 15 foreign species listed as endangered or threatened. (See Table 3.2.)

During fiscal year 2010, $90.4 million was spent by federal and state agencies on the 10 marine mammal species with the highest spending that year. (See Table 3.3.)

As of November 2011, ESA-listed endangered and threatened marine mammals fell into six main categories: whales, dolphins and porpoises, seals and sea lions, sea otters, manatees and dugongs, and polar bears.

WHALES

Whales are in the order Cetacea (along with dolphins and porpoises). Cetaceans are marine mammals that live in the water all the time and have torpedo-shaped nearly hairless bodies. (See Figure 3.1.) There are approximately 70 known whale species. The so-called great whales are the largest animals on the earth. In general, the great whale species range in size from 30 to 100 feet (9.1 to 30.5 m) in length. There are 13 whale species normally considered to be great whales. The blue whale is the largest of these species.

Whales are found throughout the world's oceans; however, many species are concentrated in cold northern waters. Even though they are warm-blooded and do not have fur, whales can survive in cold waters because they have a thick layer of dense fat and tissue known as blubber lying just beneath the skin. This blubber layer can be up to 1 foot (30.5 cm) thick in larger species.

Most whales have teeth and are in the suborder Odontoceti. By contrast, the handful of whales in the suborder Mysticeti filter their food through strong flexible plates called baleen. (See Figure 3.2.) Baleen is informally known as "whalebone." It is composed of a substance similar to human fingernails. Baleen whales

TABLE 3.1

Endangered and threatened aquatic mammals, November 1, 2011

Common name	Scientific name	Note	Listing status*	U.S. or U.S./Foreign listed
Beluga whale (1 DPS)	Delphinapterus leucas	Toothed	E	US/Foreign
Blue whale	Balaenoptera musculus	Baleen plate	E	US/Foreign
Bowhead whale	Balaena mysticetus	Baleen plate	E	US/Foreign
Caribbean monk seal	Monachus tropicalis	Earless, presumed extinct	E	US/Foreign
Finback whale	Balaenoptera physalus	Baleen plate	E	US/Foreign
Guadalupe fur seal	Arctocephalus townsendi	Eared	T	US/Foreign
Hawaiian monk seal	Monachus schauinslandi	Earless	E	US
Humpback whale	Megaptera novaeangliae	Baleen plate	E	US/Foreign
Killer whale	Orcinus orca	Toothed	E	US/Foreign
Northern sea otter	Enhydra lutris kenyoni	North Pacific Ocean stock	T	US
Polar bear	Ursus maritimus	Sea ice is primary habitat	T	US/Foreign
Right whale	Balaena glacialis (incl. australis)	Baleen plate	E	US/Foreign
Sei whale	Balaenoptera borealis	Baleen plate	E	US/Foreign
Southern sea otter	Enhydra lutris nereis	California stock	T; XN	US/Foreign
Sperm whale	Physeter catodon (=macrocephalus)	Toothed	E	US/Foreign
Steller sea-lion	Eumetopias jubatus	Eared	E; T	US/Foreign
West Indian manatee	Trichechus manatus	Florida stock	E	US/Foreign

DPS = distinct population segments.
*E = endangered; T = threatened; XN = experimental population, non-essential.

SOURCE: Adapted from "Listed U.S. Species by Taxonomic Group: Vertebrate Animals," in *USFWS Threatened and Endangered Species System (TESS)*, U.S. Department of the Interior, U.S. Fish and Wildlife Service, November 2011, http://ecos.fws.gov/tess_public/ (accessed November 1, 2011)

TABLE 3.2

Foreign endangered and threatened aquatic mammals, November 1, 2011

Common name	Scientific name	Listing status	Historic range
Amazonian manatee	Trichechus inunguis	Endangered	South America (Amazon River basin)
Cameroon clawless otter	Aonyx congicus (=congica) microdon	Endangered	Cameroon, Nigeria
Chinese River dolphin	Lipotes vexillifer	Endangered	China
Cochito	Phocoena sinus	Endangered	Mexico (Gulf of California)
Dugong	Dugong dugon	Endangered	East Africa to southern Japan, including U.S.A. (Trust Territories)
Giant otter	Pteronura brasiliensis	Endangered	South America
Gray whale	Eschrichtius robustus	Endangered	North Pacific Ocean—coastal and Bering Sea, formerly North Atlantic Ocean
Indus River dolphin	Platanista minor	Endangered	Pakistan (Indus River and tributaries)
Long-tailed otter	Lontra (=Lutra) longicaudis (incl. platensis)	Endangered	South America
Marine otter	Lontra (=Lutra) felina	Endangered	Peru south to Straits of Magellan
Mediterranean monk seal	Monachus monachus	Endangered	Mediterranean, Northwest African Coast and Black Sea
Saimaa seal	Phoca hispida saimensis	Endangered	Finland (Lake Saimaa)
Southern river otter	Lontra (=Lutra) provocax	Endangered	Chile, Argentina
Spotted seal	Phoca largha	Threatened	North Pacific Ocean
West African manatee	Trichechus senegalensis	Threatened	West Coast of Africa from Senegal River to Cuanza River

SOURCE: Adapted from "Foreign Species," in *USFWS Threatened and Endangered Species System (TESS)*, U.S. Department of the Interior, U.S. Fish and Wildlife Service, November 2011, http://ecos.fws.gov/tess_public/ (accessed November 1, 2011), and "Distribution," in *Spotted Seal (Phoca largha)*, U.S. Department of Commerce, National Oceanic and Atmospheric Administration, National Marine Fisheries Service, undated, http://www.nmfs.noaa.gov/pr/species/mammals/pinnipeds/spottedseal.htm (accessed November 1, 2011)

TABLE 3.3

The ten listed marine mammal entities with the highest expenditures under the Endangered Species Act, fiscal year 2010

Ranking	Species	Expenditure
1	Sea-lion, Steller (Eumetopias jubatus)—western pop.	$16,679,395
2	Sea-lion, Steller (Eumetopias jubatus)—eastern pop.	$16,568,504
3	Whale, right (Eubalaena glacialis (incl. australis))	$13,596,253
4	Manatee, West Indian (Trichechus manatus)	$10,353,294
5	Whale, humpback (Megaptera novaeangliae)	$7,921,457
6	Bear, polar (Ursus maritimus)	$6,804,746
7	Seal, Hawaiian monk (Monachus schauinslandi)	$5,650,652
8	Whale, sperm (Physeter catodon (=macrocephalus))	$4,432,830
9	Whale, blue (Balaenoptera musculus)	$4,198,058
10	Whale, Sei (Balaenoptera borealis)	$4,147,048

SOURCE: Adapted from "Table 2. Species Ranked in Descending Order of Total FY 2010 Reported Expenditures, Not Including Land Acquisition Costs," in *Federal and State Endangered and Threatened Species Expenditures: Fiscal Year 2010*, U.S. Department of the Interior, U.S. Fish and Wildlife Service, 2010, http://www.fws.gov/endangered/esa-library/pdf/2010.EXP.FINAL.pdf (accessed October 25, 2011)

strain large amounts of water to obtain their food, mostly zooplankton, tiny fish, and crustaceans. Nearly all the great whales are baleen whales.

Imperiled Whale Populations

As of November 2011, nine whale species had been listed for protection under the ESA in U.S. waters: beluga whales, blue whales, bowhead whales, finback whales, humpback whales, killer whales, right whales, sei whales, and sperm whales. (See Table 3.1.) All but the beluga, killer, and sperm whales have baleen plates.

The right whale is the most endangered of the great whales. It was once the "right" whale to hunt because it swims slowly, prefers shallow coastal waters, and floats when it dies. According to the NMFS, in "North Atlantic Right Whales (*Eubalaena glacialis*)" (November 18, 2011, http://www.nmfs.noaa.gov/pr/species/mammals/cetaceans/rightwhale_northatlantic.htm), right whales first received international protection from whaling during the 1930s. In 1973 the right whale was listed as endangered under the ESA. At that time two populations were recognized: northern right whales in the Northern Hemisphere and southern right whales in the Southern Hemisphere. Since that time a scientific consensus has developed that the northern right whale actually consists of two distinct species: the North Atlantic right whale (*E. glacialis*) and the North Pacific right whale (*E. japonica*). In 2008 the NMFS revised the listing for northern right whales to list these populations as two distinct endangered species under the ESA. The NMFS reports that the North Atlantic right whale can be further subdivided into two geographical populations: western and eastern. The latter is believed to be "nearly extinct." The NMFS estimates its population in the "low tens of animals." The western North Atlantic right whale population includes an estimated 300 to 400 individuals and is considered "critically" endangered.

In the annual *Stock Assessment Report*, the NMFS provides population estimates for endangered whale species in U.S. waters. Surveys of all species are not conducted every year. As of January 2012, the final 2010 reports and the draft 2011 reports were available for all three regions: the Pacific coast, the Atlantic and Gulf coasts, and the Alaskan coast.

Threats to Whales

Whale populations are imperiled due to a long history of hunting by humans. As early as the eighth century

FIGURE 3.1

Humpback whales. (© *Jan Kratochvila/Shutterstock.com.*)

FIGURE 3.2

Baleen plates

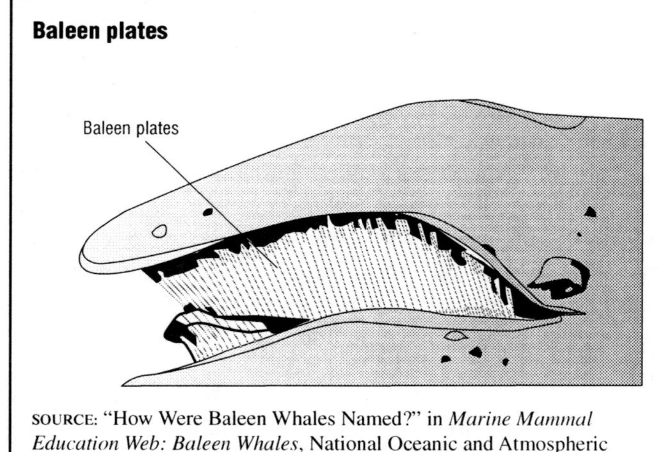

SOURCE: "How Were Baleen Whales Named?" in *Marine Mammal Education Web: Baleen Whales*, National Oceanic and Atmospheric Administration, Alaska Fisheries Science Center, National Marine Mammal Laboratory, undated, http://www.afsc.noaa.gov/nmml/education/cetaceans/baleen1.php (accessed November 2, 2011)

humans hunted whales for meat and baleen. Whales were relatively easy for fishermen to catch because the animals spend a great deal of time at the surface of the ocean and provide a large target for harpoons. Advances in shipbuilding and the invention of the steam engine allowed fishermen greater access to whale populations, even those in Arctic areas that had previously been out of reach. By the 19th century large numbers of whales were being killed for blubber and baleen. Blubber was rendered to extract whale oil, which was used to light lamps. Baleen was valued for making fans, corsets, and other consumer goods.

On December 2, 1946, the representatives of 14 nations signed the International Convention for the Regulation of Whaling (http://www.iwcoffice.org/_documents/commission/convention.pdf), which formed the International Whaling Commission (IWC). The signatory nations were Argentina, Australia, Brazil, Canada, Chile, Denmark, France, the Netherlands, New Zealand, Peru, South Africa, the Soviet Union, the United Kingdom, and the United States. The IWC was formed as a means to regulate the industry and limit the number and type of whales that could be killed. The MMPA of 1972 banned commercial whaling in U.S. waters.

Centuries of whaling severely depleted whale populations. Low birth rates and high mortality rates due to a variety of factors have prevented many species from recovering. Like other marine animals, whales are endangered by water pollution and loss or degradation of habitat. However, the biggest threats to the right whale and other whale species are believed to be entanglement in fishing gear and ship strikes. Table 3.4 lists the number of reports of whale entanglements and ship strikes between 2005 and 2009 along the northern Gulf of Mexico coast, the U.S. East Coast, and the adjacent Canadian coasts. Nearly 500 reports are recorded. A total of 308 whales were confirmed killed, including 18 by entanglement and 28 by ship strikes. Another 16 whales were killed by events that were not entanglements or ship strikes. There was insufficient information to determine the cause of death for the other 246 whales.

ENTANGLEMENT IN FISHING GEAR. The NMFS explains in "What Kinds of Fishing Gear Most Often Entangle Right Whales?" (*Right Whale News*, November 2005) that the entanglement of whales in fishing gear is a major problem. According to studies conducted in 2003 by the New England Aquarium, 71.9% of all known northern right whales have been entangled at least once in fishing gear. In 2005 researchers from Duke University investigated 31 cases of right whale entanglements and tried to identify the type of fishing gear involved in each case. They found that nearly a third of the whales had become entangled in lobster pot gear, mostly buoy lines. It was concluded "that any line rising vertically in the water column poses a significant entanglement risk" to northern right whales.

SHIP STRIKES. NOAA reports in the press release "NOAA Fisheries: North Atlantic Right Whales and Ship Strikes off the U.S. East Coast" (June 6, 2004, http://nefsc.noaa.gov/press_release/2004/advisory04.02.pdf) that between 1975 and 2002 there were 292 ship strikes on large whales off the U.S. East Coast. In *Mortality and Serious Injury Determinations for Baleen Whale Stocks along the United States Eastern Seaboard and Adjacent Canadian Maritimes, 2003–2007* (May 2009, http://www.nefsc.noaa.gov/publications/crd/crd0904/crd0904.pdf), NOAA's Northeast Fisheries Science Center (NFSC) documents 12 ship strikes on large whales between 2003 and 2004. Likewise, in *Mortality and Serious Injury Determinations for Baleen Whale Stocks along the Gulf of Mexico, United States, and Canadian Eastern Seaboards, 2005–2009* (October 2011, http://www.nefsc.noaa.gov/nefsc/publications/crd/crd1118/crd1118.pdf) the NFSC reports another 63 confirmed ship strikes between 2005 and 2009. Strikes on northern right whales are particularly troublesome because so few of the animals remain in existence. The NMFS states in "Ship Strike Reduction" (July 27, 2010, http://www.nero.noaa.gov/shipstrike/) that "collision with vessels is the leading human-caused source of mortality for the endangered North Atlantic right whale."

The Right Whale Sighting Advisory System (RWSAS) is a notification system operated by NOAA to reduce collisions between ships and right whales. Whale sightings and other detects are reported to the RWSAS and alerts are passed on to mariners in the area. Table 3.5 shows that 571 right whale sightings and detects were reported to the RWSAS in 2010. The largest number of sightings (348) was associated with aerial whale search surveys. According to Christin Khan et al. of the NMFS, in *North Atlantic Right Whale Sighting Survey (NARWSS) and Right Whale Sighting Advisory System (RWSAS) 2010 Results Summary* (March 2011, http://www.nefsc.noaa.gov/publications/crd/crd1105/1105.pdf), most detects of right whales in U.S. waters in 2010 occurred along the northeastern coast from Maine to New York. During calving season (from November through April) the NMFS performs aerial surveys and alerts ships about whales in their vicinity. In addition, federal law requires that ships remain 500 yards (457 m) from right whales. Any sightings of dead, injured, or entangled whales must be reported to authorities.

In 2004 the NMFS announced plans to propose rules requiring routing changes and speed limits for large vessels traveling in U.S. coastal waters that were frequented by northern right whales. In 2006 the NMFS published "Endangered Fish and Wildlife; Proposed Rule to Implement Speed Restrictions to Reduce the Threat of Ship Collisions with North Atlantic Right Whales" (*Federal Register*, vol. 71, no. 122, June 26, 2006). In December 2008 the rule became final and was effective through December 2013.

The NMFS and the U.S. Coast Guard have also modified shipping lanes off the coast of Boston, Massachusetts, to reduce the threat of ship collisions with whales in the area. Figure 3.3 shows areas known as right whale seasonal management areas (SMAs) that were in effect in November 2011. Speed restrictions are mandatory in SMAs and voluntary in other designated areas called dynamic management areas (DMAs). Ships are also asked to route around DMAs whenever possible.

Whale Recovery Plans

Table 3.6 lists information about population trends, recovery priority numbers, critical habitat designations, and the recovery plan status for the whale species listed as endangered under the ESA in U.S. waters. As of November 2011, critical habitat had been designated only for the beluga whale (Cook Inlet distinct population segment [DPS]), the killer whale (southern resident DPS), and the northern right whale. Draft or final recovery plans had been developed or were under development for all species except the bowhead whale and the North Pacific right whale. The populations of bowhead, humpback, and North Atlantic right whales were believed to be

TABLE 3.4

Summary of reported incidents involving whales along the northern Gulf of Mexico Coast, U.S. East Coast, and adjacent Canadian Maritimes, 2005–09

Species	Western North Atlantic right whale (Eubalaena glacialis)	Gulf of Maine humpback whale (Megaptera novaeangliae)	Western North Atlantic fin whale (Balaenoptera physalus)	Nova Scotian sei whale (B. borealis)	Western North Atlantic blue whale (B. musculus)	Canadian East Coast minke whale (B. acutorostrata)	Northern Gulf of Mexico Bryde's whale (B. edeni)	Unidentified fin/sei whale	Unidentified balaenopterid[c]	Unidentified whale spp.
Total events[a, b] (2005, 2006, 2007, 2008, 2009)	60 (14, 12, 6, 13, 15)	202 (30, 48, 35, 47, 42)	46 (10, 9, 13, 6, 8)	12 (0, 5, 1, 4, 2)	0	110 (19, 25, 27, 24, 15)	2 (0, 1, 0, 0, 1)	6 (3, 2, 1, 0, 0)	8 (4, 1, 2, 0, 1)	50 (10, 13, 8, 11, 8)
Total confirmed mortalities	20 (4, 6, 3, 3, 4)	115 (13, 33, 21, 27, 21)	34 (9, 7, 8, 5, 5)	10 (0, 4, 1, 3, 2)	0	8 (18, 18, 20, 17, 9)	2 (0, 1, 0, 0, 1)	4 (2, 1, 1, 0, 0)	8 (4, 1, 2, 0, 1)	33 (4, 10, 5, 8, 6)
Confirmed entanglement mortalities	2 (0, 1, 1, 0, 0)	6 (0, 1, 1, 2, 2)	2 (0, 0, 2, 0, 0)	1 (0, 0, 0, 1, 0)	0	7 (1, 1, 1, 4, 0)	0	0	0	0
Confirmed ship strike mortalities	6 (2, 4, 0, 0, 0)	7 (0, 3, 3, 1, 0)	9 (5, 0, 2, 1, 1)	3 (0, 1, 1, 0, 1)	0	2 (1, 0, 0, 0, 1)	1 (0, 0, 0, 0, 1)	0	0	0
Confirmed mortalities, NOT ship strike or entanglement	5 (0, 0, 1, 3, 1)	2 (1, 1, 0, 0, 0)	3 (0, 0, 0, 2, 1)	0	0	5 (2, 1, 1, 0, 1)	1 (0, 1, 0, 0, 0)	0	0	0
Confirmed mortalities, IITD[d]	7 (2, 1, 1, 0, 3)	100 (12, 28, 17,24, 19)	20 (4, 7, 4, 2, 1)	6 (0, 3, 0, 2, 1)	0	68 (14, 16, 18, 13, 7)	0	4 (2, 1, 1, 0, 0)	8 (4, 1, 2, 0, 1)	33 (4, 10, 5, 8, 6)

[a]Includes all types of events : entanglements, ship strikes, natural causes, and unconfirmed origin or fate.
[b]Excludes resights of previously reported individuals unless a new injury was documented.
[c]Described as having throat grooves (rorqual pleats).
[d]IITD = insufficient information to determine cause of death or if the injury was serious and likely lethal.

SOURCE: Allison G. Henry et al., "Table 1. Summary of All Unique Whale Events and Mortalities along the Gulf of Mexico Coast, U.S. East Coast and Adjacent Canadian Maritimes, 2005–2009," in *Mortality and Serious Injury Determinations for Baleen Whale Stocks along the Gulf of Mexico, United States and Canadian Eastern Seaboards, 2005–2009*, U.S. Department of Commerce, National Oceanic and Atmospheric Administration, Northeast Fisheries Science Center, October 2011, http://www.nefsc.noaa.gov/nefsc/publications/crd/crd1118/crd1118.pdf (accessed November 2, 2011)

TABLE 3.5

Number and breakdown by source of right whale sightings in the Northeast reported to the Right Whale Sighting Advisory System, 2010

	Jan	Feb	Mar	Apr	May	Jun	Jul	Aug	Sep	Oct	Nov	Dec	Total #	%
Aerial	16	16	55	161	54	1		3	2	6	20	14	348	61%
Shipboard		1	3	31	11			4	6	19	11	2	88	15%
Whale watch				7		2	7	12	15	9			52	9%
Coast Guard			1	5	6	5			2				19	3%
Commercial/fishing	1	1			6		1	3					12	2%
Opportunistic	3	3	8	20	2	5	3	2	1	3	2		52	9%
Total #	**20**	**21**	**67**	**224**	**79**	**13**	**11**	**24**	**26**	**37**	**33**	**16**	**571**	**100%**

SOURCE: Adapted from Christin Khan et al., "Table 3. The Total Number of Right Whale Sighting Reports to the RWSAS in 2010 by Reporting Source and Month within the (A) Northeast Region—Maine through New York, (B) Mid-Atlantic Region—New Jersey through North Carolina, and (C) Canadian Waters," in *North Atlantic Right Whale Sighting Survey (NARWSS) and Right Whale Sighting Advisory System (RWSAS) 2010 Results Summary*, U.S. Department of Commerce, National Oceanic and Atmospheric Administration, Northeast Fisheries Science Center, March 2011, http://www.nefsc.noaa.gov/publications/crd/crd1105/1105.pdf (accessed November 3, 2011)

increasing, whereas the populations of the beluga whale (Cook Inlet DPS) and the killer whale (southern resident DPS) were believed to be decreasing. Information was inadequate to determine population trends for the other whale species.

Table 3.6 also shows the recovery priority numbers assigned by the NMFS to each endangered whale species. Priority numbers can range from a value of 1 (highest priority) to 12 (lowest priority). The North Atlantic right whale has a priority level of 1, indicating strong concern about its abundance and chances for survival as a species.

In *Recovery Plan for the North Atlantic Right Whale* (August 2004, http://ecos.fws.gov/docs/recovery_plan/whale_right_northatlantic.pdf), the NMFS lists five goals for recovering the species. In order of importance, the goals are:

- Significantly reduce sources of human-caused death, injury and disturbance
- Develop demographically-based recovery criteria
- Identify, characterize, protect and monitor important habitats
- Monitor the status and trends of abundance and distribution of the western North Atlantic right whale population
- Coordinate federal, state, local, international and private efforts to implement the recovery plan

Imperiled Whales around the World

The International Union for Conservation of Nature (IUCN) indicates in *Red List of Threatened Species Version 2011.2* (2011, http://www.iucnredlist.org/) that in 2011 the following whale species were threatened:

- North Atlantic right whale
- North Pacific right whale
- Beluga whale
- Narwhal
- Sperm whale
- Blue whale
- Fin whale
- Sei whale

The IUCN notes that it lacked sufficient data to determine the status of more than two dozen other whale species.

As shown in Table 3.2, there were no completely foreign whale species listed under the ESA as of November 2011. However, all the species deemed threatened by the IUCN (except the narwhal) were listed under the ESA. (See Table 3.1.) All the species are found in both U.S. and foreign waters.

INTERNATIONAL WHALING CONTROVERSIES. As noted earlier, the IWC was formed in 1946 to regulate the commercial whaling industry and limit the number and type of whales that can be killed. In 1986 the IWC banned commercial whaling after most whale populations were placed under Appendix I of the Convention on International Trade in Endangered Species of Wild Fauna and Flora (CITES) agreement. In 1991 Iceland dropped out of the IWC over the commercial whaling ban. It rejoined in 2002, but as of January 2012 it refused to abide by the ban. Norway also does not adhere to the ban. Commercial whalers and some scientists argue that some whale species are not imperiled and thus can be hunted, assuming that reasonable catch limits are employed.

The IWC (http://www.iwcoffice.org/commission/members.htm) indicates that as of January 2012 it had 89 member nations. The commission allows whaling for "subsistence" purposes by native peoples, for example, Alaskan Eskimos. It also allows whaling for "scientific purposes." Conservation and wildlife groups have complained for decades that some IWC member countries, particularly Japan, kill many whales under this loophole and sell the meat commercially. In addition, Iceland and

FIGURE 3.3

Management areas for right whales off New England Coast, November 2011

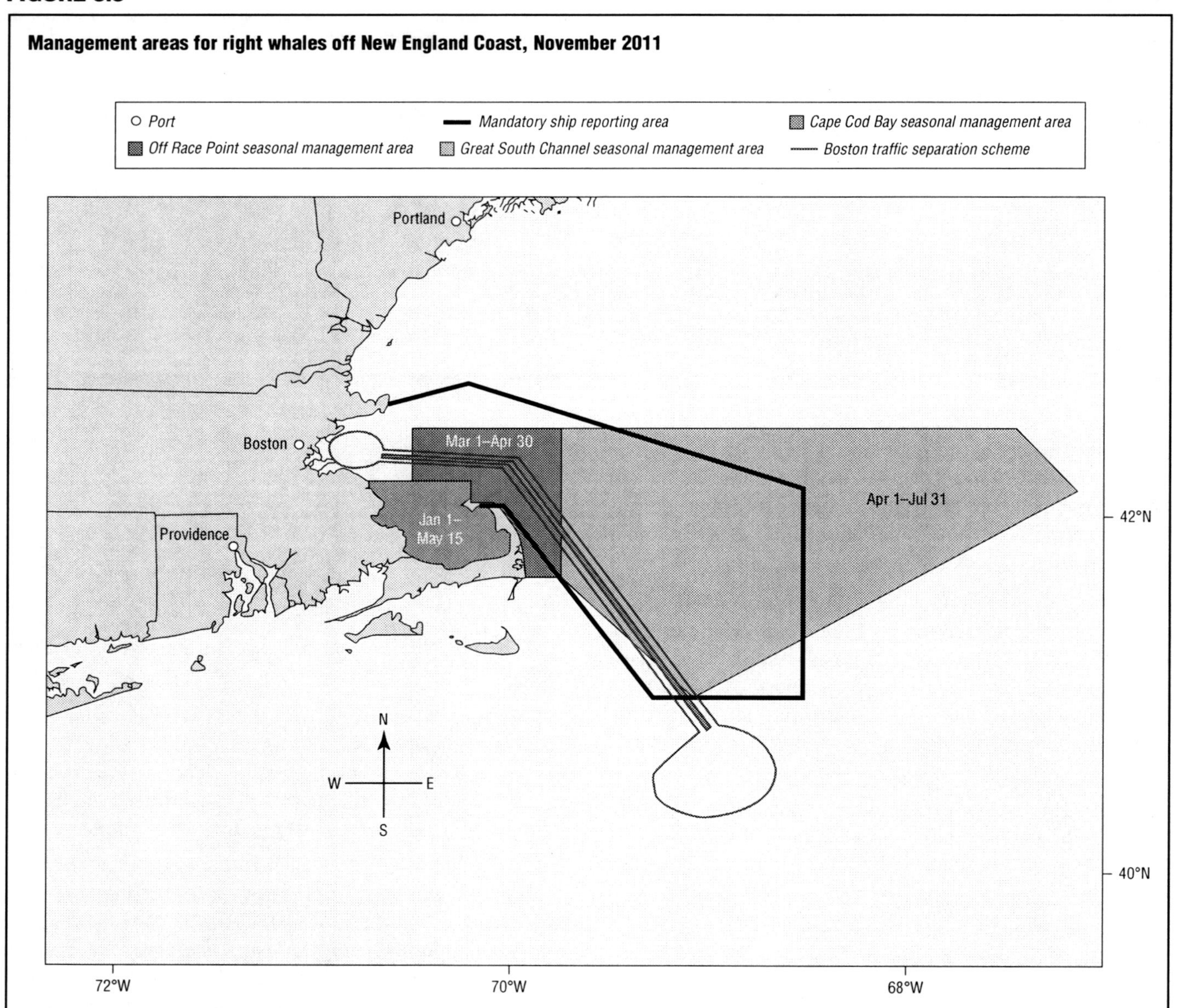

SOURCE: "Northeast U.S. Seasonal Management Areas," in *Compliance Guide for Right Whale Ship Strike Reduction Rule (50 CFR 224.105)*, U.S. Department of Commerce, National Oceanic and Atmospheric Administration, National Marine Fisheries Service, undated, http://www.nmfs.noaa.gov/pr/pdfs/shipstrike/compliance_guide.pdf (accessed November 3, 2011)

Norway openly hunt whales for commercial purposes and sell the meat primarily in Japanese markets.

Whaling vessels, particularly those from Japan, are often aggressively confronted at sea by conservation and wildlife groups that are staunchly opposed to whale hunting for any reason. In 2010 the Japanese whaling fleet was forced to cut its season short because of clashes with boats that were sponsored by the Sea Shepherd Conservation Society, a nonprofit organization based in the United States. Also in 2010 the anti-whaling government of Australia initiated legal action against Japan in the International Court of Justice for continuing to conduct whaling. As of January 2012, no decision in that case had been reached.

Together, the three whaling countries of Japan, Norway, and Iceland typically kill between 1,300 and 1,600 whales per year, with Japan accounting for the majority of the total. According to the IWC, in "Catches under Objection since 1985" (January 26, 2010, http://www.iwcoffice.org/conservation/table_objection.htm), commercial capture during the 2008–09 season was 536 whales by Norway and 38 whales by Iceland. All of the whales were minke whales. Likewise, the IWC notes in "Special Permit Catches since 1985" (January 26, 2010, http://www.iwcoffice.org/conservation/table_permit.htm) that Japan killed 1,004 whales during the 2008–09 season under scientific permit. Most (851) of the whales were minke whales. Other captured species included sei (100),

TABLE 3.6

Endangered U.S. whale species, 2011

Species (DPS)[a]	Date listed or reclassified	Endangered Species Act status	Population trend	Recovery priority number[b]	Status of recovery plan	Critical habitat
Beluga whale (Cook Inlet DPS)	10/22/2008	Endangered	Declining	2	Under development	Final
Blue whale	6/2/1970	Endangered	Unknown	5	Completed 07/1998	None
Bowhead whale	6/2/1970	Endangered	Increasing	7	None	None
Fin whale	6/2/1970	Endangered	Unknown	9	Completed 07/2010	None
Humpback whale	6/2/1970	Endangered	Increasing	5	Completed 11/1991	None
Killer whale (southern resident DPS)	11/18/2005	Endangered	Declining	3	Completed 01/2008	Final
North Atlantic right whale	6/2/1970; 3/6/2008	Endangered	Increasing	1	Completed 05/2005	Final
North Pacific right whale	6/2/1970; 3/6/2008	Endangered	Unknown	4	None	Final
Sei whale	6/2/1970	Endangered	Unknown	11	Draft completed 08/2011	None
Sperm whale	6/2/1970	Endangered	Unknown	5	Completed 12/2010	None

[a]DPS = distinct population segment.
[b]Recovery priority numbers are designated according to guidelines published by the National Marine Fisheries Service on June 15, 1990 (55 FR 24296).
Priorities are designated from 1 (high) to 12 (low) based on the following factors: degree of threat, recovery potential, and conflict with development projects or other economic activity.

SOURCE: Adapted from "Table 1. ESA-Listed Species under NMFS Jurisdiction Including Listing Status, Trends, Priority Numbers, and Recovery Plan Status," in *Biennial Report to Congress on the Recovery Program for Threatened and Endangered Species October 1, 2008–September 30, 2010*, U.S. Department of Commerce, National Oceanic and Atmospheric Administration, National Marine Fisheries Service, 2011, http://www.nmfs.noaa.gov/pr/pdfs/laws/esabiennial2010.pdf (accessed November 3, 2011), and "List of Mammal Species under NMFS' Jurisdiction," in *Marine Mammal Species under the Endangered Species Act (ESA)*, U.S. Department of Commerce, National Oceanic and Atmospheric Administration, National Marine Fisheries Service, Office of Protected Resources, undated, http://www.nmfs.noaa.gov/pr/species/esa/mammals.htm (accessed November 3, 2011)

Bryde's (50), sperm (2), and fin (1) whales. Sei, sperm, and fin whales are designated as threatened by the IUCN and are listed as endangered or threatened under the ESA.

According to the IWC, in "Aboriginal Subsistence Whaling Catches since 1985" (March 7, 2011, http://www.iwcoffice.org/conservation/table_aboriginal.htm), 336 whales were killed for "aboriginal subsistence" purposes in 2009, including 38 bowhead whales killed by Alaskan Eskimos. Other captured species included minke (168), gray (116), bowhead (41), fin (10), and humpback (1) whales. Excluding the fin whale, none of these species is considered threatened by the IUCN. All but the minke and gray whales are listed as endangered or threatened under the ESA.

DOLPHINS AND PORPOISES

Dolphins and porpoises are toothed cetaceans. They are similar in shape; however, dolphins are generally larger than porpoises and prefer shallower, warmer waters. Dolphins tend to have long bottlenoses and cone-shaped teeth, whereas porpoises have flatter noses and teeth. Dolphins are members of the Delphinidae family, a large family containing at least 30 known species. Porpoises are members of the Phocoenidae family, which includes only six existing species. As of November 2011, there were no U.S. species of dolphins or porpoises listed under the ESA.

Imperiled Foreign Dolphins and Porpoises

Most dolphin and porpoise populations around the world are hardy and not in danger of extinction. However, there are several species that are in trouble due to limited geographical distribution. According to the IUCN, in *Red List of Threatened Species Version 2011.2*, nine dolphin species and three porpoise species were considered threatened in 2011. The most imperiled (i.e., those with a critically endangered rating) were the baiji (also known as the Chinese River dolphin) and the cochito (or vaquita).

As of November 2011, there were three foreign dolphin and porpoise species listed under the ESA: the Chinese River dolphin, the Indus River dolphin, and the cochito. (See Table 3.2.) All were listed as endangered. The Chinese River and Indus River dolphins live in freshwater rivers in China and Pakistan, respectively. Their numbers are considered to be extremely small. In both cases extensive river damming, water drawdown due to human consumption, fishing, and pollution are blamed for the declines.

The cochito is a kind of porpoise found only in the Gulf of California, a narrow body of water that separates the western Mexican mainland from the Baja California peninsula. This stretch of water is known in the United States as the Sea of Cortez and contains a great diversity of sea life. Cochitos are among the rarest of all marine mammals. According to the NMFS, the cochito species has been nearly eliminated because so many of the animals have become entangled in fishing lines and drowned.

Protection of Prevalent Dolphins

Even though they are not considered endangered or threatened, dolphins receive special consideration under U.S. law because of public concern about them. Dolphins are believed to be highly intelligent. In addition, many people have been exposed to dolphins through marine entertainment parks, movies, television shows, and even personal encounters and sightings at tourist beaches. As a result, there is widespread public fondness for the animals.

Dolphins are protected by the MMPA and by laws that are designed to limit their capture during tuna fishing. In 1990 large U.S. tuna canning companies announced they would no longer purchase tuna caught in a manner that endangered dolphins. The companies began labeling their products "Dolphin Safe" if their practices met specific standards established by the U.S. government. The International Dolphin Conservation Program Act, passed in 1992, reduced the number of legally permitted dolphin deaths. This act also made the United States a dolphin-safe zone in 1994, when it became illegal to sell, buy, or ship tuna products obtained using methods that kill dolphins.

SEALS AND SEA LIONS

Seals and sea lions are considered pinnipeds. This designation comes from the Latin word *pinnipedia*, which means "feather or fin foot." Pinnipeds have finlike flippers. Even though they spend most of their time in the ocean, pinnipeds come on shore to rest, breed, give birth, and nurse their young. Areas preferred for breeding, birthing, and nursing are called rookeries. Pinnipeds not yet of reproductive age congregate at shore areas known as haul-outs.

Seals and sea lions were hunted extensively during the 1800s and early 1900s for their blubber, fur, and meat. They continue to be imperiled by human encroachment of haul-out beaches, entanglement in marine debris and fishing nets, incidental catches, disease, and lack of food due to competition from humans for prey species.

Imperiled Seal and Sea Lion Populations

As of November 2011, there were three U.S. species of seals and sea lions listed under the ESA: the Guadalupe fur seal (also found in Mexico), the Hawaiian monk seal, and the Steller sea lion. (See Table 3.1.) Another species, the Caribbean monk seal, was formerly listed, but is now believed to be extinct.

GUADALUPE FUR SEALS. The Guadalupe fur seal breeds along the eastern coast of Isla de Guadalupe, Mexico. The island is approximately 400 miles (644 km) west of Baja California. The Seal Conservation Society (SCS) notes in "Guadalupe Fur Seal (*Arctocephalus townsendi*)" (2011, http://www.pinnipeds.org/seal-information/species-information-pages/sea-lions-and-fur-seals/guadalupe-fur-seal) that even though populations once included as many as 20,000 to 100,000 individuals, decline and endangerment resulted from extensive fur hunting during the 1700s and 1800s. The species was believed to be extinct by the early 20th century, but a small population was discovered in 1954. According to the NMFS, in "Guadalupe Fur Seal (*Arctocephalus townsendi*)" (January 2012, http://www.nmfs.noaa.gov/pr/species/mammals/pinnipeds/guadalupefurseal.htm), the species "is slowly recovering from the brink of extinction." Scientists believe the population numbers about 10,000 individuals and is increasing each year.

HAWAIIAN MONK SEALS. In "Mixed News for the Hawaiian Monk Seal" (*Endangered Species Bulletin*, vol. 34, no. 2, Summer 2009), T. David Schofield of the NMFS calls the Hawaiian monk seal a "living fossil." The species originated 14 million to 15 million years ago and is endemic (limited) to Hawaii. Schofield estimates the seal population at around 1,100 individuals. Ninety percent of them are found in the northwestern Hawaiian Islands (NWHI), a string of remote and nearly uninhabited small islands and atolls to the northwest of the main Hawaiian Islands. The NWHI population has decreased from 1,300 to 1,400 individuals during the late 1990s to less than 1,000 individuals in 2007. The decline is blamed on a number of factors, including entanglement in fishing gear and other debris, low rates of survival for young females, predation on the seals by sharks, loss of beach habitat due to increasing sea levels, and loss of food fish and other prey.

Approximately 100 Hawaiian monk seals inhabit the main Hawaiian Islands, and this population appears to be thriving. Schofield reports that annual birth numbers are increasing and that the pups appear larger and healthier than those in the NWHI area. However, seals in the main Hawaiian Islands are increasingly imperiled by human interaction. Three of the animals were moved to other areas after they became too used to humans. One of the seals was "adopted" by a local community and became so "friendly" that people were swimming with it, feeding it, and petting it. When the seal became a nuisance by jumping aboard kayaks and surfboards, it was moved to another island. The NMFS explains in "Hawaiian Monk Seal (*Monachus schauinslandi*)" (http://www.nmfs.noaa.gov/pr/species/mammals/pinnipeds/hawaiianmonkseal.htm) that as of January 2012 it was conducting outreach programs in Hawaii to educate people about the problems that are associated with human interaction with the seals.

STELLER SEA LIONS. Steller sea lions are large animals, with males reaching a length of about 11 feet (3.4 m) and a weight of 2,500 pounds (1,100 kg). Females are significantly smaller. Steller sea lions are found in Pacific waters from Japan to central California, but most populations breed near Alaska and the Aleutian Islands. The breeding season is from May through July. The species was named after Georg Wilhelm Steller (1709–1746), a German naturalist who studied the animals when he accompanied the Danish explorer Vitus Jonassen Bering (1681–1741) on an expedition to Alaska in 1741.

The Steller sea lion population is divided into two stocks. (See Figure 3.4.) The eastern stock inhabits the area east of 144 degrees West longitude (near Cape Suckling, Alaska) and extends down the western coast of Canada and the U.S. mainland. The western stock is found west of 144 degrees West longitude and extends across the Aleutian Islands to Russia and Japan.

According to the NMFS, in *Recovery Plan for Steller Sea Lion: Eastern and Western Distinct Population Segments (*Eumetopias jubatus*)* (March 2008, http://www.nmfs.noaa.gov/pr/pdfs/recovery/stellersealion.pdf), the western stock numbered between 220,000 and 265,000 individuals during the mid-1970s. The population declined by 72% between 1976 and 1990. In April 1990 the Steller sea lion was listed under the ESA as threatened. Over the following decade the western stock continued to decline, dropping to 18,325 individuals. This stock was declared endangered in 1997. Between 2000 and 2004 the western population increased by 12% to 20,533 individuals. The eastern stock has been increasing at a rate of approximately 3% per year since the 1970s. This stock remains classified as threatened.

Steller sea lion populations have declined for a variety of reasons including bycatch, legal and illegal hunting, predation, and disease. In addition, scientists believe the animal has experienced reduced productivity due to the indirect effects of climate change and competition from humans for prey species (food fish).

In fiscal year 2010 species-specific expenditures under the ESA for the western and eastern populations of Steller sea lions totaled $33.2 million, accounting for approximately one-third of expenditures on all marine mammals. (See Table 3.3.) The Steller sea lion ranked fifth in spending among all species covered by the ESA. (See Table 2.16 in Chapter 2.)

Recovery Plans for Seals and Sea Lions

Table 3.7 lists information about population trends, recovery priority numbers, critical habitat designations, and the recovery plan status for the Guadalupe fur seal, the Hawaiian monk seal, and the Steller sea lion (eastern and western DPSs). As of December 2011, critical habitat and recovery plans had not been developed for the Guadalupe fur seal. Critical habitats and recovery plans had

FIGURE 3.4

Range and rookeries of Steller sea lions

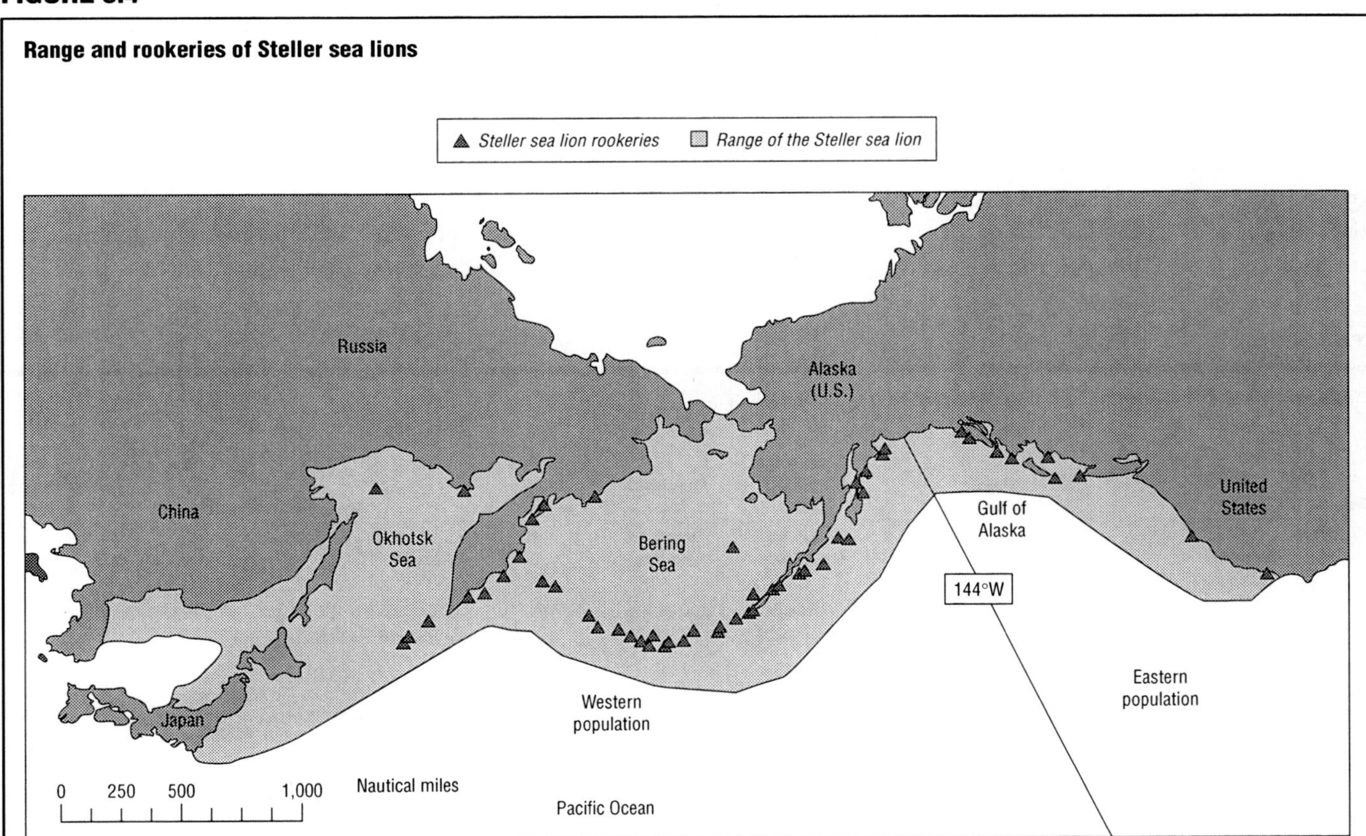

SOURCE: "Figure 13. Steller Sea Lion Range and Rookeries," in *Biennial Report to Congress on the Recovery Program for Threatened and Endangered Species October 1, 2008–September 30, 2010*, U.S. Department of Commerce, National Oceanic and Atmospheric Administration, National Marine Fisheries Service, 2011, http://www.nmfs.noaa.gov/pr/pdfs/laws/esabiennial2010.pdf (accessed November 3, 2011)

been developed for the Hawaiian monk seal and both Steller sea lion DPSs.

Table 3.7 also shows the recovery priority numbers assigned by the NMFS to each endangered seal and sea lion species. Priority numbers can range from a value of 1 (highest priority) to 12 (lowest priority). The Hawaiian monk seal has a priority level of 1, indicating strong concern about its abundance and chances for survival as a species.

Imperiled Foreign Seals and Sea Lions

According to the IUCN, in *Red List of Threatened Species Version 2011.2*, eight seal species and three sea lion species were considered threatened in 2011. The most imperiled (i.e., those with a critically endangered rating) were the Mediterranean monk seal and the Hawaiian monk seal. As of November 2011, three foreign seal and sea lion species were listed under the ESA: the Mediterranean monk seal, the Saimaa seal, and the spotted seal. (See Table 3.2.)

Mediterranean monk seals inhabit remote areas around the Mediterranean Sea and the northwestern African coast. Most are found off the coasts of Mauritania–Western Sahara, Greece, and Turkey. The species has been driven to the edge of extinction by a variety of factors including sickness and overhunting. Saimaa seals are found only in the cold waters of the Saimaa Lake system in eastern Finland. Their numbers were decimated by hunting over the centuries; however, protection measures and fishing restrictions have allowed some measure of recovery. Spotted seals inhabit the cold waters of the North Pacific Ocean from Alaska westward across the coasts of Russia and China. The NMFS explains in "Spotted Seal (*Phoca largha*)" (January 2012, http://www.nmfs.noaa.gov/pr/species/mammals/pinnipeds/spottedseal.htm) that the loss of sea ice is the primary threat facing the species.

SEA OTTERS

Sea otters are the smallest marine mammals in North America. They are furry creatures that grow to be about 4 feet (1.2 m) in length and weigh up to 65 pounds (30 kg). Otters are related to weasels and mink and are members of the Mustelidae family. Sea otters are almost entirely aquatic and inhabit relatively shallow waters along the rocky coasts of the North Pacific Ocean. They eat a wide variety of marine invertebrate. They use rocks and other objects to smash open the hard shells of clams and crabs to get the meat inside.

Even though they inhabit cold waters, sea otters do not have a blubber layer to keep them warm. Instead, they have extremely dense fur coats and high metabolism rates. Their fur coats are waterproof, but only if kept clean. This makes sea otters susceptible to water contaminants, such as oil.

Imperiled Otter Populations

At one time sea otters were populous along the entire U.S. West Coast from Southern California to Alaska. However, their thick and lustrous fur made them a target of intensive hunting for many centuries. By the dawn of the 20th century sea otters were on the brink of extinction. In 1911 they became protected under the International Fur Seal Treaty, and their numbers began to increase.

Biologists recognize two distinct populations. The northern sea otter extends from Russia across the Aleu-

TABLE 3.7

Endangered and threatened U.S. seal and sea lion species, 2011

Species (DPS)[a]	Date listed or reclassified	Endangered Species Act status	Population trend	Recovery priority number[b]	Status of recovery plan	Critical habitat
Guadalupe fur seal	12/16/1985	Threatened	Increasing	10	None	None
Hawaiian monk seal	11/23/1976	Endangered	Declining	1	Completed 03/1983; revision completed 08/2007	Final
Steller sea lion (2 DPSs)						
Eastern	4/5/1990; 11/26/1990; 5/5/1997[c]	Threatened	Increasing	10	Completed 12/1992; revision completed 03/2008	Final
Western	4/5/1990; 11/26/1990; 5/5/1997[c]	Endangered	Mixed	7	Completed 12/1992; revision completed 03/2008	Final

[a]DPS = distinct population segment.
[b]Recovery priority numbers are designated according to guidelines published by the National Marine Fisheries Service (NMFS) on June 15, 1990 (55 FR 24296).
Priorities are designated from 1 (high) to 12 (low) based on the following factors: degree of threat, recovery potential, and conflict with development projects or other economic activity.
[c]This species was first listed as threatened via a 240-day emergency rule on 4/5/1990, then officially listed as threatened in a final rule on 11/26/1990. NMFS separated the species into western and eastern DPSs via final rule on 5/5/1997, which maintained the eastern DPS as threatened and reclassified the western DPS as endangered.

SOURCE: Adapted from "Table 1. ESA-Listed Species under NMFS Jurisdiction Including Listing Status, Trends, Priority Numbers, and Recovery Plan Status," in *Biennial Report to Congress on the Recovery Program for Threatened and Endangered Species October 1, 2008–September 30, 2010*, U.S. Department of Commerce, National Oceanic and Atmospheric Administration, National Marine Fisheries Service, 2011, http://www.nmfs.noaa.gov/pr/pdfs/laws/esabiennial2010.pdf (accessed November 3, 2011), and "List of Mammal Species under NMFS' Jurisdiction," in *Marine Mammal Species under the Endangered Species Act (ESA)*, U.S. Department of Commerce, National Oceanic and Atmospheric Administration, National Marine Fisheries Service, Office of Protected Resources, undated, http://www.nmfs.noaa.gov/pr/species/esa/mammals.htm (accessed November 3, 2011)

tian Islands and the coast of Alaska south to the state of Washington. The southern sea otter is found only off the California coast. As November 2011, both populations were listed as threatened under the ESA. (See Table 3.1.)

NORTHERN SEA OTTERS. Robyn P. Angliss and Bernadette M. Allen of the NMFS Alaska Fisheries Science Center report in *Alaska Marine Mammal Stock Assessments, 2010* (May 2011, http://www.nmfs.noaa.gov/pr/pdfs/sars/ak2010.pdf) that sea otters have "recolonized much of their historic range in Alaska." There are three distinct stocks in this area: southwest Alaska stock, south-central Alaska stock, and southeast Alaska stock. (See Figure 3.5.) According to Angliss and Allen, the southeastern stock contains about 10,000 to 11,000 individuals, and the south-central stock contains about 15,000 to 16,000 individuals. Current population trends of these stocks are "believed to be stable." The southwestern stock includes about 47,000 individuals. Even though this number has increased slightly in recent years, the long-term trend is negative, as this stock has declined by more than 50% since the 1980s.

SOUTHERN SEA OTTERS. Southern (or California) sea otters were designated as a threatened species in 1977. (See Figure 3.6.) At that time the animals inhabited a small stretch of coastline in central California. Scientists feared that this isolated population was in grave danger of being wiped out by a single catastrophe, such as an oil spill. In 1987 the USFWS decided to establish an "experimental population" of sea otters at another location. Over the next few years more than 100 sea otters were moved, a few at a time, to San Nicolas Island. (See Figure 3.7.) It was hoped that these translocated animals would thrive and develop an independent growing colony. However, the venture achieved only limited success. Many of the otters swam back to their original habitat; others died, apparently from the stress of moving. During the early 1990s the transport effort was abandoned.

FIGURE 3.5

Distribution of northern sea otters in Alaska waters, 2008

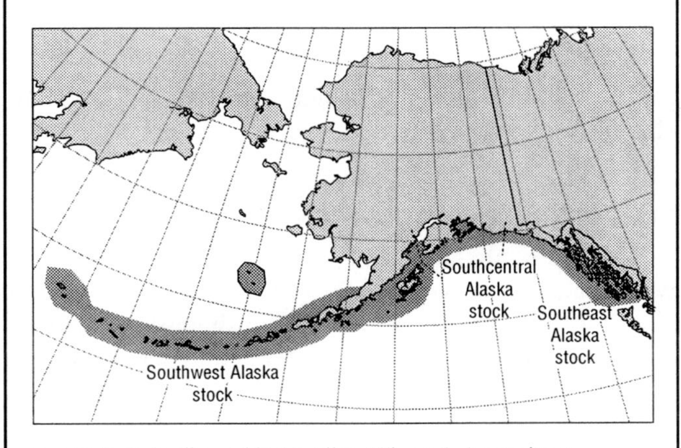

SOURCE: R. P. Angliss and B. M. Allen, "Figure 1. Approximate Distribution of Northern Sea Otters in Alaska Waters (Shaded Area)," in *Alaska Marine Mammal Stock Assessments, 2010*, U.S. Department of Commerce, National Oceanic and Atmospheric Administration, National Marine Fisheries Service, Alaska Fisheries Science Center, May 2011, http://www.nmfs.noaa.gov/pr/pdfs/sars/ak2010.pdf (accessed November 5, 2011)

FIGURE 3.6

Southern sea otter. (© *U.S. Fish and Wildlife Service.*)

In August 2011 the USFWS (http://www.gpo.gov/fdsys/pkg/FR-2011-08-26/pdf/2011-21556.pdf) proposed officially ending the translocation program and removing the designation of "experimental population" for the San Nicolas Island otters. The USFWS noted that as of December 2010 "up to 46 independent southern sea otters have been counted at San Nicolas Island." If the proposal is finalized, these animals would be considered threatened under the ESA, just like their fellow southern sea otters. As of January 2012, a final decision had not been reached on this proposal.

Figure 3.8 shows annual survey results between 1983 and 2010. These surveys were conducted during the springtime and count both independent otters and pups. As indicated in Figure 3.8, the populations have been generally increasing. During the spring of 2010 more than 2,500 otters were counted in the mainland population and San Nicolas Island colony.

The USFWS indicates in *Final Revised Recovery Plan for the Southern Sea Otter (*Enhydra lutis nereis*)* (February 24, 2003, http://ecos.fws.gov/docs/recovery_plans/2003/030403.pdf) that the primary recovery objective is the management of human activities (e.g., oil spills) that can damage or destroy habitat. The USFWS indicates that southern sea otters can be considered for delisting under the ESA when the average population level over a three-year period exceeds 3,090 animals. As shown in Figure 3.8, population trends through 2010 were encouraging. Absent any catastrophic events,

FIGURE 3.7

Location map of San Nicolas Island

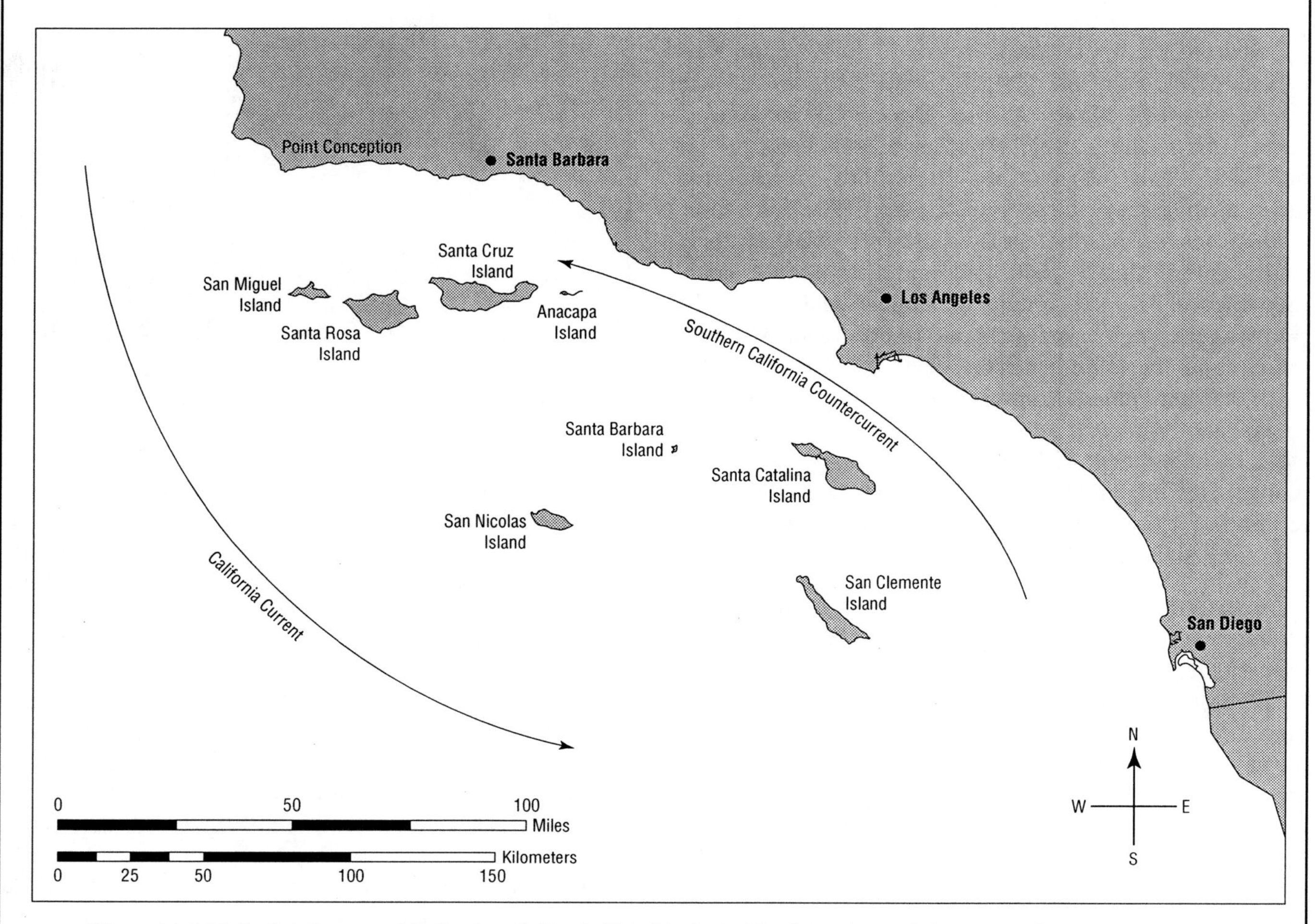

SOURCE: "Figure 4-1. Main Surface Currents of the Southern California Bight," in *Revised Draft Supplemental Environmental Impact Statement on the Translocation of Southern Sea Otters*, U.S. Department of the Interior, U.S. Fish and Wildlife Service, Ventura Fish and Wildlife Office, August 2011, http://www.fws.gov/ventura/species_information/so_sea_otter/rdseis/RDSEIS%20VOL%201.pdf (accessed November 5, 2011)

the southern sea otter could achieve delisting within the next decade.

Foreign Species of Sea Otters

According to the IUCN, in *Red List of Threatened Species Version 2011.2*, two sea otter species—the marine otter (*Lontra felina*) and the sea otter (*Enhydra lutris*)—were considered threatened in 2011.

As of November 2011, five foreign otter species were listed as endangered under the ESA. (See Table 3.2.) They populate areas of Africa and South America. Even though technically only the marine otter (*Lontra* [=*Lutra*] *felina*) could be considered a true sea-dwelling otter, the other species are sometimes found in waters in and near coastal areas and face the same types of threats as sea otters. All species are imperiled by illegal hunting for meat and fur. Loss of habitat and water pollution are also threats to their survival.

MANATEES AND DUGONGS

Manatees are large stout mammals that inhabit freshwaters and coastal waterways. (See Figure 3.9.) They are from the Sirenian order, along with dugongs. There are only five Sirenian species, and all are endangered or extinct. Scientists believe Steller's sea cow, the only species of cold-water manatee, was hunted to extinction during the 1700s.

The West Indian manatee, also known as the Florida manatee, primarily swims in the rivers, bays, and estuaries of Florida and surrounding states. As November 2011, this species was listed as endangered under the ESA. (See Table 3.1.)

FIGURE 3.8

Spring population counts of southern sea otters, 1983–2010

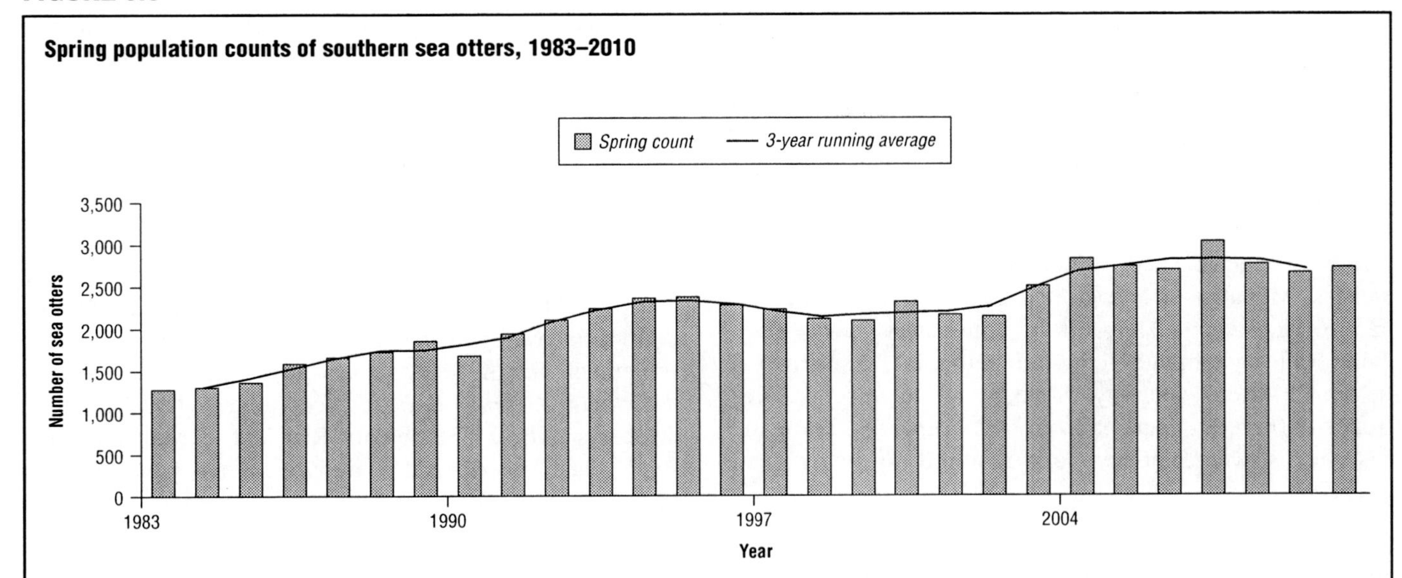

SOURCE: "Figure 4-19. Southern Sea Otter Counts 1983–2010 (Mainland Population)," in *Revised Draft Supplemental Environmental Impact Statement on the Translocation of Southern Sea Otters*, U.S. Department of the Interior, U.S. Fish and Wildlife Service, Ventura Fish and Wildlife Office, August 2011, http://www.fws.gov/ventura/species_information/so_sea_otter/rdseis/RDSEIS%20VOL%201.pdf (accessed November 5, 2011)

Manatees are often called "sea cows" and can weigh up to 2,000 pounds (900 kg). They swim just below the surface of the water and feed on vegetation. West Indian manatees migrate north during the summer, though generally no farther than the North Carolina coast. In 1995 a manatee nicknamed Chessie made headlines by swimming all the way to Chesapeake Bay. Eventually, biologists, concerned about his health in cooler waters, had him airlifted back to Florida. During the winter many manatees huddle around warm-water discharges from power plants and other industrial facilities. Even though this can keep them warm, scientists worry that overcrowding in small areas makes the animals more susceptible to sickness.

Imperiled Manatee Populations

Each year during cold weather biologists conduct surveys to determine the number of Florida manatees remaining in the wild. The numbers are estimates based on surveys conducted at known wintering habitats. The Florida Fish and Wildlife Research Institute states in "Manatee Synoptic Surveys" (2012, http://myfwc.com/research/manatee/projects/population-monitoring/synoptic-surveys/) that the January 2011 survey found 4,834 manatees living along the Florida coast. This number compares with 1,267 reported in 1991 (the first year of the survey) and 5,076 reported in 2010 (the previous high count). Many manatees have scars on their backs from motorboat propellers—these allow individual manatees to be recognized.

FIGURE 3.9

West Indian manatee (Florida manatee). (© *Kipling Brock/Shutterstock.com.*)

Threats to Manatees

Manatees are imperiled for a variety of reasons. Even though they can live for 50 or 60 years, their birth rate is low. Mature females bear a single offspring only every three to five years. Many baby manatees die in the womb or soon after birth for unknown reasons. These are called perinatal fatalities. Disease, natural pathogens, and cold-water temperatures are also deadly. However, motorboat strikes are the major documented cause of manatee mortalities. Manatees are large and swim slowly at the surface of the water. They often cannot move away from boats quickly enough to avoid being hit. As a result, several Florida waterways have been declared boat-free zones to protect manatees from boat collisions. There are also areas where boaters are required to lower their speed.

The Florida Fish and Wildlife Research Institute maintains an online database (http://research.myfwc.com/manatees/search_summary.asp) that summarizes causes of death reported for manatees. As of December 31, 2011, the database included cause of death and county of death for 7,991 manatees found dead between April 1974 and December 2010. The following is a summary of mortality causes:

- Undetermined—2,335 deaths
- Watercraft—1,820 deaths
- Perinatal (the period shortly before, during, and soon after birth)—1,640 deaths
- Natural causes—1,098 deaths
- Cold stress—725 deaths
- Gates/locks—200 deaths
- Other human causes—173 deaths

Regulating Water Activities

The Florida manatee was on the first list of endangered species published in 1967. (See Table 2.1 in Chapter 2.) According to the USFWS (June 22, 2011, http://www.fws.gov/northflorida/Manatee/Documents/MPARules/June11_KB_Proposed_Rule/20110621_frn_Federal_Register_Notice_for_Proposed_Kings_Bay_Manatee_Rule.html), in 1979 the agency "adopted a regulatory process to provide a means for establishing manatee protection areas in waters under the jurisdiction of the United States where manatees were taken by waterborne activities." The regulatory process is codified at 50 Code of Federal Regulations (http://ecfr.gpoaccess.gov/), part 17, subpart J.

As noted earlier, the MMPA and the ESA prohibit taking of marine mammals and listed species, respectively. Between 1980 and 1998 the USFWS designated seven manatee protection areas in and around Crystal River/Kings Bay on the western coast of Florida. The bay is fed by warm spring waters and is a major manatee habitat, both during cold and warm weather. However, the bay is also used for commercial boating and is a popular recreation area that features an active "manatee viewing industry." As a result, the USFWS reports that the bay has suffered increasing problems with manatee harassment and fatal boat strikes.

In 2000 a coalition of 18 environmental and wildlife groups and three individuals sued the USFWS and the state of Florida over manatee protection measures. In 2001 a settlement with the federal government was reached in *Save the Manatee Club v. Ballard*. According to the USFWS, in the press release "Settlement Reached in Manatee Lawsuit" (January 4, 2001, http://www.fws.gov/southeast/news/2001/r01-001.html), the agency agreed to a time table for completing the following actions:

- Designate additional manatee protection areas
- Revise the manatee recovery plan
- Develop regulations allowing manatee taking under the MMPA
- Coordinate with the U.S. Army Corps of Engineers (the Corps) to improve procedures for reviewing permit applications for the construction of boating facilities in manatee habitat and improve the related consultation process required under the ESA

As of November 2010, the USFWS (http://www.fws.gov/northflorida/Manatee/Documents/MPARules/index-federal-mpa-maps.htm) had established 19 manatee protection areas (five sanctuaries and 14 refuges) in 12 counties along the central western coast of Florida. In addition, the Florida Fish and WildlifeConservation Commission (http://myfwc.com/wildlifehabitats/managed/manatee/data-and-maps/) had designated dozens of "manatee protection zones" with specific watercraft speed limit restrictions.

In 2010 the USFWS published an emergency rule that temporarily designated a new manatee protection area, the Kings Bay Manatee Refuge, covering the entire Kings Bay (i.e., encompassing the seven previously designated small areas within the bay). The emergency rule, which was effective for 120 days, was issued because manatees faced "imminent" taking within the bay. In June 2011 the USFWS (http://www.gpo.gov/fdsys/pkg/FR-2011-06-22/pdf/2011-15603.pdf) issued a proposed rule to permanently establish the Kings Bay Manatee Refuge. As of January 2012, the rule had not been finalized.

In October 2001 the USFWS published a revised recovery plan for the manatee. In *Florida Manatee Recovery Plan (*Trichechus manatus latirostris*) Third Revision* (http://ecos.fws.gov/docs/recovery_plan/011030.pdf), the agency identifies the threats to the species and lists specific population benchmarks that must be reached before the manatee can be delisted under the ESA. Full recovery was not anticipated for at least two decades.

Critical Habitat

In 1975 the USFWS designated critical habitat for the species in 17 Florida counties. In December 2008 the Wildlife Advocacy Project, Save the Manatee Club, the Center for Biological Diversity, and the Defenders of Wildlife petitioned the agency to expand the critical habitat based on new information, including a 2007 status review performed by the USFWS. In September 2009 the USFWS (http://frwebgate.access.gpo.gov/cgi-bin/getdoc.cgi?dbname=2009_register&docid=fr29se09-23) published a 90-day finding indicating its intention to revise the critical habitat designation for the Florida manatee. The USFWS acknowledged that numerous factors have changed since critical habitat was designated in 1975. In particular, the state has seen an enormous increase in its human population accompanied by expanded urban and beachfront development. In January 2010 the USFWS (http://www.gpo.gov/fdsys/search/citation.result.FR.action?federalRegister.volume=2010&federalRegister.page=1574&publication=FR) concluded that the expanded critical habitat is warranted. However, the agency noted that "sufficient funds are not available due to higher priority actions such as court-ordered listing-related actions and judicially approved settlement agreements."

Foreign Manatee and Dugong Species

According to the IUCN, in *Red List of Threatened Species Version 2011.2*, three manatee species—the South American manatee, the West Indian manatee, and the West African manatee—and the only remaining dugong species were considered threatened in 2011.

As of November 2011, the species found in western Africa and in and around the Amazon River in South America were listed under the ESA and were in grave danger of extinction due to illegal hunting, deforestation, habitat destruction, and water pollution. (See Table 3.2.) The only remaining dugongs live in the coastal waters of the Indian Ocean and the Pacific Ocean. Their populations are also considered imperiled. Dugongs around the tiny island of Palau in the western Pacific Ocean are listed as endangered under the ESA.

POLAR BEARS

Polar bears are the largest of the bear species. They are believed to have evolved from grizzly bears hundreds of thousands of years ago. Polar bears have stocky bodies and can weigh up to 1,400 pounds (635 kg) when fully grown. They are relatively long lived and have small litters, typically only a cub or two per litter. Their fur includes water-repellent guard hairs and a dense undercoat that is white to pale yellow in color. They have large paddle-like paws that help make them excellent swimmers.

Polar bears are considered a marine mammal because sea ice is their primary habitat. They are found throughout the Arctic and near-Arctic regions of the Northern Hemisphere. The bears prefer coastal sea ice and other areas in which water conditions are conducive to providing prey. Seals are their primary food source, particularly ringed seals. The USFWS notes in the fact sheet "Polar Bear *Ursus maritimus*" (October 2009, http://www.fws.gov/home/feature/2009/pdf/polar_bearfactsheet1009.pdf) that there are 20,000 to 25,000 bears worldwide and 3,500 bears in the United States. There are two stocks of polar bears in the United States; both are in and around Alaska: the southern Beaufort Sea stock and the Chukchi/Bering Sea stock. (See Figure 3.10.) There is extensive overlap between the two stocks.

Population Estimates and Trends

As noted earlier, the MMPA requires periodic surveys of marine mammals by the NMFS and the USFWS to estimate populations and predict population trends. The results are published annually in the *Stock Assessment Report*. As of January 2012, the most recent reports for the two polar bear stocks were published in January 2010. According to the USFWS, in *Polar Bear (*Ursus maritimus*): Southern Beaufort Sea Stock* (January 1, 2010, http://alaska.fws.gov/fisheries/mmm/stock/final_sbs_polar_bear_sar.pdf), the southern Beaufort Sea stock has been extensively studied since the 1960s. Scientists estimate that the population consists approximately of 1,500 individuals. The bears eat primarily ringed and bearded seals and thus are concentrated in relatively shallow and biologically rich waters over the continental shelf that provides sustenance for the seals. The bears are found mostly in areas that have at least 50% ice cover.

The USFWS notes that after passage of the MMPA in 1972 the populations of both stocks increased, due to hunting restrictions. However, the southern Beaufort Sea population showed "little or no growth" during the 1990s and declined by approximately 3% per year between 2001 and 2005. Reduced sea ice, particularly during the summer and fall, is believed to be a factor in the decline.

In *Polar Bear (*Ursus maritimus*): Chukchi/Bering Seas Stock* (January 1, 2010, http://alaska.fws.gov/fisheries/mmm/stock/final_cbs_polar_bear_sar.pdf), the USFWS indicates that the Chukchi/Bering Sea stock is "widely distributed" and ranges from Alaska in the east to Russia in the west. The population of the stock is roughly estimated at 2,000 bears. However, the agency states that it does not have a "reliable" population estimate for the stock due to logistical factors (i.e., the vast and difficult terrain) and the passage of the animals into Russian territory. The USFWS notes that this stock faces "different stressors" than those

FIGURE 3.10

Range of polar bear populations, 2010

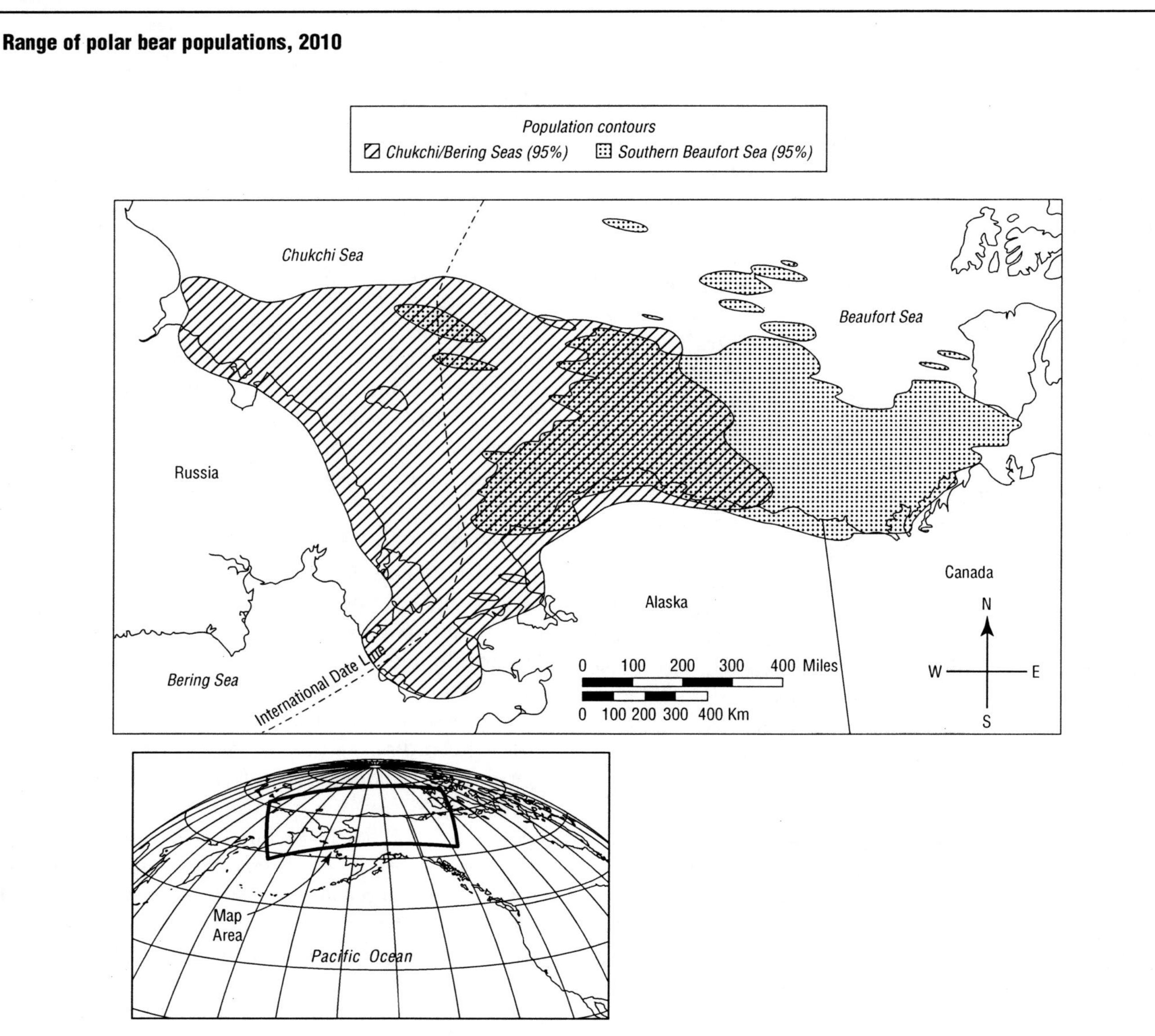

SOURCE: "Ranges of the Chukchi/Bering Seas and Southern Beaufort Sea Polar Bear (Ursus maritimus) Populations," in "Endangered and Threatened Wildlife and Plants; Designation of Critical Habitat for the Polar Bear (Ursus maritimus) in the United States," *Federal Register*, vol. 75, no. 234, December 7, 2010, http://alaska.fws.gov/fisheries/mmm/polarbear/pdf/federal_register_notice.pdf (accessed November 5, 2011)

affecting the southern Beaufort Sea stock. In particular, the Chukchi/Bering Sea stock endures "increased harvest in Russia" and a greater loss of sea ice during the summer. The population of the stock is believed to be declining.

Overall, the USFWS reports that the "sea ice is rapidly diminishing throughout the Arctic." A similar assessment is provided by the National Snow and Ice Data Center (NSIDC; http://nsidc.org/about/), a research organization sponsored by NOAA, the National Aeronautics and Space Administration, the National Science Foundation, and the University of Colorado, Boulder. The NSIDC collects data and supports research that is related to the world's frozen environments. In "Positively Arctic: Arctic Oscillation Switches Phase" (January 5, 2012, http://nsidc.org/arcticseaicenews/2012/01/positively-arctic-arctic-oscillation-switches-phase/), the organization presents data showing that the extent of Arctic ice present during the month of December declined by 3.5% per decade between 1979 and 2011. The seasonal minimum extent occurs each September. According to the NSIDC, the minimum extent in September 2011 was the second lowest on record between 1979 and 2011. The lowest minimum extent occurred during September 2007.

Listing Controversy

In 2005 the USFWS was petitioned by the Center for Biological Diversity (later joined by the Natural Resources Defense Council and Greenpeace) to list the polar bear as threatened throughout its range. Due to a heavy workload, the agency failed to issue a 90-day finding to the petition. The organizations sued in the U.S. District Court. In February 2006 the agency issued its 90-day finding, noting that there was sufficient scientific information to warrant the listing. In January 2007 the polar bear was officially proposed for listing as a threatened species.

In March 2008 the original petitioners filed a lawsuit in the U.S. District Court against the USFWS for failing to make a listing decision in a timely manner. The court ruled in favor of the plaintiffs, forcing the USFWS to make a decision by May 15, 2008. On that date the agency officially listed the polar bear as threatened. The USFWS also issued a "special rule" under section 4(d) of the ESA adopting the existing conservation regulatory requirements for polar bears under the MMPA and the CITES agreement. In other words, activities already authorized in these regulations would take precedence over the general prohibitions under U.S. law that apply to threatened species. Such activities include take, import and export, and shipment in interstate or foreign commerce for commercial purposes.

The polar bear's situation is unique among other endangered and threatened species in the United States, because the principal and most significant threat to its survival is believed to be climate change. The USFWS notes in "Endangered and Threatened Wildlife and Plants; 12-Month Petition Finding and Proposed Rule to List the Polar Bear (*Ursus maritimus*) as Threatened throughout Its Range" (*Federal Register*, vol. 72, no. 5, January 9, 2007) that melting and thinning sea ice have already stressed polar bear populations. Computer models indicate these conditions are expected to worsen in the future as temperatures continue to warm. Polar bears have evolved to move on ice. They have special suckers on their paws to help them walk on the ice. They are poorly suited to walking on ground and expend a great deal of energy when they are forced to do so, because of a lack of sea ice. In addition, warmer temperatures are degrading the snowy birthing dens of ringed seals, the major prey of the polar bears, and endangering seal pup survival. Thus, both populations face problems due to the warming of their icy ecosystems.

Environmentalists hoped the listing status granted to polar bears would facilitate legal restrictions on oil and gas exploration in Alaska and on U.S. emissions of carbon dioxide (a suspected cause of global warming). However, according to the press release "Secretary Kempthorne Announces Decision to Protect Polar Bears under Endangered Species Act" (May 14, 2008, http://www.doi.gov/archive/news/08_News_Releases/080514a.html), the U.S. secretary of the interior Dirk Kempthorne (1951–) said that "listing the polar bear as threatened can reduce avoidable losses of polar bears. But it should not open the door to use...the ESA to regulate greenhouse gas emissions from automobiles, power plants, and other sources." Kempthorne vowed "to make certain the ESA isn't abused to make global warming policies." In addition, the 4(d) rule issued for the polar bear specifically noted that the incidental take of the species due to oil and gas exploration is already allowed under regulations of the MMPA.

Environmentalists were further disappointed in May 2009, when the Obama administration announced that it would retain the 4(d) rule. In the press release "Salazar Retains Conservation Rule for Polar Bears Underlines Need for Comprehensive Energy and Climate Change Legislation" (May 8, 2009, http://www.fws.gov/news/NewsReleases/showNews.cfm?newsId=20FB90B6-A188-DB01-04788E0892D91701), the U.S. secretary of the interior Ken Salazar (1955–) said the melting of sea ice due to climate change is the biggest threat facing polar bears. He stated, "However, the Endangered Species Act is not the proper mechanism for controlling our nation's carbon emissions. Instead, we need a comprehensive energy and climate strategy that curbs climate change and its impacts—including the loss of sea ice. Both President Obama and I are committed to achieving that goal."

Following the USFWS listing of the polar bear in 2008, the state of Alaska sued the USFWS arguing that the listing was not justified. The article "Alaska Sues to Overturn Polar Bear Protection" (*Washington Times*, November 16, 2009) explains that Alaska feared the listing would threaten oil development in the state, particularly of offshore reserves. In October 2009 the Alaska governor Sean Parnell (1962–) announced that new briefs had been filed in the case. According to the article, Parnell complained that the ESA is being used as a "land-use planning tool" and that the polar bear listing is not warranted based on existing scientific information. In June 2011 a federal judge rejected the suit and ruled that the USFWS followed proper procedures in listing the polar bear.

In "Alaska Joins Appeal of Ruling on Polar Bear 'Threatened' Status" (*Anchorage Daily News*, August 27, 2011), Richard Mauer reports that dozens of plaintiffs (including Alaska) and intervenors filed appeal notices following the decision. According to Mauer, the governor's position was that "polar bears survived prehistoric periods of global warming and there's no reason to believe they wouldn't again." Parnell also claimed the ESA listing was improper because it was based on speculation, rather than on actual evidence of serious population declines. Mauer notes that state officials were worried that the polar bear listing will impede oil and gas exploration in Alaska. As of January 2012, courts had not rendered final decisions on the appeals.

In August 2008 scientists from the University of Wyoming, the U.S. Geological Survey, and the USFWS began a study to determine the effects of longer ice-free seasons on polar bears in the southern Beaufort Sea region. The USFWS explains in the fact sheet "A New Study to Understand How Polar Bears Cope with Longer Ice-Free Seasons" (November 14, 2008, http://www.uwyo.edu/polarbear/project%20fact%20sheet.pdf) that 29 polar bears were captured and sedated for collection of physical data and breath and blood samples. Specially designed radio collars were placed on 12 of the bears so their movements could be followed. A similar study was begun in 2008 on polar bears in the Chukchi Sea. Radio collars were placed on 11 of those bears for collection of location information for up to two years. In January 2009 Canadian researchers announced their findings that polar bears in the Beaufort Sea region are increasingly malnourished. According to Kate Ravilious, in "More Polar Bears Going Hungry" (*New Scientist*, January 1, 2009), blood tests performed on bears in 1985 and 1986 revealed that 9.6% and 10.5%, respectively, were fasting. Similar tests performed in 2005 and 2006 indicated that 21.4% and 29.3%, respectively, of the tested bears were fasting.

During the spring months between 2008 and 2010 the USFWS performed polar bear surveys in both the Bering and Chukchi Seas. Figure 3.11 shows the locations at which 140 of the bears were captured and sedated for biological specimen collection. More than 50 bears were equipped with radio collars or radio ear tags so that scientists can track the bears' movements. In *Polar Bear News 2010* (November 23, 2010, http://alaska.fws.gov/fisheries/mmm/polarbear/pdf/news_11_23_10.pdf), the USFWS notes that it plans to continue the surveys through at least 2012.

In December 2010 the USFWS (http://www.gpo.gov/fdsys/search/citation.result.FR.action?federalRegister.volume=2010&federalRegister.page=76086&publication=FR) designated 187,157 square miles (484,734 sq km) in Alaska and adjacent U.S. and territorial waters as critical habitat for the polar bear. The designation encompasses three types of habitat: barrier island habitat, sea ice habitat, and terrestrial denning habitat.

FIGURE 3.11

Locations of polar bears captured as part of monitoring program, 2008–10

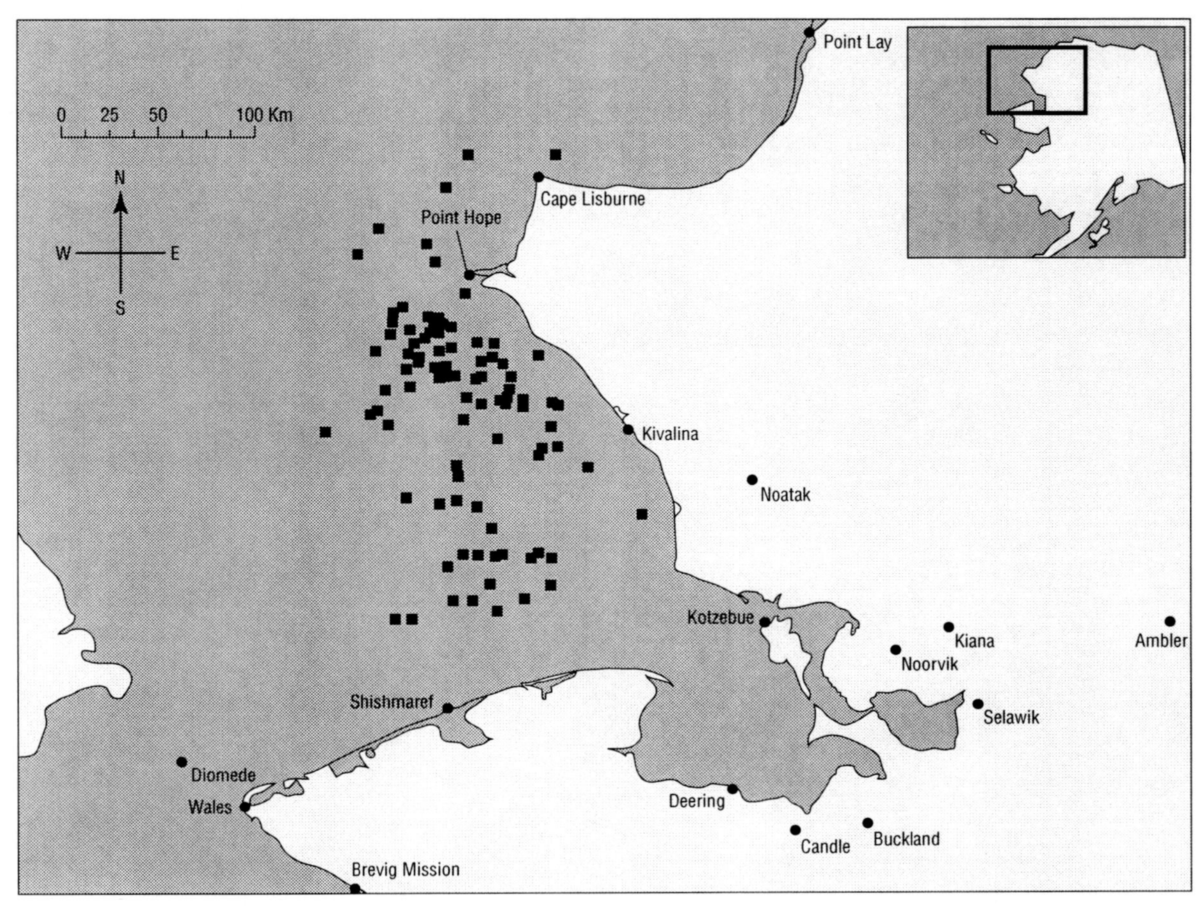

SOURCE: "Fig. 5. Locations of Captured Polar Bears at Time of Capture between 2008 and 2010," in *Polar Bear News 2010*, U.S. Department of the Interior, U.S. Fish & Wildlife Service, Marine Mammals Management Office, November 23, 2010, http://alaska.fws.gov/fisheries/mmm/polarbear/pdf/news_11_23_10.pdf (accessed November 5, 2011)

CHAPTER 4
FISH

Fish are cold-blooded vertebrates with fins. They occur in nearly all permanent water environments, from deep oceans to remote alpine lakes and desert springs. Marine fish inhabit the salty waters of oceans and seas, whereas freshwater fish inhabit inland rivers, lakes, and ponds. Some fish species migrate between fresh and marine waters. These include species called anadromous fish that are born in freshwater, migrate to the ocean to spend their adulthood, and then return to freshwater to spawn.

Fish are the most diverse vertebrate group on the planet and include thousands of different species. The largest known fish are the whale sharks, which can grow to be more than 50 feet (15.2 m) long and weigh several tons. At the other end of the spectrum is *Paedocypris progenetica*, a tiny fish discovered in Sumatra, Indonesia, that is less than 0.3 of an inch (0.8 cm) in length.

FishBase (http://www.fishbase.org/search.php) is a comprehensive online database of scientific information about fish. It was developed by the WorldFish Center of Malaysia in collaboration with the United Nations Food and Agriculture Organization and is supported by many government and research institutions. As of October 2011, FishBase contained information on 32,100 fish species around the world. Scientists report that only a small fraction of these species have been assessed for their conservation status.

As of November 2011, the U.S. Fish and Wildlife Service (USFWS) listed a total of 145 endangered and threatened fish species in the United States. (See Table 1.2 in Chapter 1.) In *Federal and State Endangered and Threatened Species Expenditures: Fiscal Year 2010* (September 2011, http://www.fws.gov/endangered/esa-library/pdf/2010.EXP.FINAL.pdf), the USFWS indicates that more than $670 million was spent under the Endangered Species Act (ESA) during fiscal year (FY) 2010 on imperiled fish.

GENERAL THREATS TO FISH

Fish species have become endangered and threatened in the United States for a variety of reasons, both natural and anthropogenic (caused by humans). Most fish are not imperiled by a single threat to their survival, but by multiple threats that combine to produce daunting challenges to recovery. Some scientists believe natural threats, such as disease, have been aggravated by human actions that stress fish populations. Dams and other structures that are used for power generation, flood control, irrigation, and navigation have dramatically changed water flow patterns in many rivers. These impediments disrupt migration patterns and affect water temperature and quality. Likewise, the dredging of river and stream beds to produce channels and the filling of wetlands and swamps have changed water habitats.

The Problem with Dams

Dams affect rivers, the lands abutting them, the water bodies they join, and aquatic wildlife throughout the United States. Water flow is reduced or stopped altogether downstream of dams, altering aquatic habitats and drying wetlands. Arthur C. Benke and Colbert E. Cushing, the editors of *Rivers of North America* (2005), note that it is difficult to find any river in the United States that has not been dammed or channeled. According to Benke and Cushing, "All human alterations of rivers, regardless of whether they provide services such as power or drinking water supply, result in degradation."

The U.S. Army Corps of Engineers (the Corps) maintains the National Inventory of Dams (NID; https://nid.usace.army.mil/). As of December 2011, the inventory included more than 84,000 dams throughout the country. To be included in this inventory, dams have to be at least 6 feet (1.8 m) tall or hold back a minimum of 15 acre-feet (4.9 million gallons [18.5 million L]) of water. Dams are built for a variety of purposes. The most frequent purposes listed in the NID are recreation, fire protection, stock or small farm pond creation, flood control, and storm water management.

Even though only a small percentage of the dams listed with the NID produce hydroelectric power, these dams tend to be the largest in size and affect large watersheds. These structures provide many challenges to aquatic species, besides impeding water flow and migration paths. Turbines operate like massive underwater fans. Passage through running turbine blades can result in the death of many small aquatic creatures that are unable to escape their path. Some modern hydroelectric dams include stairlike structures called fish ladders that provide migrating fish a watery path to climb up and over the dams.

SNAIL DARTERS. The snail darter, a small fish species related to perch, was at the center of a dam-building controversy during the 1970s. The USFWS listed the snail darter as endangered in 1975. At the time it was believed to exist only in the Little Tennessee River, and this area was designated as critical habitat for the species. That same year the Tellico Dam was near completion on the Little Tennessee River, and the filling of the Tellico Reservoir would have destroyed the entire habitat of the snail darter. A lawsuit was filed to prevent this from happening. The case went all the way to the U.S. Supreme Court, which ruled in *Tennessee Valley Authority v. Hill* (437 U.S. 153 [1978]) that under the ESA species protection must take priority over economic and developmental concerns. One month after this ruling Congress amended the ESA to allow for exemptions under certain circumstances.

In late 1979 the Tellico Dam received an exemption, and the Tellico Reservoir was filled. The snail darter is now extinct in that habitat. However, snail darter populations were later discovered in other river systems. In addition, the species has been introduced into several other habitats. Because of an increase in numbers, the snail darter was reclassified as threatened in 1984.

Entrainment and Impingement

Entrainment occurs when fish are pulled or diverted away from their natural habitat by mechanical equipment or other nonnatural structures in water bodies. A prime example is a freshwater intake structure in a river or lake. Small fish, in particular, are susceptible to being sucked into pipes through which water is being pumped out of a water body. Even if the pipe end is screened, the force of the suction can impinge (crush) fish against the screen, causing serious harm. In a more general sense entrainment refers to diversions that occur when fish accidentally feed through artificial water structures, such as gates, locks, or dams, and cannot return to their original location.

Excessive Sediment

Many river and stream banks and adjacent lands have been stripped of vegetation by timber harvesting, crop growing, and excessive grazing of livestock. This eliminates habitat for insects and other tiny creatures that serve as foodstuff for fish. It also aggravates erosion problems and allows large amounts of dirt to enter water bodies. Once in the water, this dirt is known as silt or sediment. Most of these particles settle to the bottom. However, sediment is easily stirred up by the movement of fish and other aquatic creatures, many of which spawn or lay eggs at the bed of their habitat. The dirt that remains in suspension in the water is said to make water turbid. The measure of the dirtiness (lack of clarity) of a water body is called its turbidity.

Freshwater aquatic creatures are sensitive to turbidity levels and choose their habitats based in part on their sediment preferences. Some fish prefer waters with large amounts of sediment. It provides cover that prevents predator fish from seeing them. Other species prefer clean waters with low turbidity levels. Excessive sediment may clog their gills or smother their eggs. (See Figure 4.1.)

Forestry and agricultural practices can drastically affect the sediment levels in a water system through the deforestation of banks and nearby lands. Excessive grazing of livestock along riverbanks can strip vegetation and permit large amounts of dirt to enter the water. Likewise, timber harvesting and crop production can expose loosened dirt to wind and rain that carry it into water bodies. Dams and diversion structures trap sediments behind them, interrupting the natural downstream flow of sediments that takes place in moving waters.

Chemical and Biological Pollutants

Water pollution poses a considerable threat to many aquatic species. Industrial pollution introduces metal and organic chemicals to water bodies. In agricultural areas there is runoff of manure, fertilizers, and pesticides. U.S. pesticide use during the 1960s centered around chlorine-containing organic compounds, such as dichlorodiphenyltrichloroethane (DDT) and chlordane. Scientists eventually learned that these chemicals are extremely persistent in the environment and have damaging effects on wildlife, particularly fish and bird species. DDT was banned in the United States in 1972 and chlordane in 1983; however, nearly four decades later both pesticides continue to show up in water, sediment, and fish samples.

In general, aquatic creatures are not killed outright by water contamination. A major exception is an oil spill, which can kill many creatures through direct contact. The more widespread and common threat is overall degradation of water quality and habitats due to pollutants. Exposure to contaminants can weaken the immune systems of aquatic

FIGURE 4.1

Effects of siltation on aquatic life

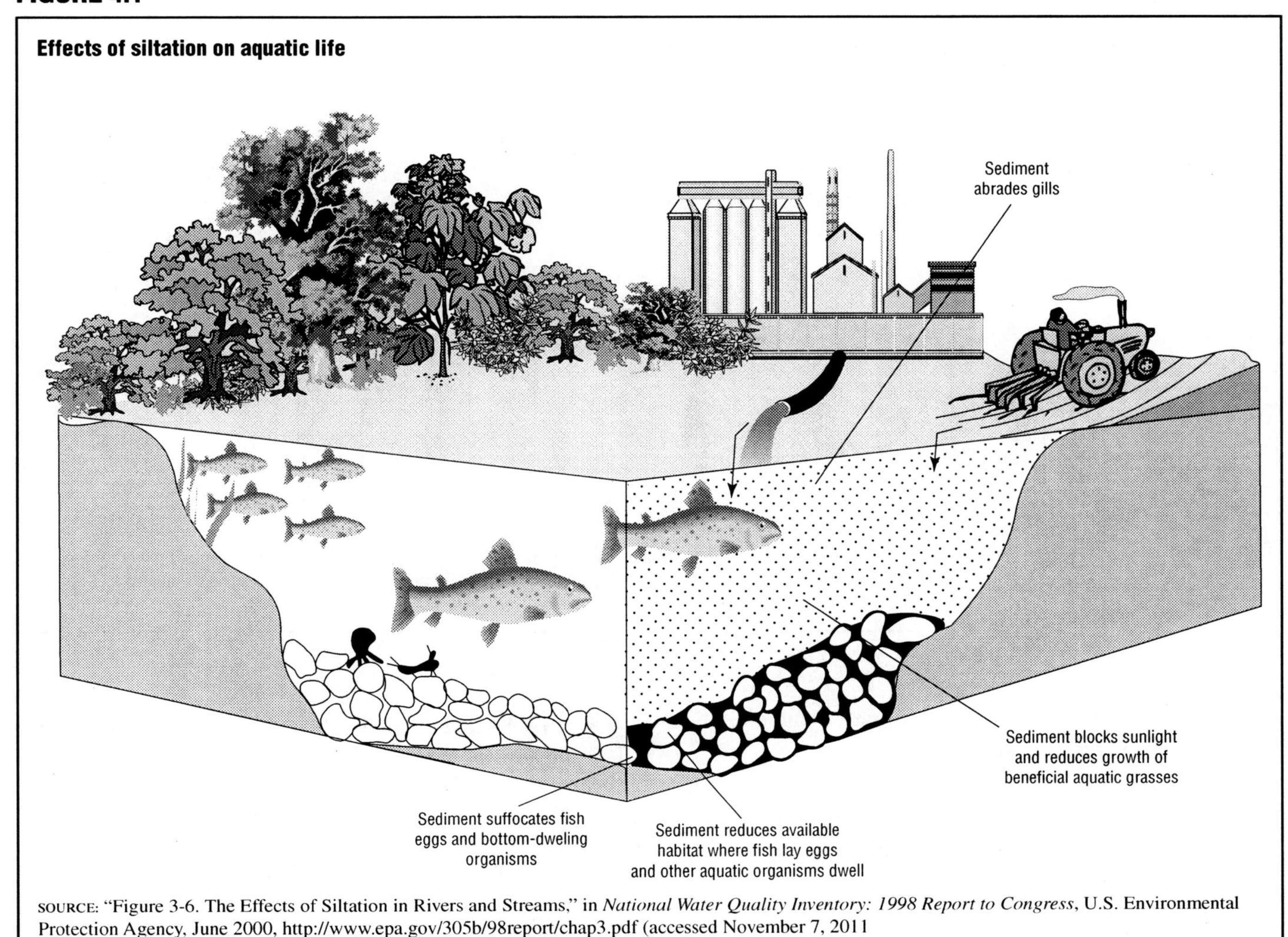

SOURCE: "Figure 3-6. The Effects of Siltation in Rivers and Streams," in *National Water Quality Inventory: 1998 Report to Congress*, U.S. Environmental Protection Agency, June 2000, http://www.epa.gov/305b/98report/chap3.pdf (accessed November 7, 2011

animals and make them more susceptible to disease and to other health and reproductive problems.

Bioaccumulative contaminants are those that accumulate in the tissues of aquatic organisms at much higher concentrations than are found in the water body itself. This biomagnification effect occurs with mercury (a metal), the pesticides DDT and chlordane, and dioxins. Dioxins are a category of several hundred chlorinated organic compounds. Polychlorinated biphenyls (PCBs) are dioxins that were widely used to cool and lubricate electrical equipment before a 1977 ban on their manufacture. Bioaccumulative contaminants are a particular concern for fish at the higher end of the aquatic food chain, such as salmon and large freshwater species.

TOXIC POLLUTANT ADVISORIES. The states issue advisories to protect residents from the adverse health risk of eating fish that are contaminated with certain pollutants. In *2010 National Listing of Fish Advisories: Biennial National Listing of Fish Advisories* (September 2011, http://water.epa.gov/scitech/swguidance/fishshellfish/fishadvisories/upload/National-Listing-of-Fish-Advisories-Technical-Fact-sheet-2010.pdf), the U.S. Environmental Protection Agency states that 4,600 advisories were in effect in 2010. The advisories covered approximately 1.5 million miles (2.4 million km) of river and 18.3 million acres (7.4 million ha) of lakes. These values represented 41% of the total river miles and 44% of the total lake acres that were assessed for contamination in 2010. As shown in Figure 4.2, the percentage of total river miles and lake acres under advisory has increased dramatically since 1993. The vast majority of the advisories issued in 2010 were due to mercury contamination (3,711). (See Table 4.1.) PCBs accounted for 1,086 of the advisories. Dioxins, DDT, and chlordane each accounted for much smaller numbers. All these chemicals are bioaccumulative. As shown in Table 4.1, the number of advisories issued due to mercury and PCB contamination and the number of lake acres and river miles contaminated by mercury and PCBs increased between 2008 and 2010. The number of advisories issued due to dioxin contamination increased only slightly between 2008 and 2010,

FIGURE 4.2

Percentage of lake acres and river miles under fish advisories, 1993–2010

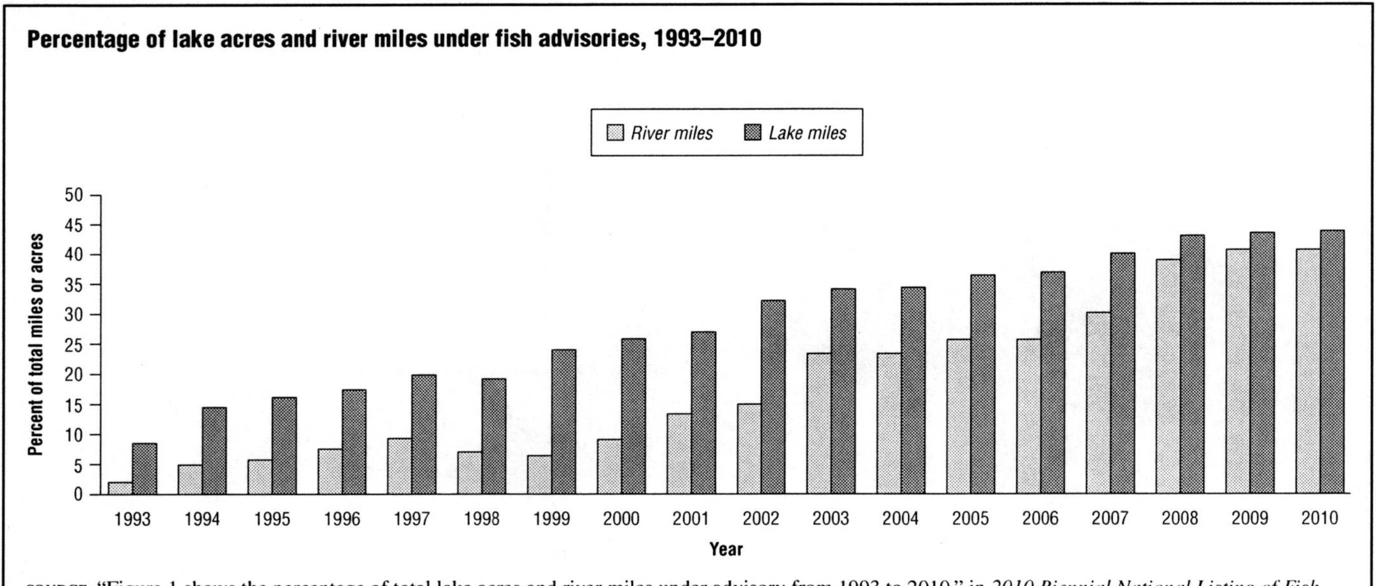

SOURCE: "Figure 1 shows the percentage of total lake acres and river miles under advisory from 1993 to 2010," in *2010 Biennial National Listing of Fish Advisories*, U.S. Environmental Protection Agency, Office of Science and Technology, September 2011, http://water.epa.gov/scitech/swguidance/fishshellfish/fishadvisories/upload/National-Listing-of-Fish-Advisories-Technical-Fact-sheet-2010.pdf (accessed November 7, 2011)

whereas the number of advisories for chlordane and DDT decreased over this two-year period.

Even though the purpose of fish advisories is to protect human health, the underlying data illustrate that toxic pollutants pose a worrisome threat to the nation's fish species.

Unwelcome Guests: Injurious Invasive Fish

Historically, fish were generally geographically limited in their habitats, such as to streams, rivers, or lakes in a particular area. However, human intervention has allowed species to move into new habitats, sometimes thousands of miles from where they originated. Many fish species have been introduced purposely to new water bodies to improve sport and recreational fishing. Some species that are popular in aquariums have been intentionally dumped into the environment. Other species have been introduced unintentionally by migrating through human-built canals and locks, stowing away in the ballast water of ships, or escaping from research, breeding, and aquaculture (fish farming) facilities. Some introduced species are not problematic in their new habitats, but others cause ecological or economic harm by competing with native species for habitat and food, preying on them, or breeding with them. Injurious nonnative species are commonly called invasive species.

The Nonindigenous Aquatic Nuisance Prevention and Control Act of 1990 and the National Invasive Species Act of 1996 are intended to help prevent unintentional introductions of aquatic invasive species. In 2005 the U.S. Department of Agriculture's National Agricultural Library established the National Invasive Species Information Center (NISIC; http://www.invasivespeciesinfo.gov/), a web-based source for educational information about common invasive species. As of November 2011, the NISIC (http://www.invasivespeciesinfo.gov/aquatics/main.shtml) listed the following invasive fish species and their current areas of distribution in the United States:

- Alewife—Great Lakes and inland waters throughout the eastern United States
- Asian swamp eel—Florida, Georgia, and Hawaii
- Bighead carp (Asian carp)—Midwest and south-central United States
- Black carp (Asian carp)—Mississippi River basin
- Eurasian ruffe—Great Lakes
- Flathead catfish—western United States and East Coast
- Grass carp (Asian carp)—widely distributed across much of the United States
- Lionfish—East Coast and Gulf of Mexico
- Northern snakehead—inland waters of the upper Atlantic, scattered populations across the United States
- Round goby—Great Lakes
- Sea lamprey—Great Lakes
- Silver carp (Asian carp)—Midwest and south-central United States

TABLE 4.1

Number of fish advisories and lake acres and river miles under fish advisories, by contaminant, 2008 and 2010

	Number of advisories		Lake acres		River miles	
Contaminant	2008	2010	2008	2010	2008	2010
Mercury	3,361	3,711	16,808,032	16,998,664	1,225,016	1,257,689
PCBs	1,025	1,086	6,049,506	7,513,434	130,372	295,773
Chlordane	67	60	842,913	824,290	54,029	53,893
Dioxins	123	128	35,400	35,400	2,055	2,333
DDT	76	58	876,520	876,520	69,198	68,884

PCB = polychlorinated biphenyls.
DDT = dichlorodiphenyltrichloroethane.

SOURCE: "Table 1 shows the number of advisories and size of waters under advisory, by contaminant," in *2010 Biennial National Listing of Fish Advisories*, U.S. Environmental Protection Agency, Office of Science and Technology, September 2011, http://water.epa.gov/scitech/swguidance/fishshellfish/fishadvisories/upload/National-Listing-of-Fish-Advisories-Technical-Fact-sheet-2010.pdf (accessed November 7, 2011)

It should be noted that even though these fish are definitely injurious from a general environmental and economic standpoint, they are not necessarily harmful to species that are designated as endangered or threatened under the ESA. For example, several of the invasive fish listed by the NISIC inhabit the Great Lakes. However, the USFWS (http://www.fws.gov/midwest/endangered/fishes/index.html) indicated that as of October 2011 the Great Lakes did not contain any endangered or threatened fish species. Even so, other nonnative fish species do harm imperiled species. For example, the threatened bull trout has become imperiled, in part, due to the introduction of nonnative brook trout to bull trout habitats. Because brook trout are an extremely popular sport and food fish, they are not considered by the public to be injurious.

Overcrowding

The overcrowding of stressed fish populations into smaller and smaller areas has contributed to hybridization (uncharacteristic mating between closely related species resulting in hybrid offspring). According to the USFWS, environmental degradation appears to inhibit natural reproductive instincts that historically prevented fish from mating outside their species. In addition, a shortage of suitable space for spawning has resulted in more mating between species. Cross-mating can be extremely detrimental to imperiled species, because the offspring can be sterile.

IMPERILED FRESHWATER FISH

Freshwater fish listed under the ESA fall within the jurisdiction of the USFWS. They include a wide variety of species and are found all over the country. As of November 2011, there were 113 freshwater fish species listed under the ESA. (See Table 4.2.) In addition, there were 16 experimental populations considered nonessential. As noted in Chapter 2, an experimental population is achieved by introducing a species to an area in which it is not currently found. The nonessential designation means that the survival of the experimental population is not believed essential to the survival of the species as a whole. Thus, the experimental population receives less protection under the ESA than does the species as a whole. Most of the imperiled fish species had recovery plans in place. In general, imperiled freshwater fish are small in size and are associated with flowing (lotic) waters, such as rivers and streams, rather than with still (lentic) waters, such as lakes and ponds. Nearly half of the listed freshwater fish fall into four species groups: chubs, daces, darters, and shiners.

Table 4.3 shows the 10 freshwater species with the highest expenditures under the ESA during FY 2010. Approximately $66.5 million was spent on only two of the fish: bull trout ($43.8 million) and pallid sturgeon ($22.7 million).

Bull Trout

Bull trout are relatively large fish that live in streams, lakes, and rivers. They can weigh more than 20 pounds (9.1 kg); however, those that inhabit small streams seldom exceed 4 pounds (1.8 kg) in weight. Bull trout are members of the char subgroup of the salmon family (Salmonidae). (See Figure 4.3.) Their backs are dark in color (green to brown) with small light-colored spots (crimson to yellow), and their undersides are pale. The fish prefer cold and clean inland waters in the Northwest.

Historically, bull trout were found throughout much of the northwestern United States and as far north as Alaska. Large populations have disappeared from major rivers, leaving mostly isolated pockets of smaller-sized fish in headwater streams. A variety of factors have contributed to the decline of the bull trout. The species is extremely sensitive to changes in water temperature and purity. Its survival is threatened by water pollution, degraded habitat, and dams and other diversion structures. In addition, the introduction of a nonnative game

TABLE 4.2

Endangered and threatened freshwater fish species, November 2011

Common name	Scientific name	Listing status[a]	U.S. or U.S./foreign listed	Recovery plan date	Recovery plan stage[b]
Alabama cavefish	Speoplatyrhinus poulsoni	E	US	10/25/1990	RF(2)
Alabama sturgeon	Scaphirhynchus suttkusi	E	US	None	—
Amber darter	Percina antesella	E	US	6/20/1986	F
Apache trout	Oncorhynchus apache	T	US	9/3/2009	RF(2)
Arkansas River shiner	Notropis girardi	T	US	None	—
Ash Meadows Amargosa pupfish	Cyprinodon nevadensis mionectes	E	US	9/28/1990	F
Ash Meadows speckled dace	Rhinichthys osculus nevadensis	E	US	9/28/1990	F
Bayou darter	Etheostoma rubrum	T	US	7/10/1990	RF(1)
Beautiful shiner	Cyprinella formosa	T	US/foreign	3/29/1995	F
Big Bend gambusia	Gambusia gaigei	E	US	9/19/1984	F
Big Spring spinedace	Lepidomeda mollispinis pratensis	T	US	1/20/1994	F
Blackside dace	Phoxinus cumberlandensis	T	US	8/17/1988	F
Bluemask (=jewel) darter	Etheostoma sp.	E	US	7/25/1997	F
Blue shiner	Cyprinella caerulea	T	US	8/30/1995	F
Bonytail chub	Gila elegans	E	US	None	—
Borax Lake chub	Gila boraxobius	E	US	2/4/1987	F
Boulder darter	Etheostoma wapiti	E (1), EXPN (1)	US	7/27/1989	F
Bull trout	Salvelinus confluentus	T (1), EXPN (1)	US	None	—
Cahaba shiner	Notropis cahabae	E	US	4/23/1992	F
Cape Fear shiner	Notropis mekistocholas	E	US	10/7/1988	F
Cherokee darter	Etheostoma scotti	T	US	11/17/2000	F
Chihuahua chub	Gila nigrescens	T	US/foreign	4/14/1986	F
Chucky Madtom	Noturus crypticus	E	US	10/11/2011	0
Clear Creek gambusia	Gambusia heterochir	E	US	1/14/1982	F
Clover Valley speckled dace	Rhinichthys osculus oligoporus	E	US	5/12/1998	F
Colorado pikeminnow (=squawfish)	Ptychocheilus lucius	E (1), EXPN (1)	US	None	—
Comanche Springs pupfish	Cyprinodon elegans	E	US	9/2/1981	F
Conasauga logperch	Percina jenkinsi	E	US	6/20/1986	F
Cui-ui	Chasmistes cujus	E	US	5/15/1992	RF(2)
Cumberland darter	Etheostoma susanae	E	US	10/13/2011	0
Delta smelt	Hypomesus transpacificus	T	US	11/26/1996	F
Desert dace	Eremichthys acros	T	US	5/27/1997	F
Desert pupfish	Cyprinodon macularius	E	US/foreign	12/8/1993	F
Devils Hole pupfish	Cyprinodon diabolis	E	US	9/28/1990	F
Devils River minnow	Dionda diaboli	T	US/foreign	9/13/2005	F
Duskytail darter	Etheostoma percnurum	E (1), EXPN (2)	US	None	—
Etowah darter	Etheostoma etowahae	E	US	11/17/2000	F
Foskett speckled dace	Rhinichthys osculus ssp.	T	US	None	—
Fountain darter	Etheostoma fonticola	E	US	2/14/1996	RF(1)
Gila chub	Gila intermedia	E	US/foreign	None	—
Gila topminnow (incl. Yaqui)	Poeciliopsis occidentalis	E	US	None	—
Gila trout	Oncorhynchus gilae	T	US	9/10/2003	RF(3)
Goldline darter	Percina aurolineata	T	US	11/17/2000	F
Greenback cutthroat trout	Oncorhynchus clarki stomias	T	US	3/1/1998	RF(2)
Hiko White River springfish	Crenichthys baileyi grandis	E	US	5/26/1998	F
Humpback chub	Gila cypha	E	US	None	—
Hutton tui chub	Gila bicolor ssp.	T	US	None	—
Independence Valley speckled dace	Rhinichthys osculus lethoporus	E	US	5/12/1998	F
June sucker	Chasmistes liorus	E	US	6/25/1999	F
Kendall Warm Springs dace	Rhinichthys osculus thermalis	E	US	7/12/1982	F
Lahontan cutthroat trout	Oncorhynchus clarki henshawi	T	US	1/30/1995	F
Laurel dace	Phoxinus saylori	E	US	10/11/2011	0
Leon Springs pupfish	Cyprinodon bovinus	E	US	8/14/1985	F
Leopard darter	Percina pantherina	T	US	5/3/1993	RD(1)
Little Colorado spinedace	Lepidomeda vittata	T	US	1/9/1998	F
Little Kern golden trout	Oncorhynchus aguabonita whitei	T	US	Exempt	—
Loach minnow	Tiaroga cobitis	T	US/foreign	9/30/1991	F
Lost River sucker	Deltistes luxatus	E	US	10/12/2011	RD(1)
Maryland darter	Etheostoma sellare	E	US	10/17/1985	RF(1)
Moapa dace	Moapa coriacea	E	US	5/16/1996	RF(1)
Modoc sucker	Catostomus microps	E	US	Exempt	—

TABLE 4.2

Endangered and threatened freshwater fish species, November 2011 [CONTINUED]

Common name	Scientific name	Listing status[a]	U.S. or U.S./foreign listed	Recovery plan date	Recovery plan stage[b]
Mohave tui chub	Gila bicolor mohavensis	E	US	9/12/1984	F
Neosho madtom	Noturus placidus	T	US	9/30/1991	F
Niangua darter	Etheostoma nianguae	T	US	7/17/1989	F
Okaloosa darter	Etheostoma okaloosae	T	US	10/26/1998	RF(1)
Oregon chub	Oregonichthys crameri	T	US	9/3/1998	F
Owens pupfish	Cyprinodon radiosus	E	US	9/30/1998	F
Owens tui chub	Gila bicolor snyderi	E	US	9/30/1998	F
Ozark cavefish	Amblyopsis rosae	T	US	12/17/1986	F
Pahranagat roundtail chub	Gila robusta jordani	E	US	5/26/1998	F
Pahrump poolfish	Empetrichthys latos	E	US	3/17/1980	F
Paiute cutthroat trout	Oncorhynchus clarki seleniris	T	US	9/10/2004	RF(1)
Palezone shiner	Notropis albizonatus	E	US	7/7/1997	F
Pallid sturgeon	Scaphirhynchus albus	E	US	11/7/1993	F
Pecos bluntnose shiner	Notropis simus pecosensis	T	US	9/30/1992	F
Pecos gambusia	Gambusia nobilis	E	US	5/9/1983	F
Pygmy madtom	Noturus stanauli	E (1), EXPN (1)	US	None	—
Pygmy sculpin	Cottus paulus (=pygmaeus)	T	US	8/6/1991	F
Railroad Valley springfish	Crenichthys nevadae	T	US	3/15/1997	F
Razorback sucker	Xyrauchen texanus	E	US/foreign	None	—
Relict darter	Etheostoma chienense	E	US	7/31/1994	D
Rio Grande silvery minnow	Hybognathus amarus	E (1), EXPN (1)	US/foreign	None	—
Roanoke logperch	Percina rex	E	US	3/20/1992	F
Rush darter	Etheostoma phytophilum	E	US	10/20/2011	0
San Marcos gambusia	Gambusia georgei	E	US	2/14/1996	RF(1)
Santa Ana sucker	Catostomus santaanae	T	US	None	—
Scioto madtom	Noturus trautmani	E	US	Exempt	—
Shortnose sucker	Chasmistes brevirostris	E	US	10/12/2011	RD(1)
Shovelnose sturgeon	Scaphirhynchus platorynchus	SAT	US	None	—
Slackwater darter	Etheostoma boschungi	T	US	3/8/1984	F
Slender chub	Erimystax cahni	T (1), EXPN (1)	US	7/29/1983	F
Smoky madtom	Noturus baileyi	E (1), EXPN (1)	US	None	—
Snail darter	Percina tanasi	T	US	5/5/1983	F
Sonora chub	Gila ditaenia	T	US/foreign	9/30/1992	F
Spikedace	Meda fulgida	T	US/foreign	9/30/1991	F
Spotfin chub	Erimonax monachus	T (1), EXPN (3)	US	None	—
Tidewater goby	Eucyclogobius newberryi	E	US	None	—
Topeka shiner	Notropis topeka (=tristis)	E	US	None	—
Unarmored threespine stickleback	Gasterosteus aculeatus williamsoni	E	US	12/26/1985	RF(1)
Vermilion darter	Etheostoma chermocki	E	US	8/6/2007	F
Virgin River chub	Gila seminuda (=robusta)	E	US	4/19/1995	RF(2)
Waccamaw silverside	Menidia extensa	T	US	8/11/1993	F
Warm Springs pupfish	Cyprinodon nevadensis pectoralis	E	US	9/28/1990	F
Warner sucker	Catostomus warnerensis	T	US	4/27/1998	F
Watercress darter	Etheostoma nuchale	E	US	3/29/1993	RF(2)
White River spinedace	Lepidomeda albivallis	E	US	3/28/1994	F
White River springfish	Crenichthys baileyi baileyi	E	US	5/26/1998	F
White sturgeon	Acipenser transmontanus	E	US/foreign	None	—
Woundfin	Plagopterus argentissimus	E (1), EXPN (1)	US	None	—
Yaqui catfish	Ictalurus pricei	T	US/foreign	3/29/1995	F
Yaqui chub	Gila purpurea	E	US/foreign	3/29/1995	F
Yellowcheek darter	Etheostoma moorei	E	US	None	—
Yellowfin madtom	Noturus flavipinnis	T (1), EXPN (3)	US	None	—

[a]E = endangered; T = threatened; SAT = threatened due to similarity of appearance; EXPN = experimental population, non-essential.
[b]F = final; D = draft; RD = draft revision; RF = final revision; O = other.

SOURCE: Adapted from "Generate Species List," in *Species Reports*, U.S. Department of the Interior, U.S. Fish & Wildlife Service, November 2011, http://ecos.fws.gov/tess_public/pub/adHocSpeciesForm.jsp (accessed November 8, 2011), and "Listed FWS/Joint FWS and NMFS Species and Populations with Recovery Plans (Sorted by Listed Entity)," in *Recovery Plans Search*, U.S. Department of the Interior, U.S. Fish & Wildlife Service, November 2011, http://ecos.fws.gov/tess_public/pub/speciesRecovery.jsp?sort=1 (accessed November 8, 2011)

TABLE 4.3

Freshwater fish species with the highest expenditures under the Endangered Species Act, fiscal year 2010

Ranking	Species	Expenditure
1	Trout, bull (Salvelinus confluentus)—U.S.A., conterminous, lower 48 states	$43,809,838
2	Sturgeon, pallid (Scaphirhynchus albus)	$22,696,252
3	Minnow, Rio Grande silvery (Hybognathus amarus)—Rio Grande, from Little Box Canyon to Amistad Dam	$10,857,819
4	Sucker, razorback (Xyrauchen texanus)—entire	$8,089,950
5	Kootenai R. system	$7,768,332
6	Sucker, Santa Ana (Catostomus santaanae)—3 CA river basins	$7,437,464
7	Chub, humpback (Gila cypha)—entire	$4,907,675
8	Trout, Lahontan cutthroat (Oncorhynchus clarki henshawi)	$4,280,684
9	Chub, bonytail (Gila elegans)—entire	$4,236,525
10	Pikeminnow (=squawfish), Colorado (Ptychocheilus lucius)—except Salt and Verde R. drainages, AZ	$3,573,523

SOURCE: Adapted from "Table 2. Species Ranked in Descending Order of Total FY 2010 Reported Expenditures, Not Including Land Acquisition Costs," in *Federal and State Endangered and Threatened Species Expenditures: Fiscal Year 2010*, U.S. Department of the Interior, U.S. Fish and Wildlife Service, 2010, http://www.fws.gov/endangered/esa-library/pdf/2010.EXP.FINAL.pdf (accessed October 25, 2011)

FIGURE 4.3

Bull trout

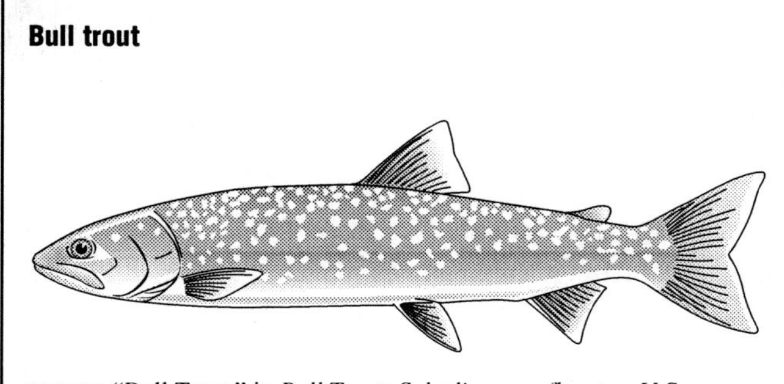

SOURCE: "Bull Trout," in *Bull Trout: Salvelinus confluentus*, U.S. Department of the Interior, U.S. Fish and Wildlife Service, January 2003, http://training.fws.gov/library/Pubs/bulltrt03.pdf (accessed November 7, 2011)

fish called brook trout has been devastating. The two species are able to mate, but produce mostly sterile offspring, which is a genetic dead end for the imperiled bull trout.

The legal history of the bull trout is extensive. In 1992 three environmental groups petitioned the USFWS to list the fish as an endangered species under the ESA. In 1993 the agency concluded that listing the species was warranted, but low in priority. This set off a long series of court battles that culminated in 1999, when all bull trout in the coterminous United States (the lower 48 states) were listed as threatened under the ESA. In 2001 the Alliance for the Wild Rockies and Friends of the Wild Swan (two of the original petitioners) filed a lawsuit against the USFWS for failing to designate critical habitat for the bull trout. A settlement was reached in 2002. In September 2005 the USFWS indicated in the press release "U.S. Fish and Wildlife Service Designates Critical Habitat for Bull Trout" (http://www.fws.gov/news/NewsReleases/) that it designated 4,812 miles (7,746 km) of stream/coastal shoreline and 143,218 acres (57,958 ha) of lakes and reservoirs as critical habitat for the bull trout in Idaho, Montana, Oregon, and Washington.

However, in 2006 the Alliance for the Wild Rockies and Friends of the Wild Swan sued over the designation. According to the USFWS (October 18, 2011, http://www.gpo.gov/fdsys/search/citation.result.FR.action?federalRegister.volume=2010&federalRegister.page=63898&publication=FR), the plaintiffs (suing parties) alleged that the agency had "failed to designate adequate critical habitat, failed to rely on the best scientific and commercial data available, failed to consider the relevant factors that led to listing, and failed to properly assess the economic benefits and costs of critical habitat designation." In October 2011 the USFWS again designated critical habitat, this time consisting of 19,729 miles (31,751 km) of stream/coastal shoreline and 488,252 acres (197,589 ha) of lakes and reservoirs in five states: Idaho, Montana, Nevada, Oregon, and Washington. In the critical habitat rule the USFWS stated its intention to revise the 2002 recovery plan for the bull trout to include six recovery units: coastal, Columbia headwaters, mid-Columbia, Saint Mary, Upper Snake, and Klamath. (See Figure 4.4.) The agency noted that "conserving each [recovery unit] is essential to conserving the listed entity as a whole." As of January 2012, the proposed revised recovery plan for the bull trout had not been issued.

In June 2011 the USFWS (http://www.gpo.gov/fdsys/pkg/FR-2011-06-21/pdf/2011-15370.pdf) listed a population of bull trout in the Clackamas River in Oregon under the ESA as a nonessential experimental population.

In 2008 the USFWS announced completion of a five-year status review of the bull trout. In the press release "Status Review of Bull Trout Completed" (April 29, 2008, http://www.fws.gov/mountain-prairie/pressrel/08-37.htm), the agency recommends that the species remain listed under the ESA as threatened throughout its range in the coterminous United States. In addition, the USFWS notes that scientists believe there are multiple distinct population segments (DPSs) of bull trout in the lower 48 states and recommends that each DPS be evaluated separately for listing under the ESA. The agency lists four requirements for the bull trout's continued survival: "cold water, clean streambed gravel, complex stream habitat features and connected habitats for migration across the landscape." Bull trout are increasingly imperiled by too-warm and sediment-laden water, degradation of habitat and migration corridors (e.g., by dams), and past introductions of nonnative competing species of trout.

FIGURE 4.4

Proposed recovery units for bull trout

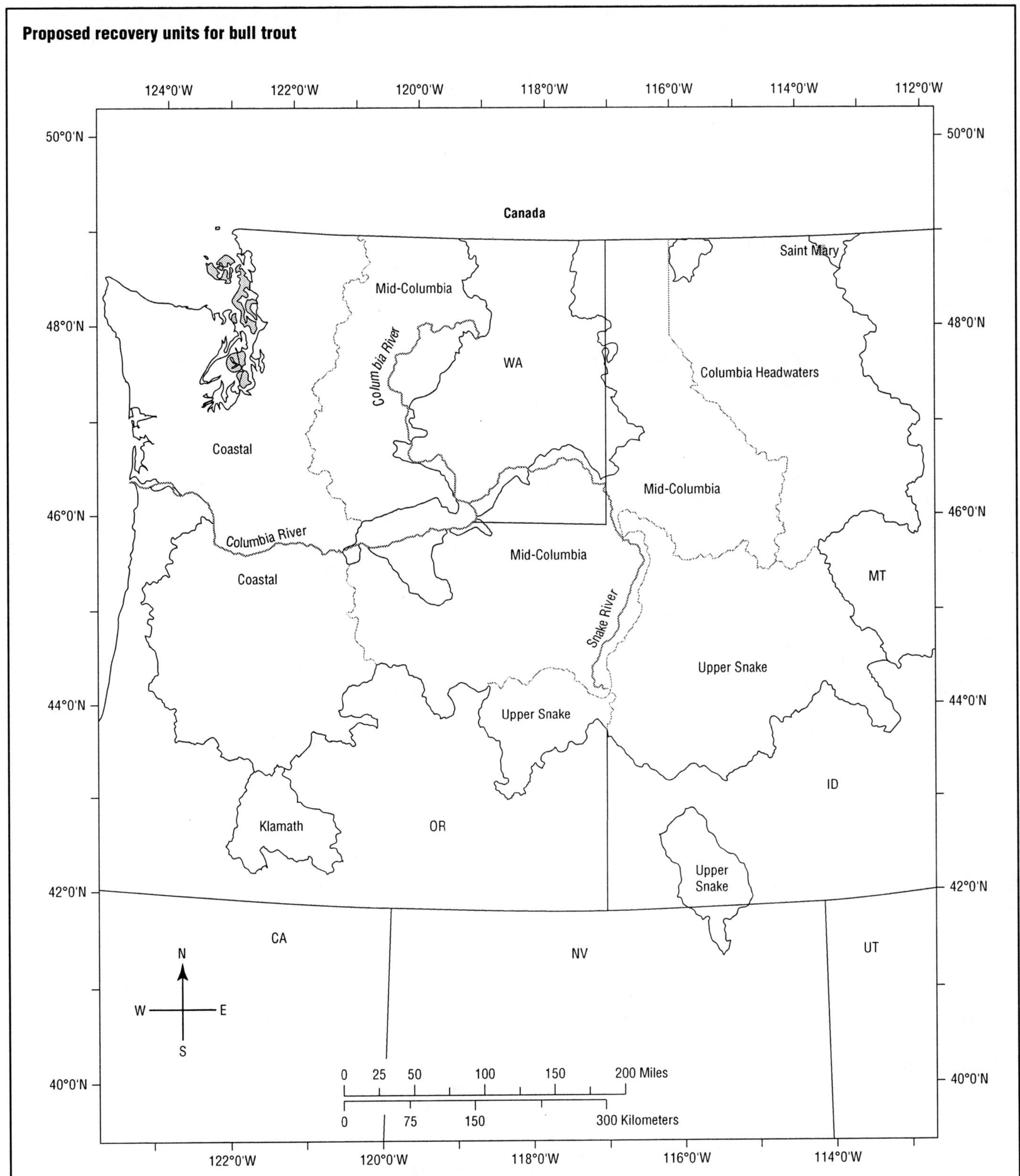

SOURCE: "Figure 1. Bull Trout Recovery Units," in "Endangered and Threatened Wildlife and Plants; Revised Designation of Critical Habitat for Bull Trout in the Coterminous United States," *Federal Register*, vol. 75, no. 200, October 18, 2010, http://www.fws.gov/pacific/bulltrout/pdf/BTCHFR101810.pdf (accessed November 7, 2011)

Pallid Sturgeon

The pallid sturgeon is a unique and rare freshwater fish that is sometimes called the "swimming dinosaur." It is descended from fish that were common more than 50 million years ago. The pallid sturgeon has a long flat snout and a slender body that ends with a pronounced tail fin. (See Figure 4.5.) Adults range in size from 3 to 5 feet (0.9 to 1.5 m) and typically weigh 25 to 50 pounds (11.3 to 22.7 kg). The fish is a bottom-feeder and prefers large rivers of relatively warm free-flowing water with high turbidity (high mud content).

Historically, the pallid sturgeon was found throughout the Mississippi and Missouri River systems from Montana and North Dakota south to the Gulf of Mexico. During the early 1900s specimens as large as 85 pounds (38.6 kg) and 6 feet (1.8 m) long were reported. Over the next century the fish virtually disappeared. In 1990 it was listed as endangered under the ESA. Three years later the USFWS published *Pallid Sturgeon Recovery Plan (*Scaphirhynchus albus*)* (1993, http://ecos.fws.gov/docs/recovery_plans/1993/931107.pdf). The agency blames human destruction and modification of habitat as the two primary causes for the pallid sturgeon's decline.

Pallid sturgeons are believed to be extremely sensitive to changes in the velocity and volume of river flows. They are nearly blind and forage along muddy river bottoms feeding on tiny fish and other creatures that prefer turbid waters. Dams and channelization have reduced the erosion of riverbank soil into the Missouri and Mississippi Rivers, their last remaining primary habitat. This has given other fish species with better eyesight an advantage over the pallid sturgeon at finding small prey. In addition, mating between the pallid sturgeon and the shovelnose sturgeon in the lower Mississippi River has produced a population of hybrid sturgeon that is thriving compared with their imperiled parents.

FIGURE 4.5

Pallid sturgeon

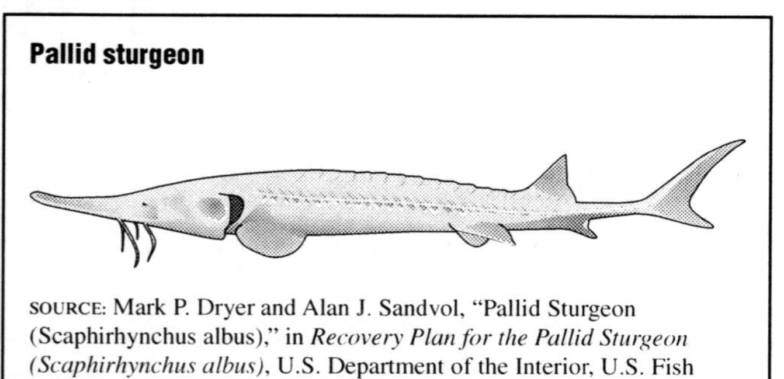

SOURCE: Mark P. Dryer and Alan J. Sandvol, "Pallid Sturgeon (Scaphirhynchus albus)," in *Recovery Plan for the Pallid Sturgeon (Scaphirhynchus albus)*, U.S. Department of the Interior, U.S. Fish and Wildlife Service, 1993, http://ecos.fws.gov/docs/recovery_plans/1993/931107.pdf (accessed November 7, 2011)

All these factors combine to provide a bleak outlook for the future of the pallid sturgeon. The USFWS recovery plan notes that "it is unlikely that successfully reproducing populations of pallid sturgeon can be recovered without restoring the habitat elements (morphology, hydrology, temperature regime, cover, and sediment/organic matter transport) of the Missouri and Mississippi Rivers necessary for the species continued survival."

In *Pallid Sturgeon (*Scaphirhynchus albus*): 5-Year Review—Summary and Evaluation* (June 15, 2007, http://ecos.fws.gov/docs/five_year_review/doc1059.pdf), the USFWS examines the scientific and commercial data that have become available since the species was listed as endangered in 1990. The agency concludes that the pallid sturgeon should remain listed as endangered with a recovery priority number of 2c. The USFWS ranks recovery priority on a scale from 1 to 18, with lower numbers indicating higher priority. The "c" designation means the recovery of this species is in conflict with economic activities in its region.

The Comprehensive Sturgeon Research Program (CSRP) is an interagency collaboration between the U.S. Geological Survey, the Nebraska Game and Parks Commission, the USFWS, and the Corps' Missouri River Recovery Integrated Science Program to study the pallid sturgeon. Aaron J. DeLonay et al. summarize the latest scientific findings from CSRP research in *Ecological Requirements for Pallid Sturgeon Reproduction and Recruitment in the Lower Missouri River: A Research Synthesis 2005–08* (September 22, 2009, http://pubs.usgs.gov/sir/2009/5201/pdf/sir2009_5201.pdf). The researchers discuss what is known about the movement, habitat use, and reproductive behavior of the pallid sturgeon and the recommended measures that may increase the survivability and reproductive success of the species. These measures include habitat restoration activities and channel projects that are designed to provide and improve rearing habitat.

RECOVERY PLANS FOR FRESHWATER FISH. As of November 2011, there were recovery plans for more than 100 populations of listed freshwater fish populations. (See Table 4.2.) Most of the plans were finalized. The USFWS provides details of the plans in "Listed FWS/Joint FWS and NMFS Species and Populations with Recovery Plans" (January 2012, http://ecos.fws.gov/tess_public/TESSWebpageRecovery?sort=1).

IMPERILED MARINE AND ANADROMOUS FISH

As noted earlier, marine fish inhabit the salty waters of seas and oceans. In reality, some species of fish thrive in both fresh and marine waters. For example, anadromous fish migrate between freshwater and the sea for mating purposes. As of November 2011, there were 15 marine and anadromous fish species listed under the

TABLE 4.4

Endangered and threatened marine and anadromous fish species, November 2011

Common name	Scientific name	Listing status*	Recovery plan date	Recovery plan stage
Atlantic salmon	Salmo salar	E	None	—
Bocaccio	Sebastes paucispinis	E	None (listed in 2010)	—
Canary rockfish	Sebastes pinniger	T	None (listed in 2010)	—
Chinook salmon	Oncorhynchus (=Salmo) tshawytscha	E (2) and T (7)	None	—
Chum salmon	Oncorhynchus (=Salmo) keta	T (2)	None	—
Coho salmon	Oncorhynchus (=Salmo) kisutch	E (1) and T (3)	None	—
Gulf sturgeon	Acipenser oxyrinchus desotoi	T	9/22/1995	Final
Largetooth sawfish	Pristis perotteti	E	None (listed in 2011)	—
North American green sturgeon	Acipenser medirostris	T	None	—
Pacific eulachon (smelt)	Thaleichthys pacificus	T	None (listed in 2010)	—
Shortnose sturgeon	Acipenser brevirostrum	E	12/17/1998	Final
Smalltooth sawfish	Pristis pectinata	E	1/21/2009	Final
Sockeye salmon	Oncorhynchus (=Salmo) nerka	E (1) and T (1)	None	—
Steelhead	Oncorhynchus (=Salmo) mykiss	E (1) and T (9)	None	—
Yelloweye rockfish	Sebastes ruberrimus	T	None (listed in 2010)	—

*E = endangered; T = threatened. Numbers in parentheses indicate separate populations.

SOURCE: Adapted from "Generate Species List," in *Species Reports*, U.S. Department of the Interior, U.S. Fish & Wildlife Service, November 2011, http://ecos.fws.gov/tess_public/pub/adHocSpeciesForm.jsp (accessed November 8, 2011), and "Listed FWS/Joint FWS and NMFS Species and Populations with Recovery Plans (Sorted by Listed Entity)," in *Recovery Plans Search*, U.S. Department of the Interior, U.S. Fish & Wildlife Service, November 2011, http://ecos.fws.gov/tess_public/pub/speciesRecovery.jsp?sort=1 (accessed November 8, 2011)

ESA. (See Table 4.4.) They are under the jurisdiction of the National Marine Fisheries Service (NMFS). The bocaccio, canary rockfish, and yelloweye rockfish are true marine species. They inhabit the Puget Sound in Washington.

The smalltooth and largetooth sawfish, gulf sturgeon, Pacific eulachon, and shortnose sturgeon prefer coastal marine and estuarine waters (areas where saltwater and freshwater meet). The North American green sturgeon is anadromous and inhabits Pacific coast waters. Five other imperiled anadromous species are found in the Pacific Northwest:

- Chinook salmon
- Chum salmon
- Coho salmon
- Sockeye salmon
- Steelhead

In addition, there is an endangered species of Atlantic salmon that migrates between the Gulf of Maine and Maine's rivers.

Table 4.5 shows that $529.9 million was spent on the 10 anadromous fish species listed under the ESA during FY 2010. As noted in Table 2.16 in Chapter 2, six of the species were among the 20 overall species with the highest spending during FY 2010: chinook salmon, steelhead, coho salmon, sockeye salmon, chum salmon, and Atlantic salmon. The chinook salmon topped the spending list with expenditures of $230.3 million in FY 2010.

TABLE 4.5

Marine and anadromous fish species with the highest expenditures under the Endangered Species Act, fiscal year 2010

Ranking	Species	Expenditure
1	Salmon, chinook	$230,268,061
2	Steelhead	$195,729,612
3	Salmon, coho	$37,596,542
4	Salmon, sockeye	$22,535,318
5	Salmon, chum	$18,562,958
6	Salmon, Atlantic	$13,145,296
7	Sturgeon, shortnose (Acipenser brevirostrum)	$4,569,131
8	Sturgeon, North American green (Acipenser medirostris)-CA-Southern DPS	$3,658,452
9	Sturgeon, gulf (Acipenser oxyrinchus desotoi)	$2,320,107
10	Sawfish, smalltooth (Pristis pectinata)	$1,555,876

DPS = distinct population segments.

SOURCE: Adapted from "Table 2. Species Ranked in Descending Order of Total FY 2010 Reported Expenditures, Not Including Land Acquisition Costs," in *Federal and State Endangered and Threatened Species Expenditures: Fiscal Year 2010*, U.S. Department of the Interior, U.S. Fish and Wildlife Service, 2010, http://www.fws.gov/endangered/esa-library/pdf/2010.EXP.FINAL.pdf (accessed October 25, 2011)

Pacific Salmonids

Pacific salmonids are found in waters of the northwestern United States and belong to the genus *Oncorhynchus*. There are five species of Pacific salmon: chinook, chum, coho, pink, and sockeye. As of November 2011, all but the pink salmon were listed under the ESA as endangered or threatened. (See Table 4.4.) A detailed discussion about Pacific salmonids is provided by the USFWS in "Pacific Salmon (*Oncorhynchus spp.*)" (2012, http://www.fws.gov/species/species_accounts/bio_salm.html).

Chinook salmon are the largest of the Pacific salmonids, averaging about 24 pounds (11 kg) in adulthood. (See Figure 4.6.) They spend two to seven years in the ocean and travel up to 2,500 miles (4,000 km) from their home streams. Chum, coho, and sockeye salmon adults average approximately 10 to 12 pounds (4.5 to 5.4 kg).

PERILS OF MIGRATION. Pacific salmon pose unique protection challenges because they are anadromous. Salmon eggs (or roe) are laid in the bottom gravel of cold freshwater streams, where they incubate for five to 10 weeks. Each egg ranges in size from 0.3 to 0.5 of an inch (0.8 to 1.3 cm) in size, depending on the species. The eggs hatch to release baby fish (or alevin) that are called fry as they mature. Once a fry reaches about 3 inches (7.6 cm) in length, it is called a fingerling. This typically takes less than a year.

FIGURE 4.6

Some Pacific salmon species

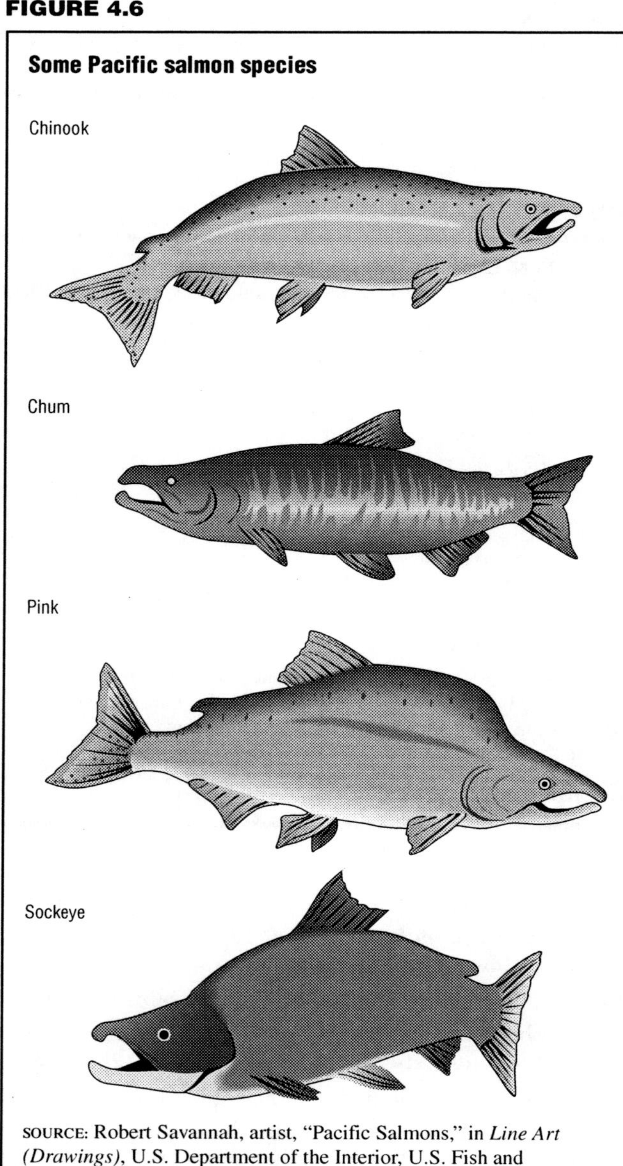

SOURCE: Robert Savannah, artist, "Pacific Salmons," in *Line Art (Drawings)*, U.S. Department of the Interior, U.S. Fish and Wildlife Service, undated, http://www.fws.gov/pictures/lineart/bobsavannah/pacificsalmons.html (accessed November 7, 2011)

At some point during their first two years the young salmon (now called smolts) migrate downstream to the ocean. There they spend several months or years of their adulthood. When they reach sexual maturity, males and females journey back to the streams where they were born to mate and deposit eggs. This is called spawning. Pacific salmon make the round-trip only once. They expend all their energy swimming back upstream and die soon after the eggs are laid and fertilized. Their upstream habitats can be hundreds and even thousands of miles away from their ocean habitats. It is a long and dangerous journey both ways.

Predator fish and birds eat salmon fry, fingerlings, and smolts as they make their way to the ocean. Bears, birds, marine mammals, and humans prey on the adult fish as they migrate upstream. Waterfalls, rapids, dams, and other water diversions pose tremendous obstacles to Pacific salmon as they try to travel across long distances.

STOCKS. Salmon heading to the same general location travel upstream in groups called stocks (or runs). Stocks migrate at different times of the year, depending on geographical and genetic factors. Figure 4.7 illustrates the life cycle of a stock that migrates upstream from late summer through early fall. During its lifetime a Pacific salmon is exposed to three different water environments: freshwater streams and rivers, estuaries (areas where freshwater and saltwater meet), and the ocean.

DECLINING POPULATIONS. Daniel L. Bottom et al. estimate in *Salmon at River's End: The Role of the Estuary in the Decline and Recovery of Columbia River Salmon* (August 2005, http://www.nwfsc.noaa.gov/assets/25/6294_09302005_153156_SARETM68Final.pdf) that between 11 million and 16 million salmon per year migrated upstream in waters of the northwestern United States before the arrival of European settlers. Extensive fishing and canning operations quickly decimated the salmon population. As early as 1893 federal officials warned that the future of salmon fisheries had a "disastrous outlook." During the 1890s hatcheries began operating and stocking rivers and streams with farm-raised salmon. Throughout the next century salmon populations were further stressed as natural river flows were dramatically altered with dams, navigational structures, and irrigation systems. By 2012 dozens of salmon hatcheries and hundreds of dams had been constructed in the Columbia River basin of the Pacific Northwest.

Endangered and threatened salmon are identified by their water of origin and, in most cases, by their upstream migration season. In 1990 the winter-run stock of chinook salmon from the Sacramento River was designated

FIGURE 4.7

Life cycle of Pacific salmon

Ecological context

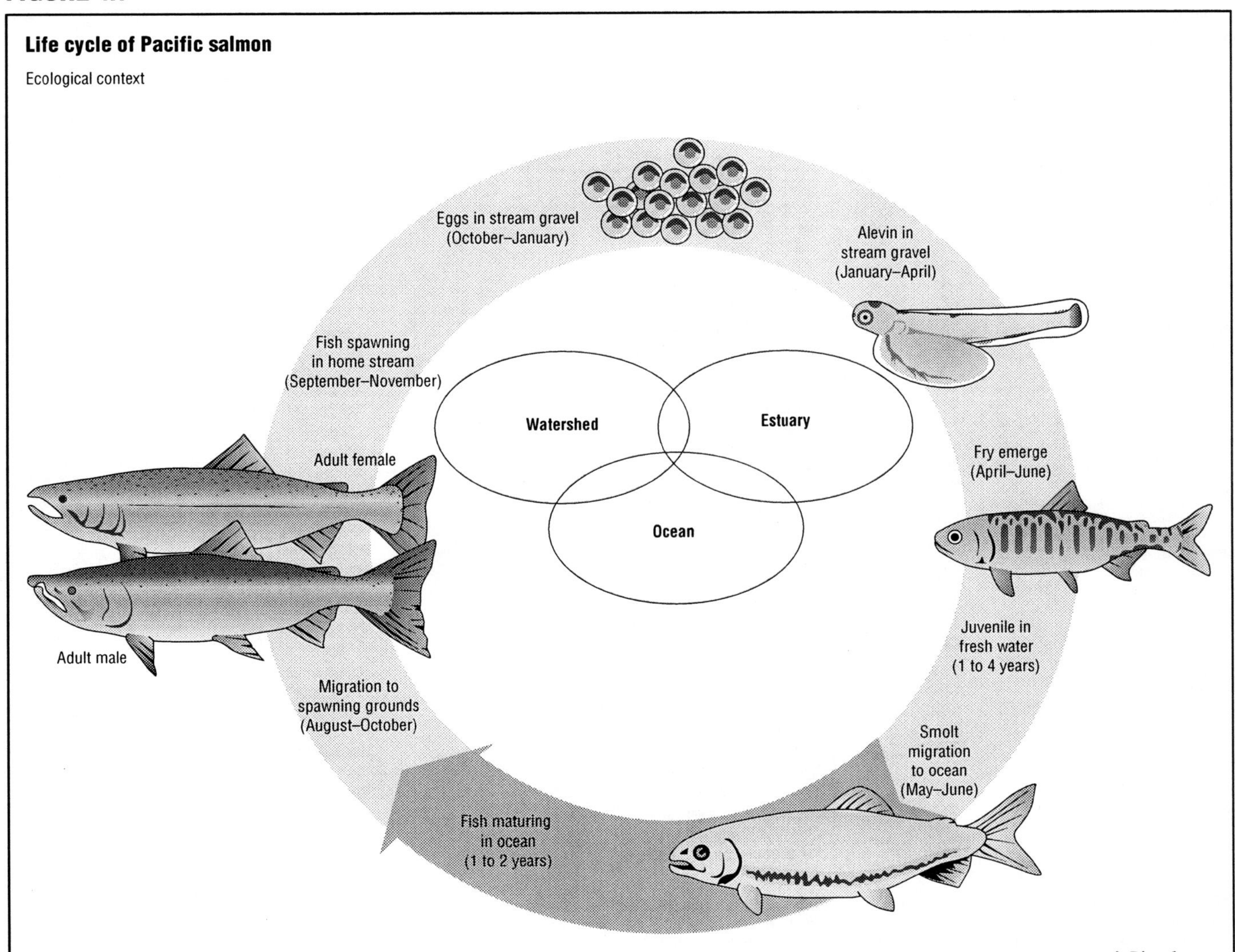

SOURCE: "Figure 1A. A Generalized Depiction of the Pacific Salmon Life Cycle," in *The Northwest Fisheries Science Center Strategic Research Plan for Salmon: Final Draft*, U.S. Department of Commerce, National Oceanic and Atmospheric Administration, National Marine Fisheries Service, June 17, 2004, http://www.nwfsc.noaa.gov/publications/researchplans/salmon_research_plan6.17.04%20.pdf (accessed November 7, 2011)

as threatened under the ESA, the first Pacific salmon to be listed. It was reclassified as endangered four years later. During the 1990s and the first decade of the 21st century the NMFS identified 35 evolutionarily significant units (ESUs) of Pacific salmonids. As of August 2011, 17 of them were listed as endangered or threatened. (See Table 4.6.) Two ESUs were "species of concern," meaning that the agency had some concerns regarding threats to these species, but lacked sufficient information indicating the need to list them under the ESA.

THREATS TO SURVIVAL. Biologists blame four main threats for the imperiled state of Pacific salmonids:

- Habitat degradation—channelization, dredging, water withdrawals for irrigation, wetland losses, and diking have changed river, stream, and estuary environments.
- Harvesting levels—overfishing for more than a century has decimated salmon populations.
- Hatcheries—biologists fear that hatchery releases overburden estuaries with too many competing fish at the same time.
- Hydropower—impassable dams have rendered some historical habitats unreachable by salmon. Most modern dams have fish ladders. However, all dams affect water temperature, flow, and quality.

Besides these threats, scientists believe climate change and the presence of nonnative aquatic species are detrimental to salmon populations.

The effects of climate change to Pacific salmon are addressed by Thomas R. Karl, Jerry M. Melillo, and Thomas C. Peterson in *Global Climate Change Impacts in the United States* (2009, http://downloads.globalchange.gov/usimpacts/pdfs/climate-impacts-report.pdf). According to the researchers, salmon are at particular risk due to warming of their cold-water habitats, the rivers and streams in the

TABLE 4.6

Listing status of West Coast salmon populations, August 2011

		Species*	Current Endangered Species Act listing status	Endangered Species Act listing actions under review
Sockeye salmon (*Oncorhynchus nerka*)	1	Snake River	Endangered	
	2	Ozette Lake	Threatened	
	3	Baker River	Not warranted	
	4	Okanogan River	Not warranted	
	5	Lake Wenatchee	Not warranted	
	6	Quinalt Lake	Not warranted	
	7	Lake Pleasant	Not warranted	
Chinook salmon (*O. tshawytscha*)	8	Sacramento River Winter-run	Endangered	
	9	Upper Columbia River Spring-run	Endangered	
	10	Snake River Spring/Summer-run	Threatened	
	11	Snake River Fall-run	Threatened	
	12	Puget Sound	Threatened	
	13	Lower Columbia River	Threatened	
	14	Upper Willamette River	Threatened	
	15	Central Valley Spring-run	Threatened	
	16	California Coastal	Threatened	
	17	Central Valley Fall and Late Fall-run	Species of concern	
	18	Upper Klamath-Trinity Rivers	Not warranted	
	19	Oregon Coast	Not warranted	
	20	Washington Coast	Not warranted	
	21	Middle Columbia River spring-run	Not warranted	
	22	Upper Columbia River summer/fall-run	Not warranted	
	23	Southern Oregon and Northern California Coast	Not warranted	
	24	Deschutes River summer/fall-run	Not warranted	
Coho salmon (*O. kisutch*)	25	Central California Coast	Endangered	
	26	Southern Oregon/Northern California	Threatened	
	27	Lower Columbia River	Threatened	• Critical habitat
	28	Oregon Coast	Threatened	
	29	Southwest Washington	Undetermined	
	30	Puget Sound/Strait of Georgia	Species of concern	
	31	Olympic Peninsula	Not warranted	
Chum salmon (*O. keta*)	32	Hood Canal Summer-run	Threatened	
	33	Columbia River	Threatened	
	34	Puget Sound/Strait of Georgia	Not warranted	
	35	Pacific Coast	Not warranted	

*The Endangered Species Act (ESA) defines a "species" to include any distinct population segment of any species of vertebrate fish or wildlife. For Pacific salmon, National Oceanic and Atmospheric Administration (NOAA) Fisheries Service considers an evolutionarily significant unit, or "ESU," a "species" under the ESA. For Pacific steelhead, NOAA Fisheries Service has delinated distinct population segments (DPSs) for consideration as "species" under the ESA.

SOURCE: Adapted from "Endangered Species Act Status of West Coast Salmon & Steelhead," in *ESA Salmon Listings*, U.S. Department of Commerce, National Oceanic and Atmospheric Administration, National Marine Fisheries Service, August 11, 2011, http://www.nwr.noaa.gov/ESA-Salmon-Listings/upload/1-pgr-8-11.pdf (accessed November 8, 2011)

Pacific Northwest. Rising temperatures are expected to lead to more rainfall (as opposed to snowfall), which could increase flooding that could dislodge eggs from the streambeds. The eggs may begin to hatch earlier in the spring. This could be problematic if the insects that the young fish commonly eat are not active at this earlier time. It may also expose the fish to different predators. Karl, Melillo, and Peterson also suggest that warming temperatures will generally increase salmon metabolism (and hunger), forcing them to work harder to find sufficient food. Furthermore, warmer waters encourage the growth of diseases and parasites that could further decimate salmon stocks.

In recent years wildlife officials have grown increasingly concerned about predation on imperiled Pacific salmon by sea lions in the lower Columbia River. The problem is particularly severe on the downstream side of the Bonneville Dam. The dam spans the lower Columbia River between Oregon and Washington and is located approximately 145 miles (233 km) upstream of the river's mouth at the Pacific Ocean. In "Columbia River Sea Lion Management" (2012, http://wdfw.wa.gov/conservation/sealions/questions.html), the Washington Department of Fish and Wildlife (WDFW) notes that sea lions have consumed thousands of fish, some of which are listed as threatened or endangered under the ESA. The fish congregate below the dam during their migration upstream; the dam is equipped with fish ladders that provide the fish passage to upstream habitat. Each spring since 2002 trained observers at the dam have counted the numbers of sea lions preying on the fish and the number of fish eaten by the sea lions. The latter have been mostly California sea lions, a thriving species that is not listed under the ESA, but is protected under the MMPA. However, in recent years observers have counted increasing numbers of Steller sea lions (a threatened species) at the dam.

This situation raises the question of what to do when one threatened or endangered species is being harmed by another protected species. According to the WDFW, hazing (the use of boats, underwater firecrackers, and rubber buckshot) is regularly conducted in an attempt to scare the sea lions away, but it only temporarily interrupts the feeding. In 2008 Idaho, Oregon, and Washington received permission from the NMFS to remove or kill identifiable California sea lions that habitually prey on salmon at the dam. This decision was challenged in court by the Humane Society of the United States (HSUS), which argued that the animals are being used as scapegoats by authorities who do not want to reduce fishing limits. A long legal battle ensued, which is described by the HSUS in *Bonneville Dam Sea Lions under Siege* (September 13, 2011, http://www.humanesociety.org/issues/fisheries/timelines/bonneville_dam_sea_lions_under_siege.html). According to the HSUS, between 2009 and 2010 a total of 24 sea lions captured at the dam were killed by lethal injection. Oregon and Washington authorities have repeatedly received approval from the NMFS to remove and kill sea lions, and the HSUS has repeatedly filed lawsuits against the activities. Some of the lawsuits have been temporarily successful in ceasing the removal/kill program. However, in early 2011 Oregon and Washington again asked the NMFS for permission to remove and kill sea lions at the dam. As of January 2012, a final decision had not been made on the application.

In *Evaluation of Pinniped Predation on Adult Salmonids and Other Fish in the Bonneville Dam Tailrace, 2008–2010* (October 14, 2010, http://www.mediate.com/DSConsulting/docs/PINNIPED%202008-2010%20USACE%20REPORT.pdf), Robert J. Stansell, Karrie M. Gibbons, and William T. Nagy of the Corps report the results of a two-year observation program at the dam. Observers estimate that pinnipeds (seals and sea lions) consumed about 4,000 to 6,000 salmonids per year between 2008 and 2010. The number of Steller sea lions observed at the dam increased from an average of 5 per year between 2002 and 2007 to 46.7 per year between 2008 and 2010. Stansell, Gibbons, and Nagy note that various "deterrents" were used by agencies to prevent predation between 2008 and 2010. This included physical barriers and floating gates that were reportedly effective at blocking sea lion passageway. Harassment of the sea lions has also continued. In addition, 38 California sea lions were removed from the dam between 2008 and 2010.

THE KLAMATH RIVER BASIN CONTROVERSY. The Klamath River originates in southern Oregon and flows southwest for more than 200 miles (320 km) across the southern part of Oregon and the northern part of California. (See Figure 4.8.) According to the NMFS, in *Klamath River Basin: 2009 Report to Congress* (June 30, 2009, http://www.nmfs.noaa.gov/pr/pdfs/klamath2009.pdf), the basin covers over 10 million acres (4 million ha) and was once the third-largest producer of salmon on the West Coast. Over the past 100 years the salmon populations have been reduced to a fraction of what they once were. Demand for water in the basin has grown dramatically and sparked bitter battles among fishing, agricultural, timber, mining, and environmental interests. These battles are complicated by the many stakeholders involved. The NMFS notes that approximately 6 million acres (2.4 million ha) in the basin are public lands. These include national wildlife refuges, national forests, national parks and monuments, and wild and scenic river designations—all managed by a variety of federal agencies and programs. In addition, approximately 96,000 acres (39,000 ha) are trust lands of six Native American tribes. The remaining 4 million acres (1.6 million ha) are in private hands. The basin is home to several anadromous species including stocks of chinook and coho salmon and steelhead. As shown in Table 4.6, the Southern Oregon–Northern California stock of coho salmon are listed as threatened under the ESA.

In 2007 President George W. Bush (1946–) signed into law the Magnuson-Steven Fishery Conservation and Management Reauthorization Act of 2006. The act requires the NMFS to issue a recovery plan for the basin's coho salmon and to submit an annual report to Congress beginning in 2009 detailing actions that are taken to restore salmon habitat and water conditions in the basin.

In *Magnuson-Stevens Reauthorization Act Klamath River Coho Salmon Recovery Plan* (July 10, 2007, http://swr.nmfs.noaa.gov/salmon/MSRA_RecoveryPlan_FINAL.pdf), the NMFS provides historical estimates of salmon and steelhead population trends in the basin and describes the human-related factors that have severely depleted those populations. The latter are divided into three broad categories:

- Water management activities—these include dams and impoundments that block off hundreds of miles of historical salmon habitat from present populations, alter water flows, impair water quality, and increase water temperatures. In addition, domestic, agricultural, and municipal users have reduced basin stream flows by diverting water for competing uses. These diversions also increase water temperatures, reduce habitat availability and water quality, and entrain young fish in unscreened pipes. Even though some of this diverted water is later returned to the river, such as from irrigated lands, its quality is degraded compared with its original condition.

FIGURE 4.8

Klamath River basin

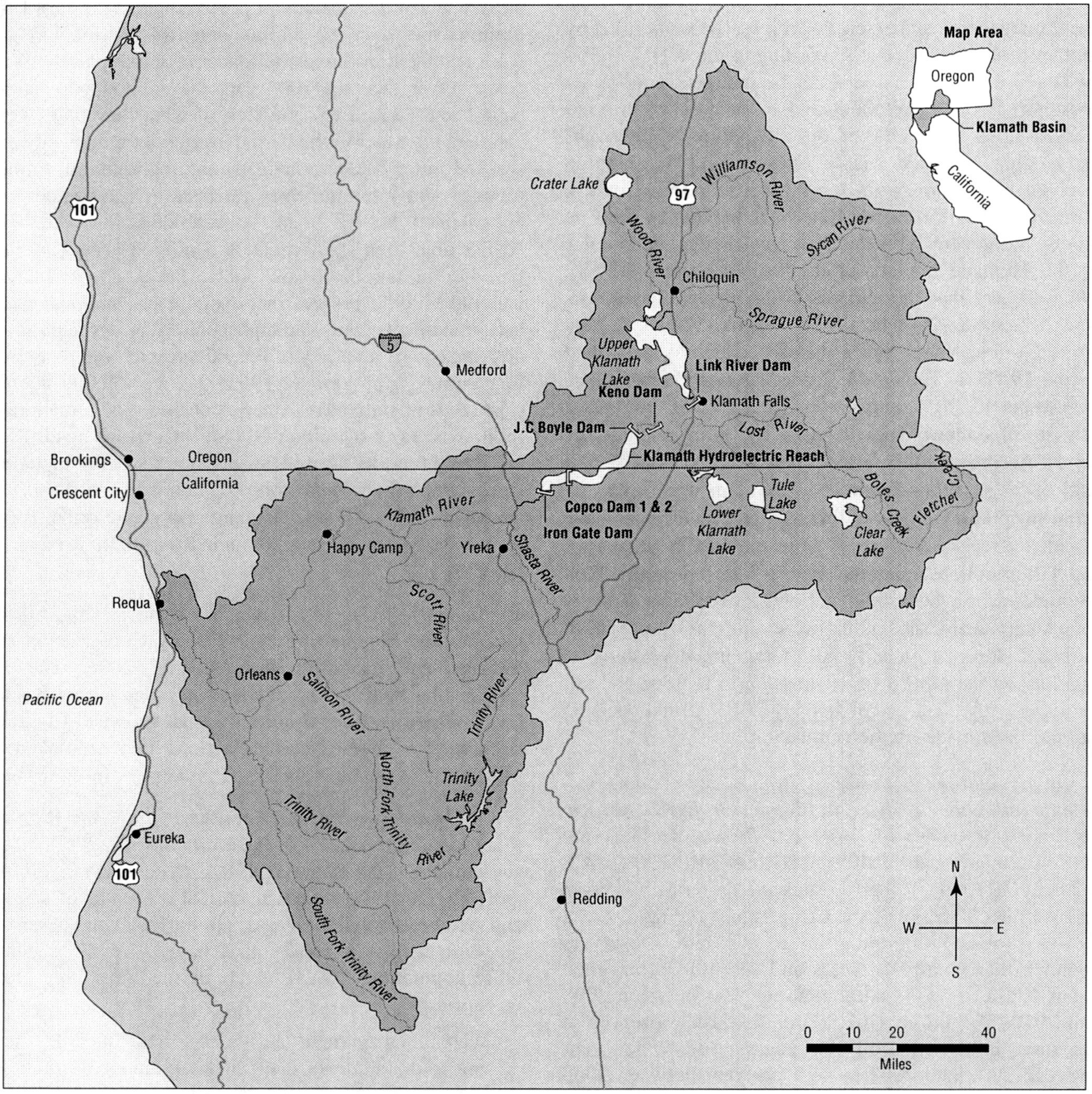

SOURCE: "Figure ES-1. The Klamath Basin," in *Klamath Facilities Removal Public Draft Environmental Impact Statement/Environmental Impact Report*, U.S. Department of the Interior and California Department of Fish and Game, September 2011, http://klamathrestoration.gov/sites/klamathrestoration.gov/files/KlamathFacilitiesRemoval_EISEIR_09222011.pdf (accessed December 10, 2011)

- Land management activities—timber harvesting and associated road building have affected large areas of public and private forestlands in the basin. These activities cause sedimentation of the waters and increase their temperatures. The same negative effects result from gold mining, which involves dredging and suction activities in the rivers and streams.
- Fish management activities—overharvesting by the commercial salmon canning industry during the early 20th century dramatically decreased the basin's

salmon populations. Since the 1980s federal and state regulations have sharply curtailed commercial salmon fishing in the region.

Freshwater resource management in the West is largely handled by the Bureau of Reclamation (BOR) under the U.S. Department of the Interior (DOI). Until 2001 the BOR managed to balance competing water demands from users in the basin and maintain river and stream flows for fish. In 2001 the region was struck by a drought, and the BOR allowed large water diversions for irrigation purposes. The following year tens of thousands of dead salmon were found in the basin, apparently the victims of low water flow. The BOR responded with drastic water cuts to farmers. This led to a full-fledged water war among Native American tribes, commercial fishing interests, conservation groups, farmers, and local water districts. Ever since, BOR-developed operating plans for water flows in the Klamath River basin have been continually challenged in court.

Meanwhile, a separate legal battle has been waged over efforts by PacifiCorp, an Oregon-based utility company, to renew a federal license for four dams it owns on the Klamath River. According to the NMFS, the license expired in 2006, but PacifiCorp can continue to operate the dams under the terms and conditions of its old license until it obtains a new license. One of the dams is called Iron Gate, and it blocks upstream fish passage to the northern reaches of the basin. The old license did not require the company to provide fish ladders. However, the Federal Energy Regulatory Commission, which oversees power companies, has included fish passage requirements as a condition of relicensing the dam. The NMFS notes that the fish passages could open up more than 350 miles (560 km) of historical habitat for salmon and other migrating fish species.

In 2008 a two-year collaborative effort between dozens of stakeholders—federal, state, and local government agencies; irrigation districts; property owners; Native American tribes; and private conservation organizations—resulted in the proposed Klamath Basin Restoration Agreement (KBRA; http://www.edsheets.com/Klamathdocs.html). The KBRA seeks to restore thriving fish populations in the river and to satisfy local needs for irrigation water and power generation. It calls on PacifiCorp to remove its four dams from the Klamath River. In November 2008 PacifiCorp reached an Agreement in Principle with the federal government and the states of California and Oregon to remove the dams beginning in 2020. In September 2009 the DOI announced in the press release "Secretary Salazar Announces Draft Agreement on Klamath Dam Removal Proposal" (http://www.doi.gov/news/pressreleases/2009_09_30_release.cfm) that PacifiCorp had reached a draft agreement with local, state, tribal, and federal agencies; Klamath basin water users; fishing organizations; and environmental groups to remove the four dams. The DOI indicated that the so-called Klamath Hydroelectric Settlement Agreement (KHSA) will "work in tandem" with the proposed KBRA.

According to the NMFS, in *Klamath River Basin: 2011 Report to Congress* (October 3, 2011, http://swr.nmfs.noaa.gov/klamath/klamath_basin_2011_final.pdf), in early 2010 the KBRA and the KHSA were finalized. The agency notes that environmental studies on the effects of dam removal were expected to be completed by early 2012.

Steelhead

Steelhead are members of the *Oncorhynchus* genus and have the scientific name *Oncorhynchus mykiss*. Freshwater steelhead are called rainbow trout. Anadromous steelhead are also trout, but they are associated with salmon due to similarities in habitat and behavior. Steelhead are found in the Pacific Northwest and are anadromous like salmon but have two major differences: steelhead migrate individually, rather than in groups, and can spawn many times, not just once.

As of August 2011, the NMFS had identified 15 DPSs of steelhead. (See Table 4.7.) Eleven of these DPSs were listed under the ESA as endangered or threatened. In addition, there was one DPS classified as a "species of concern." The USFWS indicates that $195.7 million was spent on steelhead under the ESA during FY 2010. (See Table 4.5.)

Steelhead face the same threats as Pacific salmon: habitat loss and alteration, overharvesting, dams and other water obstacles, and competition with hatchery fish.

Recovery Plans for Anadromous Fish

As of January 2012, the NMFS (http://www.nmfs.noaa.gov/pr/recovery/plans.htm) had published recovery plans for several anadromous fish species, including the Atlantic salmon, the gulf sturgeon, and various salmon and steelhead populations.

Imperiled Fish around the World

In *Red List of Threatened Species Version 2011.2* (2011, http://www.iucnredlist.org/documents/summarystatistics/2011_2_RL_Stats_Table1.pdf), the International Union for Conservation of Nature (IUCN) indicates that in 2011, 2,028 out of 9,554 species of evaluated fish were threatened. The IUCN notes that there are 32,100 known fish species, so it is expected that many more fish species will be listed in the future as more evaluations are completed.

As of November 2011, the USFWS listed 12 foreign species of fish as endangered or threatened. (See Table 4.8.) All but one of the species were endangered. Most of the species are found in Asia.

TABLE 4.7

Listing status of West Coast steelhead populations, August 2011

		Species*	Current Endangered Species Act listing status	Endangered Species Act listing actions under review
Steelhead	36	Southern California	Endangered	
(*O. mykiss*)	37	Upper Columbia River	Threatened	
	38	Central California Coast	Threatened	
	39	South Central California Coast	Threatened	
	40	Snake River Basin	Threatened	
	41	Lower Columbia River	Threatened	
	42	California Central Valley	Threatened	
	43	Upper Willamette River	Threatened	
	44	Middle Columbia River	Threatened	
	45	Northern California	Threatened	
	46	Oregon Coast	Species of concern	
	47	Southwest Washington	Not warranted	
	48	Olympic Peninsula	Not warranted	
	49	Puget Sound	Threatened	• Critical habitat
	50	Klamath Mountains Province	Not warranted	

*The Endangered Species Act (ESA) defines a "species" to include any distinct population segment of any species of vertebrate fish or wildlife. For Pacific salmon, National Oceanic and Atmospheric Administration (NOAA) Fisheries Service considers an evolutionarily significant unit, or "ESU," a "species" under the ESA. For Pacific steelhead, NOAA Fisheries Service has delinated distinct population segments (DPSs) for consideration as "species" under the ESA.

SOURCE: Adapted from "Endangered Species Act Status of West Coast Salmon & Steelhead," in *ESA Salmon Listings*, U.S. Department of Commerce, National Oceanic and Atmospheric Administration, National Marine Fisheries Service, August 11, 2011, http://www.nwr.noaa.gov/ESA-Salmon-Listings/upload/1-pgr-8-11.pdf (accessed November 8, 2011)

TABLE 4.8

Foreign endangered and threatened fish species, November 2011

Common name	Scientific name	Historic range	Listing status*
Ala balik (trout)	Salmo platycephalus	Turkey	E
Asian bonytongue	Scleropages formosus	Thailand, Indonesia, Malaysia	E
Ayumodoki (loach)	Hymenophysa curta	Japan	E
Beluga sturgeon	Huso huso	Black Sea, Caspian Sea, Adriatic Sea and Sea of Azov .	T
Catfish	Pangasius sanitwongsei	Thailand	E
Cicek (minnow)	Acanthorutilus handlirschi	Turkey	E
Ikan temoleh (minnow)	Probarbus jullieni	Thailand, Cambodia, Vietnam, Malaysia, Laos	E
Mexican blindcat (catfish)	Prietella phreatophila	Mexico	E
Miyako tango (=Toyko bitterling)	Tanakia tanago	Japan	E
Nekogigi (catfish)	Coreobagrus ichikawai	Japan	E
Thailand giant catfish	Pangasianodon gigas	Thailand	E
Totoaba (seatrout or weakfish)	Cynoscion macdonaldi	Mexico (Gulf of California)	E

*E = endangered; T = threatened.

SOURCE: Adapted from "Generate Species List," in *Species Reports*, U.S. Department of the Interior, U.S. Fish & Wildlife Service, November 2011, http://ecos.fws.gov/tess_public/pub/adHocSpeciesForm.jsp (accessed November 8, 2011)

CHAPTER 5
CLAMS, SNAILS, CRUSTACEANS, AND CORALS

CLAMS, SNAILS, AND CRUSTACEANS

Clams, snails, and crustaceans are small aquatic creatures. They are invertebrates, meaning they lack an internal skeleton made of bone or cartilage. Clams and snails are in the phylum Mollusca. Mollusks have soft bodies that are usually enclosed in a thin hard shell made of calcium. The U.S. Fish and Wildlife Service (USFWS) uses the generic term *clam* to refer to clams and mussels, but there are physical and reproductive differences between the two creatures. In general, mussels are larger than clams and have an oblong lopsided shell, as opposed to the round symmetrical shell of the clam.

Crustaceans are a large class of creatures with a hard exoskeleton (external skeleton), appendages, and antennae. This class includes lobsters, shrimps, and crabs.

As of November 2011, there were 134 U.S. and foreign species of clams (including mussels), snails, crustaceans, and corals listed under the Endangered Species Act (ESA) as endangered or threatened. (See Table 1.2 in Chapter 1.) Table 5.1 shows the 10 clam, snail, crustacean, and coral species with the highest expenditures under the ESA during fiscal year (FY) 2010. Nearly $13.1 million was spent on activities to conserve these 10 species.

CLAMS AND MUSSELS

There were 72 U.S. species of clams and mussels listed under the ESA as of November 2011. (See Table 5.2.) The vast majority of imperiled clams and mussels in the United States are freshwater species that inhabit inland rivers, primarily in the Southeast.

Mussels are bivalved (two-shelled) creatures encased in hard hinged shells made of calcium. The freshwater species can grow to be up to 6 inches (15.2 cm) in length. The United States, with nearly 300 species, has the greatest diversity of freshwater mussels in the world. According to the U.S. Geological Survey (USGS), in "Conservation of Southeastern Mussels" (March 7, 2011, http://fl.biology.usgs.gov/Southeastern_Aquatic_Fauna/Freshwater_Mussels/freshwater_mussels.html), approximately 90% of these creatures live in southeastern states. Most of them are found burrowed into the sand and gravel beds of rivers and streams making up the Mississippi River system. Mussels have a footlike appendage that acts like an anchor to hold them in place. They can use this appendage to move themselves slowly over small distances. Mussels tend to congregate in large groups called colonies.

Mussels are filter-feeders. They have a siphoning system that sucks in food and oxygen from the water. Their gills can filter impurities out of the water. Thus, mussels are tiny natural water purifiers.

Most mussel species have a unique way of spreading their offspring. A female mussel can produce several thousand eggs in a year. After the eggs are fertilized, they develop into larva and are released. The larva latch onto the fins or gills of passing fish and stay there until they have grown into baby clams. At that point they turn loose of the fish and drop to the river bottom. The larvae are called glochidia. It is believed that glochidia are harmless to the fish on which they hitchhike. This parasitic relationship allows mussels to spread and distribute beyond their usual range.

Mussel Declines

The decline of freshwater mussels began during the 1800s. Many of the creatures have an interior shell surface with a pearl-like sheen. These pearlymussels were in great demand as a source of buttons for clothing until the invention of plastic. Collectors also killed many mussels by prying them open looking for pearls. Until the 1990s mussel shells were ground up and used in the oyster pearl industry. Another cause for decline has been habitat disturbance, especially water pollution and the modification

TABLE 5.1

Clam, snail, crustacean, and coral species with the highest expenditures under the Endangered Species Act, fiscal year 2010

Ranking	Species	Expenditure
1	Coral, staghorn (Acropora cervicornis)	$2,560,317
2	Coral, elkhorn (Acropora palmata)	$2,460,760
3	Fairy shrimp, vernal pool (Branchinecta lynchi)	$1,978,888
4	Abalone, White (Haliotis sorenseni)—North America (West Coast from Point Conception, CA, to Punta Abreojos, Baja California, Mexico)	$1,239,899
5	Tadpole shrimp, vernal pool (Lepidurus packardi)	$1,093,686
6	Higgins eye (pearlymussel) (Lampsilis higginsii)	$998,903
7	Mucket, pink (pearlymussel) (Lampsilis abrupta)	$820,394
8	Snails, Oahu tree (Achatinella spp.)	$685,801
9	Elktoe, Appalachian (Alasmidonta raveneliana)	$637,938
10	Wedgemussel, dwarf (Alasmidonta heterodon)	$582,082

SOURCE: Adapted from "Table 2. Species Ranked in Descending Order of Total FY 2010 Reported Expenditures, Not Including Land Acquisition Costs," in *Federal and State Endangered and Threatened Species Expenditures: Fiscal Year 2010*, U.S. Department of the Interior, U.S. Fish and Wildlife Service, 2010, http://www.fws.gov/endangered/esa-library/pdf/2010.EXP.FINAL.pdf (accessed October 25, 2011)

of aquatic habitats by dams. In addition, erosion due to strip mining and agriculture causes siltation that can actually bury and suffocate mussels. The invasive zebra mussel and quagga mussel have also harmed native freshwater mussel species, such as the Higgins eye pearlymussel, by competing with them for food and other resources.

Higgins Eye Pearlymussel

As shown in Table 5.1, nearly $999,000 was spent under the ESA during FY 2010 on the Higgins eye pearlymussel. This freshwater species is native to the United States and is found in the waters of Illinois, Iowa, Minnesota, Missouri, Nebraska, and Wisconsin. The species was named after its discoverer, Frank Higgins, who found some of the mussels in the Mississippi River near Muscatine, Iowa, during the mid-1800s. Over the next few decades Muscatine developed a thriving pearl-button industry that lasted into the 1940s. Higgins eye was also harvested for the commercial pearl industry.

In 1976 the Higgins eye pearlymussel was listed as an endangered species under the ESA. More than a century of scavenging by humans had severely depleted the species. Dams, navigational structures, and water quality problems in the upper Mississippi River system were contributing factors to its decline. In 1983 the USFWS published its first recovery plan for the Higgins eye. The plan identified areas that were deemed essential habitat for the species and called for limits on construction and harvesting in those areas.

In May 2004 the USFWS published *Higgins Eye Pearlymussel (*Lampsilis higginsii*) Recovery Plan: First Revision* (http://ecos.fws.gov/docs/recovery_plans/2004/040714.pdf). The new plan examines more recent threats to the pearlymussel's survival, primarily the pervasive spread of zebra mussels. It acknowledges that there is no feasible way to eliminate zebra mussels to the extent needed to benefit the Higgins eye. Instead, the plan focuses on developing methods to prevent new zebra mussel infestations and working to lessen the impacts of already infested populations.

The USFWS indicates in "Saving the Higgins Eye Pearlymussel" (October 12, 2011, http://www.fws.gov/midwest/Endangered/clams/higginseye/propagation_fs.html) that the Genoa National Fish Hatchery in Genoa, Wisconsin, specializes in capturing female Higgins eye pearlymussels that have not yet released their glochidia. The glochidia are removed via a syringe and placed in tanks or buckets that contain suitable host fish. The fish are either released into the environment or maintained at the hatchery until the glochidia are mature enough to live on their own. These juvenile mussels are then released into natural waters that are free from invasive mussel species. In "Aquatic Species Production" (2012, http://www.fws.gov/midwest/Genoa/fish_production.html), the Genoa National Fish Hatchery indicates that it has produced the following numbers of Higgins eye pearlymussels:

- FY 2004—2,333,665
- FY 2005—1,979,480
- FY 2006—2,242,095
- FY 2007—769,040
- FY 2008—1,240,352
- FY 2009—1,373,214
- FY 2010—489,967

In September 2010 the USFWS (http://www.gpo.gov/fdsys/search/citation.result.FR.action?federalRegister.volume=2010&federalRegister.page=55820&publication=FR) initiated a five-year review for several midwestern species, including the Higgins eye pearlymussel. As explained in Chapter 2, the ESA requires all listed species to be reviewed at least every five years to determine whether they still require ESA protection. As of January 2012, the results of the review had not been issued.

ZEBRA AND QUAGGA MUSSELS: AN INFESTATION. In 1988 an unwelcomed visitor was discovered in the waters of Lake St. Clair, Michigan: a zebra mussel (*Dreissena polymorpha*). The zebra mussel is native to eastern Europe. It is smaller than the freshwater mussels found in the United States and has a different method for spreading its young. The larva of zebra mussels do not require a fish host to develop into babies. They can attach to any hard surface under the water. This allows zebra mussels to spread much easier and quicker than their American counterparts.

It is believed that the first zebra mussels migrated to the United States in the ballast water of ships. This is

TABLE 5.2

Endangered and threatened clam species, November 2011

Common name	Scientific name	Listing status[a]	U.S. or U.S./foreign	Recovery plan date	Recovery plan stage[b]
Alabama (=inflated) heelsplitter	Potamilus inflatus	T	US	4/13/1993	F
Alabama lampmussel	Lampsilis virescens	E (1), EXPN (1)	US	7/2/1985	F
Alabama moccasinshell	Medionidus acutissimus	T	US	11/17/2000	F
Altamaha spinymussel	Elliptio spinosa	E	US	None (listed in 2011)	—
Appalachian elktoe	Alasmidonta raveneliana	E	US	8/26/1996	F
Appalachian monkeyface (pearlymussel)	Quadrula sparsa	E (1), EXPN (1)	US	7/9/1984	F
Arkansas fatmucket	Lampsilis powellii	T	US	2/10/1992	F
Birdwing pearlymussel	Conradilla caelata	E (1), EXPN (2)	US	7/9/1984	F
Black clubshell	Pleurobema curtum	E	US	11/14/1989	F
Carolina heelsplitter	Lasmigona decorata	E	US	1/17/1997	F
Chipola slabshell	Elliptio chipolaensis	T	US	9/19/2003	F
Clubshell	Pleurobema clava	E (1), EXPN (1)	US	9/21/1994	F
Coosa moccasinshell	Medionidus parvulus	E	US	11/17/2000	F
Cracking pearlymussel	Hemistena lata	E (1), EXPN (2)	US	7/11/1991	F
Cumberland bean (pearlymussel)	Villosa trabalis	E (1), EXPN (2)	US	8/22/1984	F
Cumberland elktoe	Alasmidonta atropurpurea	E	US	5/24/2004	F
Cumberlandian combshell	Epioblasma brevidens	E (1), EXPN (2)	US	5/24/2004	F
Cumberland monkeyface (pearlymussel)	Quadrula intermedia	E (1), EXPN (2)	US	7/9/1984	F
Cumberland pigtoe	Pleurobema gibberum	E	US	8/13/1992	F
Curtis pearlymussel	Epioblasma florentina curtisii	E	US	2/4/1986	F
Dark pigtoe	Pleurobema furvum	E	US	11/17/2000	F
Dromedary pearlymussel	Dromus dromas	E (1), EXPN (2)	US	7/9/1984	F
Dwarf wedgemussel	Alasmidonta heterodon	E	US/foreign	2/8/1993	F
Fanshell	Cyprogenia stegaria	E (1), EXPN (1)	US	7/9/1991	F
Fat pocketbook	Potamilus capax	E	US	11/14/1989	F
Fat three-ridge (mussel)	Amblema neislerii	E	US	9/19/2003	F
Finelined pocketbook	Lampsilis altilis	T	US	11/17/2000	F
Finerayed pigtoe	Fusconaia cuneolus	E (1), EXPN (2)	US	9/19/1984	F
Flat pigtoe	Pleurobema marshalli	E	US	11/14/1989	F
Georgia pigtoe	Pleurobema hanleyianum	E	US	None (listed in 2010)	—
Green blossom (pearlymussel)	Epioblasma torulosa gubernaculum	E	US	7/9/1984	F
Gulf moccasinshell	Medionidus penicillatus	E	US	9/19/2003	F
Heavy pigtoe	Pleurobema taitianum	E	US	11/14/1989	F
Higgins eye (pearlymussel)	Lampsilis higginsii	E	US	7/14/2004	RF(1)
James spinymussel	Pleurobema collina	E	US	9/24/1990	F
Littlewing pearlymussel	Pegias fabula	E	US	9/22/1989	F
Louisiana pearlshell	Margaritifera hembeli	T	US	12/3/1990	F
Northern riffleshell	Epioblasma torulosa rangiana	E	US	9/21/1994	F
Ochlockonee moccasinshell	Medionidus simpsonianus	E	US	9/19/2003	F
Orangefoot pimpleback (pearlymussel)	Plethobasus cooperianus	E (1), EXPN (1)	US	9/30/1984	F
Orangenacre mucket	Lampsilis perovalis	T	US	11/17/2000	F
Ouachita rock pocketbook	Arkansia wheeleri	E	US	6/2/2004	F
Oval pigtoe	Pleurobema pyriforme	E	US	9/19/2003	F
Ovate clubshell	Pleurobema perovatum	E	US	11/17/2000	F
Oyster mussel	Epioblasma capsaeformis	E (1), EXPN (2)	US	5/24/2004	F
Pale lilliput (pearlymussel)	Toxolasma cylindrellus	E	US	8/22/1984	F
Pink mucket (pearlymussel)	Lampsilis abrupta	E	US	1/24/1985	F
Purple bankclimber (mussel)	Elliptoideus sloatianus	T	US	9/19/2003	F
Purple bean	Villosa perpurpurea	E	US	5/24/2004	F
Purple cat's paw (=purple cat's paw pearlymussel)	Epioblasma obliquata obliquata	E (1), EXPN (1)	US	3/10/1992	F
Ring pink (mussel)	Obovaria retusa	E (1), EXPN (1)	US	3/25/1991	F
Rough pigtoe	Pleurobema plenum	E (1), EXPN (1)	US	8/6/1984	F
Rough rabbitsfoot	Quadrula cylindrica strigillata	E	US	5/24/2004	F
Scaleshell mussel	Leptodea leptodon	E	US	4/7/2010	F
Shiny pigtoe	Fusconaia cor	E (1), EXPN (2)	US	7/9/1984	F
Shinyrayed pocketbook	Lampsilis subangulata	E	US	9/19/2003	F
Southern acornshell	Epioblasma othcaloogensis	E	US	11/17/2000	F
Southern clubshell	Pleurobema decisum	E	US	11/17/2000	F
Southern combshell	Epioblasma penita	E	US	11/14/1989	F
Southern pigtoe	Pleurobema georgianum	E	US	11/17/2000	F
Speckled pocketbook	Lampsilis streckeri	E	US	1/2/1992	F
Stirrupshell	Quadrula stapes	E	US	11/14/1989	F
Tan riffleshell	Epioblasma florentina walkeri (=E. walkeri)	E	US	10/22/1984	F
Tar River spinymussel	Elliptio steinstansana	E	US	5/5/1992	RF(1)
Triangular kidneyshell	Ptychobranchus greenii	E	US	11/17/2000	F
Tubercled blossom (pearlymussel)	Epioblasma torulosa torulosa	E (1), EXPN (1)	US	1/25/1985	F
Turgid blossom (pearlymussel)	Epioblasma turgidula	E (1), EXPN (1)	US	1/25/1985	F
Upland combshell	Epioblasma metastriata	E	US	11/17/2000	F
White catspaw (pearlymussel)	Epioblasma obliquata perobliqua	E	US	1/25/1990	F

TABLE 5.2

Endangered and threatened clam species, November 2011 [CONTINUED]

Common name	Scientific name	Listing status[a]	U.S. or U.S./foreign	Recovery plan date	Recovery plan stage[b]
White wartyback (pearlymussel)	Plethobasus cicatricosus	E (1), EXPN (1)	US	9/19/1984	F
Winged mapleleaf	Quadrula fragosa	E (1), EXPN (1)	US	6/25/1997	F
Yellow blossom (pearlymussel)	Epioblasma florentina florentina	E (1), EXPN (1)	US	1/25/1985	F

[a]E = endangered; T = threatened; EXPN = experimental population, non-essential; numbers in parentheses indicate separate populations.
[b]F = final; RF = final revision.

SOURCE: Adapted from "Generate Species List," in *Species Reports*, U.S. Department of the Interior, U.S. Fish & Wildlife Service, November 2011, http://ecos.fws.gov/tess_public/pub/adHocSpeciesForm.jsp (accessed November 8, 2011), and "Listed FWS/Joint FWS and NMFS Species and Populations with Recovery Plans (Sorted by Listed Entity)," in *Recovery Plans Search*, U.S. Department of the Interior, U.S. Fish & Wildlife Service, November 2011, http://ecos.fws.gov/tess_public/pub/speciesRecovery.jsp?sort=1 (accessed November 8, 2011)

water held in large tanks below deck to improve the stability and control of ships. Ballast water is pumped in and out as needed during a journey. Zebra mussels have also been found clinging to the hulls of small fishing and recreation boats. These boats are hauled overland on trailers, and this allows the creatures to travel great distances between inland water bodies.

Another foreign invader of concern is the quagga mussel (*Dreissena bugensis*), a native of the Ukraine in eastern Europe. In 1989 the mussel was first sighted in the United States in Lake Erie. By the mid-1990s it had spread to other lakes in the upper Midwest. In 2007 the mussel was discovered in lakes in Nevada, Arizona, and Southern California. This finding is particularly troubling to scientists because of the large concentration of imperiled aquatic species in southwestern water bodies.

Figure 5.1 shows a USGS map of zebra and quagga mussel distribution around the country as of July 2011. These invasive species have spread from the Great Lakes south to the Gulf of Mexico and east to New England and are beginning to show up in western waters. They have been found on boat hulls as far west as California. Throughout waterways in the Midwest, colonies of zebra mussels have clogged pipes and other structures that are used for municipal and industrial water supply. In addition, the pests have significantly degraded native mussel colonies by competing for available food, space, and resources.

CLAM AND MUSSEL RECOVERY PLANS. As of November 2011, all 72 species of U.S. and U.S./foreign clams and mussels listed under the ESA had recovery plans in draft or final form. (See Table 5.2.) Conservation efforts for freshwater mussels include the captive breeding and reintroduction of some species, as well as measures to restore damaged habitats.

Snails

Snails belong to the class Gastropoda of mollusks. They typically have an external spiral-shaped shell and a distinct head that includes sensory organs. Snails inhabit terrestrial (land), marine, and freshwater environments. Most land snails prefer moist, heavily vegetated locations and survive on a diet of vegetation and algae. However, there are a few carnivorous (meat-eating) snail species. Snails are found throughout the United States. Most imperiled species are located in the West (including Hawaii) and the Southeast (primarily Alabama). Land snails are imperiled by a variety of factors including the destruction of habitat, being preyed on by rats and invasive carnivorous snails, and the spread of nonnative vegetation. Some marine (ocean-dwelling) snails have historically been prized by humans as a food source.

As of November 2011, there were 37 U.S. and U.S./foreign species of snails listed under the ESA. (See Table 5.3.) Most had an endangered listing. As shown in Table 5.1, over $1.2 million was spent under the ESA in FY 2010 to conserve the white abalone.

White Abalone

The white abalone is a marine snail found along the West Coast from Southern California to northern Mexico. According to the USFWS, in "White Abalone (*Haliotis sorenseni*)" (2012, http://www.nmfs.noaa.gov/pr/species/invertebrates/whiteabalone.htm), it is a relative newcomer to the endangered species list, having been added in 2001. In addition, it is the first marine invertebrate ever to be listed under the ESA. As a marine species, it is under the jurisdiction of the National Marine Fisheries Service (NMFS). In the listing rule, the NMFS (May 29, 2001, http://www.gpo.gov/fdsys/pkg/FR-2001-05-29/pdf/01-13430.pdf#page=1) indicated that it did not intend to designate critical habitat for the species because of fears about poaching (illegal hunting). The agency noted that the "abalone as a group continue to be highly prized and in demand as food by humans."

In October 2008 the NMFS published *Final White Abalone Recovery Plan (*Haliotis sorenseni*)* (http://ecos.fws.gov/docs/recovery_plan/whiteabalone.pdf), which describes the status of the species and the activities that are required to protect and recover it. According to the agency,

FIGURE 5.1

Distribution of zebra and quagga mussel sightings, July 2011

[Dreissena polymorpha and D. rostriformis bugensis]

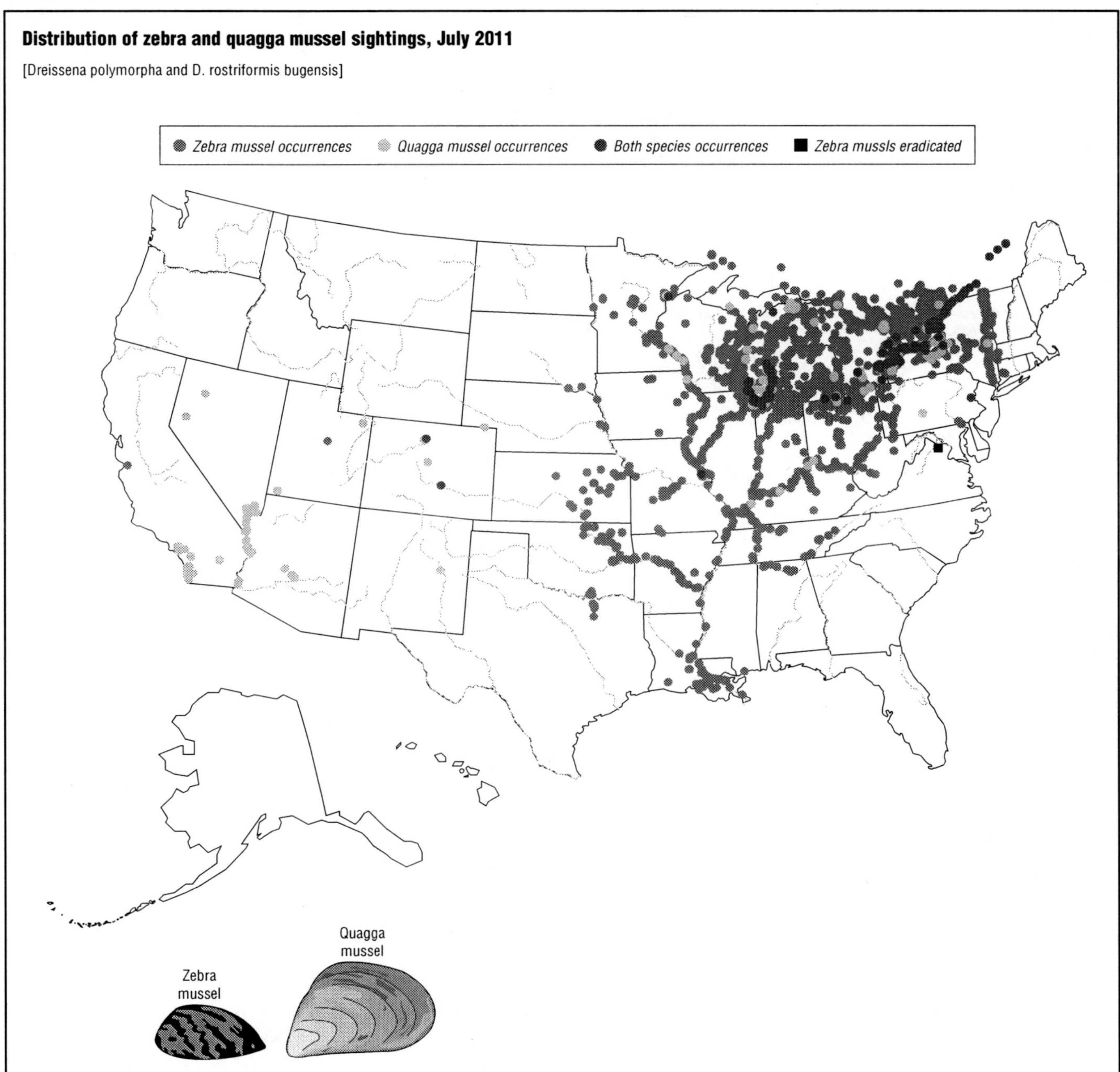

SOURCE: "Zebra and Quagga Mussel Sightings Distribution," in *Zebra and Quagga Mussel—U.S. Distribution Information*, U.S. Geological Survey, July 8, 2011, http://nas.er.usgs.gov/taxgroup/mollusks/zebramussel/maps/current_zm_quag_map.jpg (accessed November 8, 2011)

recovery will be extremely difficult, because white abalone populations have been reduced by 99% since the 1970s due to overharvesting by humans. In fact, the NMFS describes the current population size as "near zero." As such, no potential recovery date is given; however, the agency estimates that recovery will likely take "several decades and cost $20 million at a minimum."

RECOVERY PLANS FOR SNAILS. As of November 2011, most of the imperiled snail species had recovery plans. (See Table 5.3.) The black abalone, interrupted rocksnail, and rough hornsnail did not have plans because they were recently listed under the ESA in 2010 or 2011. Three other species that did not have plans—Koster's springsnail, Pecos assiminea snail, and Roswell springsnail—were listed under the ESA in 2005. In February 2009 the USFWS initiated a five-year status review for these species. As of January 2012, the results of that review and a five-year review begun in 2007 on the slender campeloma had not been published.

TABLE 5.3

Endangered and threatened snail species, November 2011

Common name	Scientific name	Listing status[a]	U.S. or U.S./foreign	Recovery plan date	Recovery plan stage[b]
Alamosa springsnail	Tryonia alamosae	E	US	8/31/1994	F
Anthony's riversnail	Athearnia anthonyi	E (1), EXPN (2)	US	8/13/1997	F
Armored snail	Pyrgulopsis (=Marstonia) pachyta	E	US	7/1/1994	D
Banbury Springs limpet	Lanx sp.	E	US	11/26/1995	F
Black Abalone	Haliotis cracherodii	E	US	None (listed in 2011)	—
Bliss Rapids snail	Taylorconcha serpenticola	T	US	11/26/1995	F
Bruneau hot springsnail	Pyrgulopsis bruneauensis	E	US	9/30/2002	F
Chittenango ovate amber snail	Succinea chittenangoensis	T	US	8/21/2006	RF(1)
Cylindrical lioplax (snail)	Lioplax cyclostomaformis	E	US	12/2/2005	F
Flat pebblesnail	Lepyrium showalteri	E	US	12/2/2005	F
Flat-spired three-toothed snail	Triodopsis platysayoides	T	US	5/9/1983	F
Interrupted (=Georgia) rocksnail	Leptoxis foremani	E	US	None (listed in 2010)	—
Iowa Pleistocene snail	Discus macclintocki	E	US	3/22/1984	F
Kanab ambersnail	Oxyloma haydeni kanabensis	E	US	10/12/1995	F
Koster's springsnail	Juturnia kosteri	E	US	None	—
Lacy elimia (snail)	Elimia crenatella	T	US	12/2/2005	F
Magazine Mountain shagreen	Mesodon magazinensis	T	US	2/1/1994	F
Morro shoulderband (=Banded dune) snail	Helminthoglypta walkeriana	E	US	9/28/1998	F
Newcomb's snail	Erinna newcombi	T	US	9/18/2006	F
Noonday snail	Mesodon clarki nantahala	T	US	9/7/1984	F
Oahu tree snails	Achatinella spp.	E	US	6/30/1992	F
Painted rocksnail	Leptoxis taeniata	T	US	12/2/2005	F
Painted snake coiled forest snail	Anguispira picta	T	US	10/14/1982	F
Pecos assiminea snail	Assiminea pecos	E	US/foreign	None	—
Plicate rocksnail	Leptoxis plicata	E	US	12/2/2005	F
Roswell springsnail	Pyrgulopsis roswellensis	E	US	None	—
Rough hornsnail	Pleurocera foremani	E	US	None (listed in 2010)	—
Round rocksnail	Leptoxis ampla	T	US	12/2/2005	F
Royal marstonia (snail)	Pyrgulopsis ogmorhaphe	E	US	8/11/1995	F
Slender campeloma	Campeloma decampi	E	US	None	—
Snake River physa snail	Physa natricina	E	US	11/26/1995	F
Socorro springsnail	Pyrgulopsis neomexicana	E	US	8/31/1994	F
Stock Island tree snail	Orthalicus reses (not incl. nesodryas)	T	US	5/18/1999	F
Tulotoma snail	Tulotoma magnifica	T	US	11/17/2000	F
Tumbling Creek cavesnail	Antrobia culveri	E	US	9/22/2003	F
Virginia fringed mountain snail	Polygyriscus virginianus	E	US	5/9/1983	F
White abalone	Haliotis sorenseni	E	US/foreign	1/12/2009	F

[a]E = endangered; T = threatened; EXPN = experimental population, non-essential; numbers in parentheses indicate separate populations.
[b]F = final; RF = final revision; D = draft.

SOURCE: Adapted from "Generate Species List," in *Species Reports*, U.S. Department of the Interior, U.S. Fish & Wildlife Service, November 2011, http://ecos.fws.gov/tess_public/pub/adHocSpeciesForm.jsp (accessed November 8, 2011), and "Listed FWS/Joint FWS and NMFS Species and Populations with Recovery Plans (Sorted by Listed Entity)," in *Recovery Plans Search*, U.S. Department of the Interior, U.S. Fish & Wildlife Service, November 2011, http://ecos.fws.gov/tess_public/pub/speciesRecovery.jsp?sort=1 (accessed November 8, 2011)

Crustaceans

Crustaceans are a large class of mandibulate (jawed) creatures in the phylum Arthropoda. They are mostly aquatic and inhabit marine and freshwaters. As of November 2011, there were 22 U.S. species listed as endangered or threatened under the ESA. (See Table 5.4.) Even though imperiled crustacean species are found throughout the United States, they are mostly located in California. Nearly $2 million was spent under the ESA during FY 2010 on one California crustacean: the vernal pool fairy shrimp. (See Table 5.1.)

VERNAL POOL FAIRY SHRIMP. The vernal pool fairy shrimp (*Branchinecta lynchi*) was listed under the ESA as threatened in 1994. It is found in California and Oregon. The term *vernal* is from the Latin word for "spring." This species inhabits temporary small ponds and pools of water that appear during the rainy season (winter or springtime) and dry up over time. The shrimp lay their eggs in these pools when they contain water. The eggs go dormant in the dirt when the pools become dry. Baby shrimp hatch only when exposed to water that is approximately 50 degrees Fahrenheit (10 degrees Celsius). Adults typically reach 0.4 to 1 inch (1 to 2.5 cm) in length. The shrimp have a life span of two to five months.

In 2003 critical habitat was designated for the vernal pool fairy shrimp along with several other species of vernal pool shrimp. In 2006 the USFWS published *Recovery Plan for Vernal Pool Ecosystems of California and Southern Oregon* (http://www.fws.gov/ecos/ajax/docs/recovery_plan/060614.pdf), a recovery plan that covers dozens of imperiled plant and animal species that inhabit vernal pool ecosystems in California and southern Oregon. The USFWS notes that vernal pool life forms are threatened by urban and agricultural development and by

TABLE 5.4

Endangered and threatened crustacean species, November 2011

Common name	Scientific name	Listing status[a]	U.S. or U.S./foreign	Recovery plan date	Recovery plan stage[b]
Alabama cave shrimp	Palaemonias alabamae	E	US	9/4/1997	F
California freshwater shrimp	Syncaris pacifica	E	US	7/31/1998	F
Cave crayfish	Cambarus zophonastes	E	US	9/26/1988	F
Cave crayfish	Cambarus aculabrum	E	US	10/30/1996	F
Conservancy fairy shrimp	Branchinecta conservatio	E	US	12/15/2005	F
Hay's Spring amphipod	Stygobromus hayi	E	US	Exempt	—
Illinois cave amphipod	Gammarus acherondytes	E	US	9/20/2002	F
Kauai cave amphipod	Spelaeorchestia koloana	E	US	7/19/2006	F
Kentucky cave shrimp	Palaemonias ganteri	E	US	10/7/1988	F
Lee County cave isopod	Lirceus usdagalun	E	US	9/30/1997	F
Longhorn fairy shrimp	Branchinecta longiantenna	E	US	12/15/2005	F
Madison Cave isopod	Antrolana lira	T	US	9/30/1996	F
Nashville crayfish	Orconectes shoupi	E	US	2/8/1989	RF(1)
Noel's Amphipod	Gammarus desperatus	E	US	None	—
Peck's Cave amphipod	Stygobromus (=Stygonectes) pecki	E	US	2/14/1996	RF(1)
Riverside fairy shrimp	Streptocephalus woottoni	E	US	9/3/1998	F
San Diego fairy shrimp	Branchinecta sandiegonensis	E	US	9/3/1998	F
Shasta crayfish	Pacifastacus fortis	E	US	8/28/1998	F
Socorro isopod	Thermosphaeroma thermophilus	E	US	2/16/1982	F
Squirrel Chimney Cave shrimp	Palaemonetes cummingi	T	US	Exempt	—
Vernal pool fairy shrimp	Branchinecta lynchi	T	US	12/15/2005	F
Vernal pool tadpole shrimp	Lepidurus packardi	E	US	12/15/2005	F

[a]E = endangered; T = threatened.
[b]F = final; RF = final revision.

SOURCE: Adapted from "Generate Species List," in *Species Reports*, U.S. Department of the Interior, U.S. Fish & Wildlife Service, November 2011, http://ecos.fws.gov/tess_public/pub/adHocSpeciesForm.jsp (accessed November 8, 2011), and "Listed FWS/Joint FWS and NMFS Species and Populations with Recovery Plans (Sorted by Listed Entity)," in *Recovery Plans Search*, U.S. Department of the Interior, U.S. Fish & Wildlife Service, November 2011, http://ecos.fws.gov/tess_public/pub/speciesRecovery.jsp?sort=1 (accessed November 8, 2011)

invasion of nonnative species. The recovery of vernal pool species will require an ecosystem-wide approach. The USFWS proposes establishing conservation areas and reserves to protect primary vernal pool habitat.

In September 2007 the USFWS published *Vernal Pool Fairy Shrimp (*Branchinecta lynchi*): 5-Year Review—Summary and Evaluation* (http://www.fws.gov/cno/es/images/Graphics/VPFS_5-yr%20review%20CNO%20FINAL%2027Sept07.pdf). Based on the collected information, the agency decided to maintain a threatened listing under the ESA for the vernal pool fairy shrimp. In May 2011 the USFWS (http://www.gpo.gov/fdsys/pkg/FR-2011-05-25/pdf/2011-12861.pdf) announced the initiation of another five-year review of the species. As of January 2012, the results of that review had not been published.

CRUSTACEAN RECOVERY PLANS. As of November 2011, nearly all endangered and threatened species of crustaceans found in the United States had recovery plans. (See Table 5.4.) Most plans were in final form.

Imperiled Mollusks and Crustaceans around the World

According to the International Union for Conservation of Nature (IUCN), in *Red List of Threatened Species Version 2011.2* (2011, http://www.iucnredlist.org/documents/summarystatistics/2011_2_RL_Stats_Table1.pdf), 1,673 species of mollusks and 596 species of crustaceans were threatened in 2011. For mollusks, this number accounted for 31% of the 5,422 species evaluated. The IUCN reports that approximately 85,000 mollusk species are known. Only 2,399 crustacean species were evaluated for the 2011 report. Threatened species accounted for 25% of this total. The IUCN notes that there are approximately 47,000 known species of crustaceans. As shown in Table 5.5, only three foreign clam and snail species were listed under the ESA as of November 2011. All had endangered listings. No foreign crustacean species were listed under the ESA at that time.

CORALS

Corals are one of the most unusual members of the animal kingdom. They are invertebrate marine creatures of the phylum Cnidaria, along with jellyfish and anemones. Many people are familiar with coral reefs—vast and colorful undersea structures that are popular with scuba divers and snorkelers. (See Figure 5.2.) Most coral reefs are composed of many hundreds or thousands of individual coral organisms called polyps. Figure 5.3 shows a coral polyp with its common parts labeled. The polyp has an opening (or mouth) surrounded by tentacles that capture sea creatures for food. Following digestion in the polyp's stomach, waste materials are expelled out through the mouth.

Most reef-building coral polyps have a symbiotic (mutually beneficial) relationship with algae (tiny plant

TABLE 5.5

Foreign endangered and threatened clam and snail species, November 2011

Common name	Scientific name	Category	Listing status	Historic range
Nicklin's pearlymussel	Megalonaias nicklineana	Clam	Endangered	Mexico
Tampico pearlymussel	Cyrtonaias tampicoensis tecomatensis	Clam	Endangered	Mexico
Manus Island tree snail	Papustyla pulcherrima	Snail	Endangered	Manus Island (Papua New Guinea)

SOURCE: Adapted from "Generate Species List," in *Species Reports*, U.S. Department of the Interior, U.S. Fish & Wildlife Service, November 2011, http://ecos.fws.gov/tess_public/pub/adHocSpeciesForm.jsp (accessed November 8, 2011)

FIGURE 5.2

Coral reefs are among the most diverse ecosystems in the world. They are also immediately threatened by global warming, which has caused unprecedented episodes of coral bleaching in recent years. *(© Dennis Sabo/Shutterstock.com.)*

cells) called zooxanthellae. The algae perform photosynthesis, which creates oxygen and provides nutrients to the coral polyps, allowing them to grow and spread. Reef-building corals secrete calcium carbonate, a hard mineral compound that forms the reef skeleton. Zooxanthellae are also responsible for the bright and varied colors found in coral reefs. Without the algae, coral polyps are naturally translucent. Scientists believe zooxanthellae serve as sort of a "sunscreen" for warm-water corals, protecting them from the harsh ultraviolet rays of the sun.

Coral reefs are the largest living structures on the earth. They are primarily found in coastal, tropical waters. These reefs are located in relatively shallow waters, making them more susceptible to human activities. The U.S. Commission on Ocean Policy states in *An Ocean Blueprint for the 21st Century: Final Report* (2004, http://www.oceancommission.gov/documents/full_color_rpt/000_ocean_full_report.pdf) that only 1% to 2% of warm-water corals are found in U.S. waters. Most warm-water corals are located in the waters of the South Pacific and around Indonesia. In addition, there are

FIGURE 5.3

A coral polyp

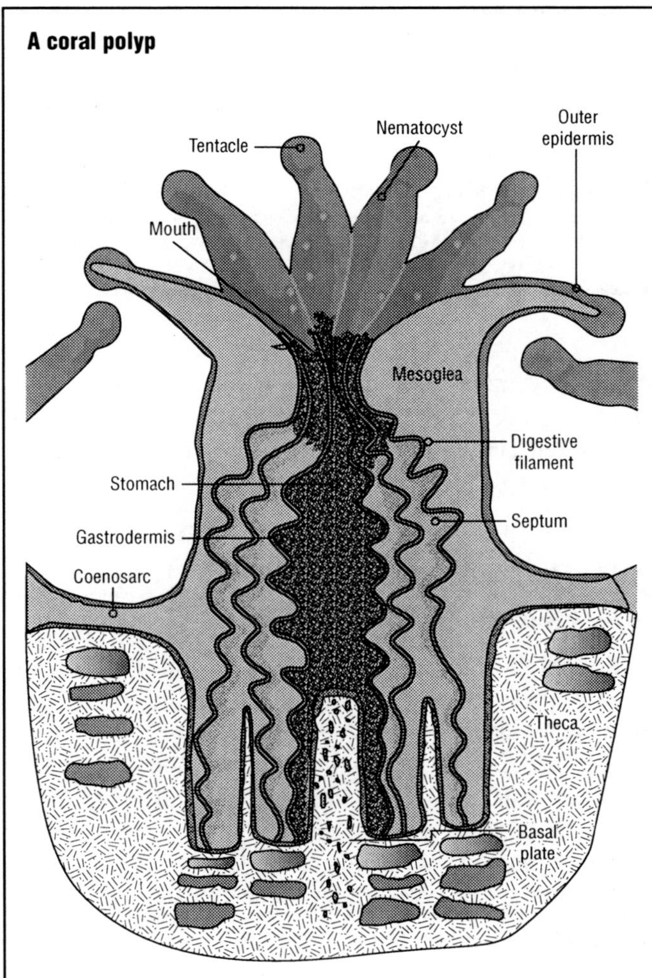

SOURCE: "Polyp," in *Education Kits: Corals*, U.S. Department of Commerce, National Oceanic and Atmospheric Administration, National Ocean Service, March 25, 2008, http://oceanservice.noaa.gov/education/kits/corals/media/supp_coral01a.html (accessed November 8, 2011)

many cold-water reefs around the world that scientists are just beginning to study. These reefs are found in cold deep waters from depths of 100 feet (30.5 m) to more than 3 miles (4.8 km).

General Threats to Corals

In *An Ocean Blueprint for the 21st Century*, the Commission on Ocean Policy explains that coral reefs are imperiled by diseases and coastal development that spur the growth of unfriendly algae. Coastal development also increases the danger of the reefs being damaged by divers and boat anchors. Other serious threats to coral reef ecosystems include marine pollution, overfishing of species found around the corals, and destructive fishing methods used around the corals. The latter include explosives or chemicals, such as cyanide, that are used to stun coral inhabitants so they are easier to capture. Such methods are illegal and are known to be practiced in areas of Southeast Asia. The collection of tropical reef specimens for the aquarium trade has also damaged a number of species.

Perhaps the greatest immediate threat to coral reefs is rising water temperature due to global climate change. Overly warm water causes zooxanthellae to leave or be expelled from coral polyps. The resulting loss of color is known as coral bleaching. Bleached coral suffers from a lack of nutrients, becoming weak and lackluster and more susceptible to disease and other environmental stressors. Long-term zooxanthellae deficiencies can cause the coral to die.

The article "Corals Can Sense What's Coming" (*ScienceDaily*, November 18, 2011) indicates that since the 1970s seven major coral bleaching events have occurred around the world. The most recent occurred in 2010 across the Indian Ocean and the Coral Triangle, a coral-rich area of the Pacific Ocean that encompasses Indonesia, the Philippine Islands, and other islands north of Australia.

As described in Chapter 1, global climate change is believed to be triggered by excessive anthropogenic (human-caused) emissions of carbon dioxide into the atmosphere. The National Oceanic and Atmospheric Administration (NOAA) explains in "What Is Ocean Acidification?" (2012, http://www.pmel.noaa.gov/co2/story/What+is+Ocean+Acidification%3F) that the increasing uptake of atmospheric carbon dioxide by the earth's oceans is lowering the pH (potential hydrogen; the level of acidity, a lower value indicates more acid) of the water and reducing the availability of carbonate ions to marine organisms that need them. The phenomenon is known as ocean acidification, and scientists fear that it poses a serious threat to the health of coral systems.

Imperiled U.S. Corals

NOAA periodically publishes a comprehensive report describing the condition of shallow-water coral reef ecosystems in the United States. As of January 2012, the most recent report was published in 2008. In *The State of Coral Reef Ecosystems of the United States and Pacific Freely Associated States: 2008* (July 2008, http://ccma.nos.noaa.gov/ecosystems/coralreef/coral2008/), NOAA's Coral Reef Conservation Program (CRCP) discusses the results of monitoring activities that were conducted by government, private, and academic entities engaged in assessing the condition of the nation's shallow-water coral reef ecosystems. According to the CRCP, approximately half the ecosystems were found to be in "poor" or "fair" condition.

As of November 2011, there were two U.S. species of coral listed as threatened under the ESA. (See Table 5.6.) The corals were Elkhorn coral (*Acropora palmata*) and Staghorn coral (*Acropora cervicornis*). Both species are branching corals found in the Caribbean, including the

TABLE 5.6

Endangered and threatened coral species, November 2011

Common name	Scientific name	Listing status	U.S. or U.S./foreign	Species group	Recovery plan date	Recovery plan stage
Elkhorn coral	Acropora palmata	Threatened	US/foreign	Corals	None	—
Staghorn coral	Acropora cervicornis	Threatened	US/foreign	Corals	None	—

SOURCE: Adapted from "Generate Species List," in *Species Reports*, U.S. Department of the Interior, U.S. Fish & Wildlife Service, November 2011, http://ecos.fws.gov/tess_public/pub/adHocSpeciesForm.jsp (accessed November 8, 2011), and "Listed FWS/Joint FWS and NMFS Species and Populations with Recovery Plans (Sorted by Listed Entity)," in *Recovery Plans Search*, U.S. Department of the Interior, U.S. Fish & Wildlife Service, November 2011, http://ecos.fws.gov/tess_public/pub/speciesRecovery.jsp?sort=1 (accessed November 8, 2011)

coastal waters of Florida, Puerto Rico, and the U.S. Virgin Islands. Their range extends to many tropical countries of Central and South America. Both species were listed under the ESA effective June 2006.

As marine creatures, they are under the jurisdiction of the NMFS. According to the NMFS (December 14, 2007, http://www.nmfs.noaa.gov/pr/pdfs/fr/fr72-71102.pdf), the corals are primarily threatened by disease, hurricanes, and elevated sea surface temperatures. The agency notes that these threats "are severe, unpredictable, [and] likely to increase in the foreseeable future." However, these corals are widely distributed. In addition, they reproduce asexually through a process called fragmentation in which branches break off and become reattached to the reef. This mechanism helps them recover from damaging events, such as hurricanes. As a result of these factors, the corals are not believed to be at risk of extinction throughout all or a significant part of their range. Thus, they are afforded a threatened, rather than an endangered ranking.

In November 2008 the NMFS (http://www.gpo.gov/fdsys/pkg/FR-2008-11-26/pdf/E8-27748.pdf#page=1) designated critical habitat in marine areas for the two species as follows:

- Puerto Rico area—1,383 square miles (3,582 sq km)
- Florida area—1,329 square miles (3,442 sq km)
- St. Croix area—126 square miles (326 sq km)
- St. John/St. Thomas area—121 square miles (313 sq km)

Imperiled Corals around the World

In *Red List of Threatened Species Version 2011.2*, the IUCN indicates that 235 species of corals were threatened in 2011. This number accounted for 27% of the 856 species evaluated. According to the IUCN, approximately 2,175 coral species are known to scientists.

As of November 2011, no foreign coral species were listed as endangered or threatened under the ESA.

CHAPTER 6
AMPHIBIANS AND REPTILES

Amphibians and reptiles are collectively known by biologists as herpetofauna. In *Red List of Threatened Species Version 2011.2* (2011, http://www.iucnredlist.org/documents/summarystatistics/2011_2_RL_Stats_Table1.pdf), the International Union for Conservation of Nature (IUCN) notes that there are 6,771 described amphibian species and 9,439 described reptile species. New species in both of these groups are being discovered every year, particularly in remote tropical regions that are only now being explored.

Most of the herpetofauna native to the United States are found in wetlands and riparian habitat (the banks and immediate areas around water bodies, such as rivers and streams). Biologists indicate that amphibians and reptiles play a crucial role in these ecosystems by controlling insects, processing dead organic matter into a form that is edible by smaller creatures, and providing an important link in the food chain.

Many herpetofauna species are under threat, primarily due to declines and degradation in their habitats in recent decades.

AMPHIBIANS

Amphibians are vertebrate animals in the taxonomic class Amphibia. They represent the most ancient group of terrestrial vertebrates. The earliest amphibians lived during the early Devonian era, some 400 million years ago. The three groups of amphibians that have survived to the present day are salamanders, frogs (and toads), and caecilians.

Salamanders belong to the orders Caudata or Urodela. They have moist smooth skin, slender bodies, four short legs, and long tails. This category includes the amphibians commonly known as newts (land-dwelling salamanders) and sirens (salamanders that have both lungs and gills).

Frogs and toads are in the order Anura. These amphibians do not have tails as adults. They have small bodies with two short front legs and two long back legs. (See Figure 6.1.) Their feet are webbed, and they are good jumpers and hoppers. True frogs belong to the family Ranidae, whereas true toads belong to the family Bufonidae. There are many other families in this order whose members are commonly described as frogs or toads. Many of the species go through a swimming tadpole stage before metamorphosing into an adult. However, in some species eggs hatch directly as juvenile froglets, which are miniature versions of the adults. Tadpoles are most often herbivorous (plant-eating), although there are some carnivorous (meat-eating) tadpoles, including cannibalistic species. Adults are carnivorous and catch prey with their sticky tongue.

Caecilians belong to the orders Gymnophiona or Apoda and share a common ancestor with the other amphibians, but look much different. They are often mistaken for worms or snakes. They have long slender bodies with no limbs and are found primarily in the tropics.

Amphi means "both," and amphibians get their name from the fact that many species occupy both aquatic and terrestrial habitats. In particular, many amphibian species undergo a dramatic change called metamorphosis, in which individuals move from an aquatic larval stage to a terrestrial adult stage. For example, in many frog species aquatic swimming tadpoles metamorphose into terrestrial jumping frogs. In the process, they lose their muscular swimming tails and acquire forelimbs and hind limbs. Many amphibian species occupy terrestrial habitats through most of the year, but migrate to ponds to breed. However, there are also species that are either entirely aquatic or entirely terrestrial. Whatever their habitat, amphibians generally require some moisture to survive. This is because amphibians pass some oxygen

FIGURE 6.1

General body type for frogs and toads

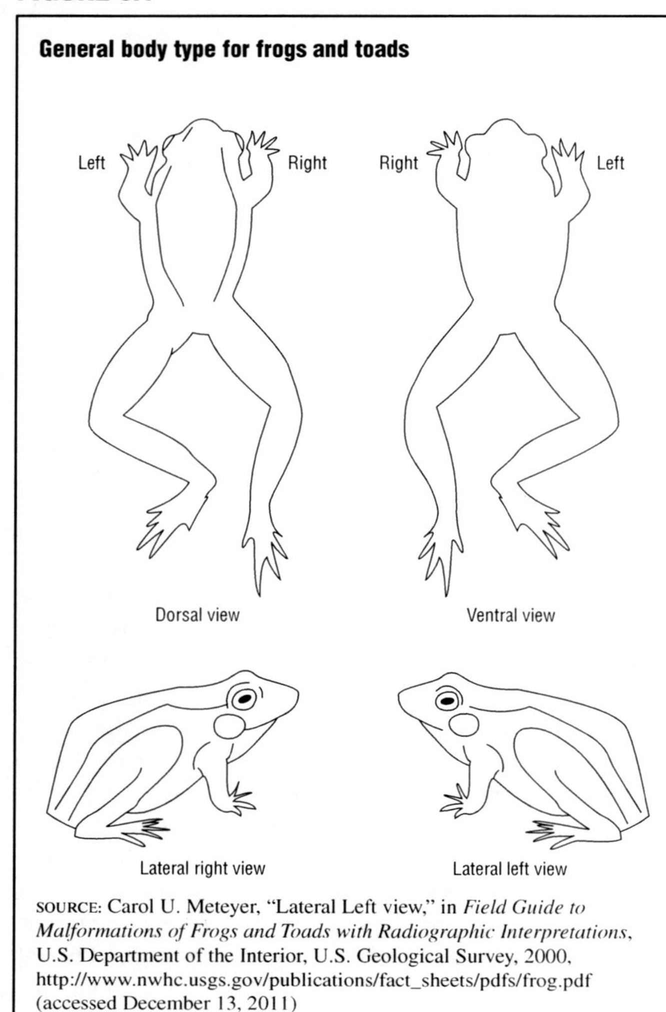

SOURCE: Carol U. Meteyer, "Lateral Left view," in *Field Guide to Malformations of Frogs and Toads with Radiographic Interpretations*, U.S. Department of the Interior, U.S. Geological Survey, 2000, http://www.nwhc.usgs.gov/publications/fact_sheets/pdfs/frog.pdf (accessed December 13, 2011)

and other chemicals in and out of their body directly through their skin, using processes that require water to function.

Many amphibian species are in serious decline due to factors such as habitat loss, pollution, and climate change. Amphibians are particularly vulnerable to pollution because their skin readily absorbs water and other substances from the environment. For this reason, amphibians are frequently considered biological indicator species, meaning that their presence, condition, and numbers are monitored as a gauge of the overall well-being of their habitat.

THREATENED AND ENDANGERED SPECIES OF AMPHIBIANS

As of November 2011, there were 23 U.S. amphibian species listed as threatened or endangered under the Endangered Species Act (ESA). (See Table 6.1.) The list contains 13 species of salamander (including the Ozark hellbender) and 10 species of frogs and toads (including the golden coqui and the guajón, which are Puerto Rican frogs). Note that no caecilian species were listed. Most of the listed species were endangered and nearly all had recovery plans in place. Geographically, the list is dominated by western and southern states, primarily California and Texas.

Table 6.2 shows the 10 amphibian species with the highest expenditures under the ESA during fiscal year (FY) 2010. Nearly $11.4 million was spent in an effort to conserve these species.

Imperiled Salamanders in the United States

As shown in Table 6.1, 13 salamanders were listed under the ESA as of November 2011.

Some endangered salamanders, including many cave species, have highly restricted habitats. For example, the Barton Springs salamander is only found in and around spring-fed pools in Zilker Park in Austin, Texas. (See Figure 6.2.) The species was first listed as endangered in 1997. Urban development has contributed to degradation of the local groundwater that feeds the spring. In addition, flows from the spring have decreased due to increasing human use of groundwater from the aquifer. Finally, the Barton Springs salamander has been the subject of contentious debate between conservationists and those who wish to expand development around the area of the pools.

CALIFORNIA TIGER SALAMANDER. As shown in Table 6.2, approximately $4.6 million was spent under the ESA during FY 2010 on three separate populations of the California tiger salamander. (See Figure 6.3.) According to the U.S. Fish and Wildlife Service (USFWS), in "California Tiger Salamander (*Ambystoma californiense*)" (January 2012, http://ecos.fws.gov/speciesProfile/profile/speciesProfile.action?spcode=D01T), the salamander lives in terrestrial habitats and is relatively large, reaching a maximum length of about 8 inches (20.3 cm).

The ESA-listed populations are as follows:

- Sonoma County, California population—endangered
- Santa Barbara County, California population—endangered
- Central California distinct population segment (DPS), excluding the Sonoma and Santa Barbara County populations—threatened

Thus, the salamander is listed across a long swath of coastal California, an area that is characterized by residential and commercial development close to the coast and ranching and farming activities inland.

In *California Tiger Salamander (*Ambystoma californiense*) Santa Barbara County Distinct Population Segment: 5-Year Review—Summary and Evaluation* (November 2009, http://ecos.fws.gov/docs/five_year_review/doc3223.pdf), the USFWS indicates that the

TABLE 6.1

Endangered and threatened amphibian species, November 2011

Common name	Scientific name	Listing status[a]	U.S. or U.S./foreign listed	Recovery plan date	Recovery plan stage[b]
Arroyo (=arroyo southwestern) toad	Bufo califomicus (=microscaphus)	E	US/foreign	7/24/1999	F
Barton Springs salamander	Eurycea sosorum	E	US	9/21/2005	F
California red-legged frog	Rana draytonii	T	US/foreign	5/28/2002	F
California tiger salamander	Ambystomacaliforniense	E (2), T (1)	US	None	—
Cheat Mountain salamander	Plethodon nettingi	T	US	7/25/1991	F
Chiricahua leopard frog	Rana chiricahuensis	T	US/foreign	6/4/2007	F
Desert slender salamander	Batrachoseps aridus	E	US	8/12/1982	F
Frosted Flatwoods salamander	Ambystomacingulatum	T	US	None	—
Golden coqui	Eleutherodactylusjasperi	T	US	4/19/1984	F
Guajon	Eleutherodactyluscooki	T	US	9/24/2004	F
Houston toad	Bufo houstonensis	E	US	9/17/1984	F
Mississippi gopher frog	Rana capito sevosa	E	US	None	—
Mountain yellow-legged frog	Rana muscosa	E	US	None	—
Ozark Hellbender	Cryptobranchusalleganiensisbishopi	E	US	None (listed in 2011)	—
Puerto Rican crested toad	Peltophrynelemur	T	US/foreign	8/7/1992	F
Red Hills salamander	Phaeognathushubrichti	T	US	11/23/1983	F
Reticulated flatwoods salamander	Ambystomabishopi	E	US	None	—
San Marcos salamander	Eurycea nana	T	US	2/14/1996	RF(1)
Santa Cruz long-toed salamander	Ambystomamacrodactylumcroceum	E	US	7/2/1999	RD(2)
Shenandoah salamander	Plethodonshenandoah	E	US	9/29/1994	F
Sonora tiger salamander	Ambystomatigrinumstebbinsi	E	US/foreign	9/24/2002	F
Texas blind salamander	Typhlomolgerathbuni	E	US	2/14/1996	RF(1)
Wyoming toad	Bufo baxteri (=hemiophrys)	E	US	9/11/1991	F

[a]E = endangered; T = threatened; numbers in parentheses indicate separate populations.
[b]F = final; D = draft; RD = draft revision; RF = final revision.

SOURCE: Adapted from "Generate Species List," in *Species Reports*, U.S. Department of the Interior, U.S. Fish & Wildlife Service, November 2011, http://ecos.fws.gov/tess_public/pub/adHocSpeciesForm.jsp (accessed November 8, 2011), and "Listed FWS/Joint FWS and NMFS Species and Populations with Recovery Plans (Sorted by Listed Entity)," in *Recovery Plans Search*, U.S. Department of the Interior, U.S. Fish & Wildlife Service, November 2011, http://ecos.fws.gov/tess_public/pub/speciesRecovery.jsp?sort=1 (accessed November 8, 2011)

TABLE 6.2

Amphibian species with the highest expenditures under the Endangered Species Act, fiscal year 2010

Ranking	Species	Expenditure
1	Frog, California red-legged (Rana draytonii)—Entire	$2,732,668
2	Salamander, California tiger (Ambystoma californiense)—CA-Sonoma County	$2,207,619
3	Toad, arroyo (=arroyo southwestern) (Bufo californicus (=microscaphus))	$1,576,429
4	Salamander, California tiger (Ambystoma californiense)—Central CA DPS	$1,392,458
5	Frog, Chiricahua leopard (Rana chiricahuensis)	$1,152,117
6	Salamander, California tiger (Ambystoma californiense)—CA-Santa Barbara County	$1,002,956
7	Frog, Mississippi gopher (Rana capito sevosa)—west of Mobile and Tombigbee Rivers in AL, MS, and LA	$446,383
8	Salamander, Reticulated flatwoods (Ambystoma bishopi)	$335,914
9	Frog, mountain yellow-legged (Rana muscosa)—southern CA DPS	$284,725
10	Salamander, Red Hills (Phaeognathus hubrichti)	$248,620

DPS = distinct population segment.

SOURCE: Adapted from "Table 2. Species Ranked in Descending Order of Total FY 2010 Reported Expenditures, Not Including Land Acquisition Costs," in *Federal and State Endangered and Threatened Species Expenditures: Fiscal Year 2010*, U.S. Department of the Interior, U.S. Fish and Wildlife Service, 2010, http://www.fws.gov/endangered/esa-library/pdf/2010.EXP.FINAL.pdf (accessed October 25, 2011)

salamander lives mostly underground, but migrates to ponds and water pools to breed. Historically, the preferred breeding sites were vernal water bodies. (As explained in Chapter 5, the term *vernal* is from the Latin word for "spring." Vernal ponds and pools appear during the rainy season—winter or springtime—and dry up over time.) The USFWS notes that the salamander is now found more often in human-made or human-modified water bodies, such as livestock ponds.

The USFWS states in "California Tiger Salamander: *Ambystoma californiense*" (July 29, 2009, http://www.fws.gov/sacramento/ES_Species/Accounts/Amphibians-Reptiles/Documents/california_tiger_salamander.pdf) that the salamander is imperiled by numerous threats, including habitat loss and fragmentation and nonnative predators and mating partners.

In August 2011 the USFWS (http://www.gpo.gov/fdsys/pkg/FR-2011-08-31/pdf/2011-21945.pdf) designated 47,383 acres (19,175 ha) of critical habitat for the Sonoma County population of the California tiger salamander. The rule making followed years of court battles between the agency and the CBD over the appropriateness of and the proper extent for critical habitat. As of January 2012, critical habitat had not been designated for the Santa

FIGURE 6.2

Location of Barton Springs in Texas

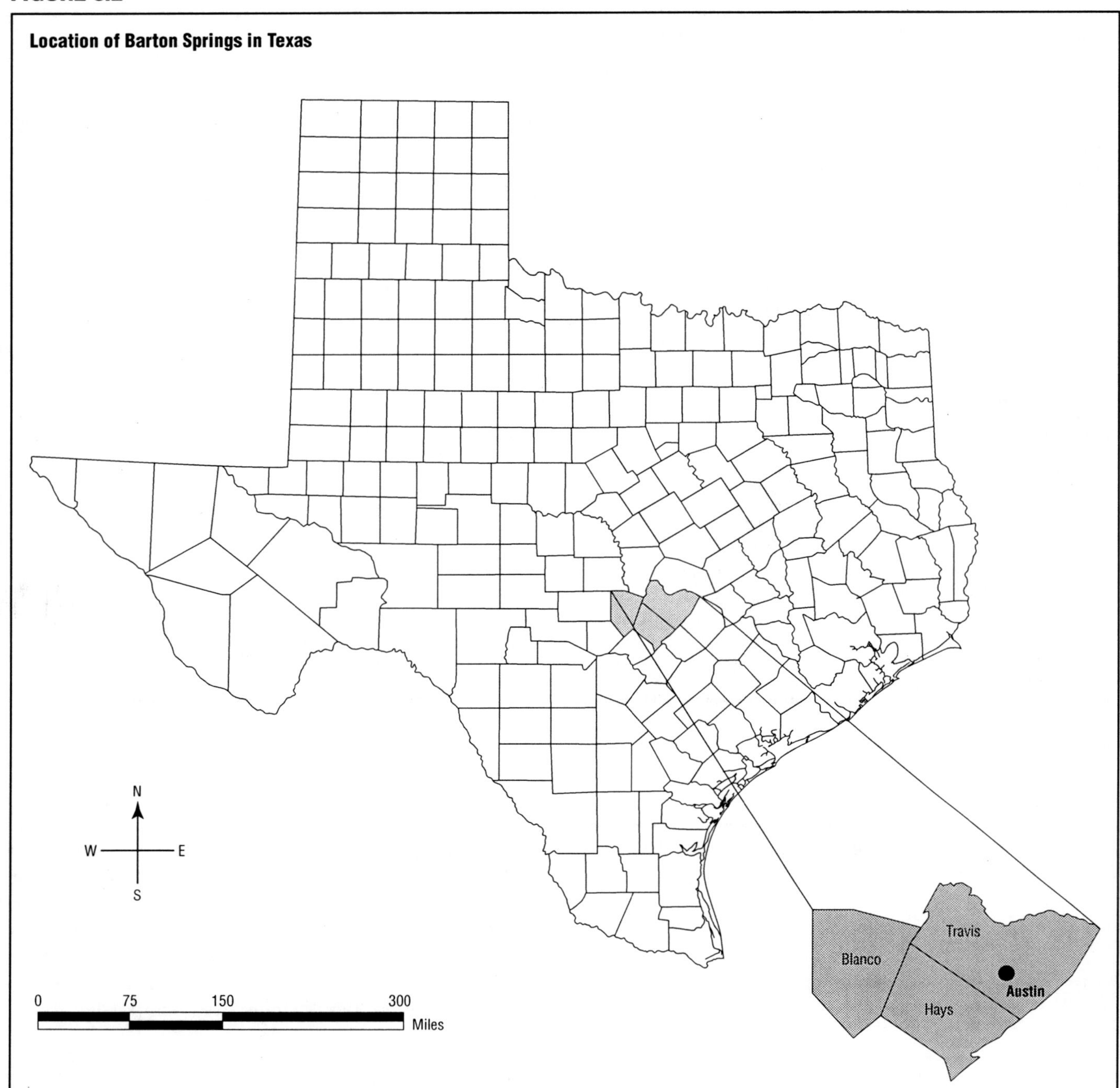

SOURCE: "Figure 3. Location of Barton Springs, Travis County, Texas," in *Barton Springs Salamander (Eurycea sosorum) Recovery Plan*, U.S. Department of the Interior, U.S. Fish and Wildlife Service, September 2005, http://www.fws.gov/contaminants/OtherDocuments/BartonSpringsSalamanderRP.pdf (accessed December 13, 2011)

Barbara County or central California DPSs. Also, recovery plans had not been developed for any of the populations.

As noted earlier, a five-year review was published for the Santa Barbara County population in 2009. The purpose of a five-year review is to examine the status of listed species and determine whether to keep or change their ESA listings. The USFWS recommended that the endangered listing be maintained for the Santa Barbara population. In May 2010 and May 2011 the agency initiated five-year reviews for the central California and Sonoma County populations, respectively. As of January 2012, the results of the reviews had not been published.

In 2007 scientists reported that the genetic distinction of the California tiger salamander as a species is in serious doubt due to long-term interbreeding with the nonnative barred tiger salamander. Benjamin M. Fitzpatrick and

FIGURE 6.3

California tiger salamander. *(© U.S. Fish and Wildlife Service.)*

H. Bradley Shaffer of the University of California, Davis, note in "Hybrid Vigor between Native and Introduced Salamanders Raises New Challenges for Conservation" (*PNAS*, vol. 104, no. 4, October 2, 2007) that the nonnative salamanders were introduced to California from Texas sometime during the late 1940s to the late 1950s as larval fish bait. Subsequent interbreeding with the native species has produced a vigorous hybrid offspring. The researchers find that the hybrids have "higher survival rates" than either of the parent species. In "Rapid Spread of Invasive Genes into a Threatened Native Species" (*PNAS*, vol. 107, no. 8, October 2, February 23, 2010), Benjamin M. Fitzpatrick et al. provide an update on their investigation, noting that introgression (gene infiltration) is continuing to occur at a rapid rate and appears to be favored by natural selection. The researchers point out that the net outcome of this genetic change on the native salamander species and its status under the ESA is uncertain.

Imperiled Frogs and Toads in the United States

The 10 species of imperiled frogs and toads found in the United States are geographically diverse. Their habitats are located in the West, Southeast, and Puerto Rico.

CALIFORNIA RED-LEGGED FROGS. The California red-legged frog is the largest native frog in the western United States. The frog was made famous by Mark Twain's (1835–1910) short story "The Celebrated Jumping Frog of Calaveras County" (1865). The species experienced a significant decline during the mid-20th century. According to the Environmental Defense Fund, in "California Red-Legged Frog" (September 29, 2009, http://apps.edf.org/page.cfm?tagID=7766), by 1960 California red-legged frogs had disappeared altogether from the state's Central Valley, probably due to the loss of most of their habitat. In 1996 the species was listed as threatened under the ESA. As noted in Table 6.2, $2.7 million was spent under the ESA during FY 2010 on conserving the California red-legged frog.

California red-legged frogs require riverside habitats that are covered by vegetation and are close to deep-water pools. They are extremely sensitive to habitat disturbance and water pollution—tadpoles are particularly sensitive to varying oxygen levels and siltation (mud and other natural impurities) during metamorphosis. The frogs require three to four years to reach maturity and have a normal life span of eight to 10 years.

Water reservoir construction and agricultural and residential development are the primary factors in the decline of this species. Biologists find that California red-legged frogs generally disappear from habitats within five years of a reservoir or water diversion project. The removal of vegetation associated with flood control and the use of herbicides and restructuring of landscapes further degrade remaining habitat. Finally, nonnative species have also attacked red-legged frog populations. These include alien fish predators as well as competing species such as bullfrogs.

In May 2002 the USFWS published *Recovery Plan for the California Red-Legged Frog (*Rana aurora draytonii*)* (http://ecos.fws.gov/docs/recovery_plans/2002/020528.pdf). The agency calls for eliminating threats in current habitats, restoring damaged habitats, and reintroducing populations into the historic range of the species. Critical habitat for the species was first set by the agency in 2001 and revised in 2006. In 2007 the CBD filed a legal challenge arguing that the critical habitat designation was not sufficient. As a result, in March 2010 the USFWS (http://www.gpo.gov/fdsys/pkg/FR-2010-03-17/pdf/2010-4656.pdf#page=1) substantially increased the critical habitat from 450,000 acres (182,000 ha) to 1.6 million acres (660,000 ha).

In May 2011 the USFWS (http://www.gpo.gov/fdsys/pkg/FR-2011-05-25/pdf/2011-12861.pdf) initiated a five-year review for dozens of western species, including the California red-legged frog. As of January 2012, the results of the review had not been published.

Foreign Amphibians in Danger

The IUCN reports in *Red List of Threatened Species Version 2011.2* that 1,917 amphibian species were threatened in 2011. This represented 30% of the 6,338 amphibian species evaluated, the highest percentage for any group of animals evaluated.

As of November 2011, there were nine foreign amphibian species listed as threatened or endangered under the ESA. (See Table 6.3.) This includes seven frog and toad species and two salamander species.

TABLE 6.3

Foreign endangered and threatened amphibian species, November 2011

Common name	Scientific name	Listing status*	Historic range
African viviparous toads	Nectophrynoides spp.	E	Tanzania, Guinea, Ivory Coast, Cameroon, Liberia, Ethiopia
Cameroon toad	Bufo superciliaris	E	Equatorial Africa
Chinese giant salamander	Andrias davidianus (=davidianusd.)	E	Western China
Goliath frog	Conraua goliath	T	Cameroon, Equatorial Guinea, Gabon
Israel painted frog	Discoglossus nigriventer	E	Israel
Japanese giant salamander	Andrias japonicus (=davidianusj.)	E	Japan
Monte Verde golden toad	Bufo periglenes	E	Costa Rica
Panamanian golden frog	Atelopus varius zeteki	E	Panama
Stephen Island frog	Leiopelma hamiltoni	E	New Zealand

*E = Endangered; T = threatened.

SOURCE: Adapted from "Generate Species List," in *Species Reports*, U.S. Department of the Interior, U.S. Fish & Wildlife Service, November 2011, http://ecos.fws.gov/tess_public/pub/adHocSpeciesForm.jsp (accessed November 8, 2011)

WORLDWIDE THREATS TO AMPHIBIANS

At the end of the 20th century biologists uncovered growing evidence of a global decline in amphibian populations. AmphibiaWeb, a conservation organization that monitors amphibian species worldwide, reports in "Worldwide Amphibian Declines: How Big Is the Problem, What Are the Causes and What Can Be Done?" (January 22, 2009, http://amphibiaweb.org/declines/declines.html) that 168 species have become extinct in recent decades and that at least 2,469 other species are declining in population. Amphibian declines have been documented worldwide, though the degree of decline varies across regions. Areas that have been hardest hit include Central America, the Caribbean, and Australia. AmphibiaWeb blames eight main factors for the declines:

- Habitat destruction, alteration, and fragmentation
- Introduction of nonnative species that prey on or compete with native amphibians
- Human overexploitation of amphibians for sale as food, pets, or medical specimens
- Climate change
- Rising levels of ultraviolet radiation
- Chemical contaminants in the environment
- Disease
- Deformities

Scientists are concerned because many amphibian species—particularly frogs—have become extinct over a very short period. Other species are either declining or showing high levels of gross deformities, such as extra limbs.

Habitat Destruction

The loss of habitat is a major factor in the decline of many amphibian species, as it is for many other species. The destruction of tropical forests and wetlands, ecosystems that are rich with amphibians, has done particular damage to populations. In the United States, deforestation is blamed for the loss or decline of salamander species in the Pacific Northwest and the Appalachian hardwood forests. In addition, some amphibians have lost appropriate aquatic breeding habitats, particularly small bodies of water such as ponds. These aquatic habitats are often developed or filled in by humans, because they appear to be less biologically valuable than larger aquatic habitats.

Nonnative Species

Many amphibian species have also been affected by the introduction of nonnative species that either compete with or prey on them. These include fish, crayfish, and other amphibians. The bullfrog, the cane toad (a large frog species), and the African clawed frog (a species often used in biological research) are some of the invasive species believed to have affected amphibian populations.

Human Collection and Consumption

Many amphibian species are vigorously hunted for food, the pet trade, or as medical research specimens. In "Over-collection and the Global Amphibian Crisis" (November 2009, http://www.defenders.org/resources/publications/programs_and_policy/international_conservation/the_amphibian_crisis_and_the_threat_of_international_trade.pdf), the Defenders of Wildlife, a nonprofit wildlife conservation organization, indicates that up to 1 billion frogs are believed to be consumed annually around the world. Consumption occurs in both undeveloped and developed countries; for example, in the United States and Europe frog legs are considered a delicacy.

Lisa M. Schloegel et al. report in "Magnitude of the US Trade in Amphibians and Presence of *Batrachochytrium dendrobatidis* and Ranavirus Infection in Imported North American Bullfrogs (*Rana catesbeiana*)" (*Biological*

Conservation, vol. 142, no. 7, 2009) on government importation data they collected for Los Angeles, San Francisco, and New York between 2000 and 2005 for live amphibians and amphibian parts. The total was nearly 30.5 million individual live amphibians and approximately 13.1 million pounds (6 million kg) of amphibian parts. Nearly all (28 million) of the live amphibians were for human consumption. The others were imported for the pet trade or other purposes.

Climate Change

As noted in Chapter 1, continual warming of the earth's atmosphere is bringing about climate change, which has repercussions for many ecosystems. In "Repercussions of Global Change" (Michael Lannoo, ed., *Amphibian Declines: The Conservation Status of United States Species*, 2005), Jamie K. Reaser and Andrew Blaustein note that amphibians will likely be among the first species to show "broad-scale" changes due to global warming because they are so sensitive to moisture and temperature variations.

According to AmpibiaWeb, in *Climate Change* (May 14, 2008, http://amphibiaweb.org/declines/ClimateChange.html), some amphibian species are breeding earlier in the year due to unseasonably warm temperatures. This can expose eggs and young to adverse environmental conditions, such as floods due to snowmelt or a sudden cold snap. Warmer temperatures may also spur outbreaks of parasites and other diseases that are detrimental to amphibian species.

Ultraviolet Radiation

Air pollution by substances such as chlorofluorocarbons has reduced the amount of protective ozone in the earth's atmosphere. This has resulted in increased levels of ultraviolet (UV) radiation striking the earth's surface. UV radiation has wavelengths of 290 to 400 nanometers (nm). Wavelengths of between 290 and 315 nm are called UV-B radiation and are the most dangerous, because they can damage deoxyribonucleic acid by producing chemicals called cyclobutane pyrimidine dimers. Exposure to UV-B radiation causes genetic mutations that can prevent normal development or kill eggs. Increased UV-B levels particularly affect the many frog species whose eggs lack shells and float on the exposed surfaces of ponds. Tadpoles and adults are also at risk because of their thin delicate skins.

Chemical Contaminants

Pollution is believed to be a major factor in global amphibian declines. As noted in Chapter 4, in 2010 U.S. states issued 4,600 health advisories due to concerns about the risks of humans eating fish that were known to be contaminated with certain pollutants, mainly mercury, polychlorinated biphenyls, dioxins, and the pesticides dichlorodiphenyltrichloroethane (DDT) and chlordane. Because amphibians absorb water directly through skin and into their body, they are particularly vulnerable to water that is polluted with these chemicals and pesticides.

Disease

Amphibian diseases caused by bacteria, viruses, and fungi have devastated certain populations. Of particular importance in recent years is the chytrid fungus. It attacks skin, and even though there are often no symptoms initially, affected individuals eventually begin to shed their skin and die. The chytrid fungus is believed to be responsible for the demise of many species in Australia and Panama. In "Is the Frog-Killing Chytrid Fungus Fueled by Climate Fluctuations?" (*Scientific American*, August 11, 2009), Brendan Borrell reports that scientists believe the fungus originated in South Africa and began to spread during the 1930s as African clawed frogs were exported around the world for use in pregnancy tests. Scientists are split in their opinion on whether warming global temperatures and climate change play a significant role in the spread of the fungus.

Amphibian Deformities

Amphibian deformities first hit the spotlight during the mid-1990s, when reports about frogs with missing, extra, or misshapen limbs or other physical abnormalities began to surface around the country. In *Field Guide to Malformationsof Frogs and Toads* (April 2001, http://www.nwhc.usgs.gov/publications/fact_sheets/pdfs/frog.pdf), Carol U. Meteyer of the U.S. Geological Survey (USGS) indicates that as of 2000 frog malformations had been reported in more than 50 species of frog and toads in 44 states.

The high incidence of amphibian deformities in some U.S. species in some areas appears to have multiple causes, as no single hypothesis accounts for all the different types of deformities seen. Probable causes include chemical contaminants, nutrient deficiencies, injuries (e.g., from predators), UV-B radiation, and parasites.

In April 2009 the Public Broadcasting Service (PBS) began televising the documentary *Frogs: The Thin Green Line* (http://www.pbs.org/wnet/nature/episodes/frogs-the-thin-green-line/video-full-episode/4882/), which focuses on the worldwide decline of frog species and the resulting effects on ecosystems and humans. Frogs are particularly important from an ecological standpoint, because they lie at the "middle" of the food chain. Tadpoles eat algae, a process that helps maintain clean water quality. Adult frogs eat insects, including insects such as mosquitoes and flies that can spread diseases to animals and humans. Frog eggs provide food for wasps and spiders. Fish and other aquatic creatures eat tadpoles, while birds and reptiles feed on adult frogs. Furthermore, PBS notes that scientists have discovered hundreds of chemicals in the skins of frogs, particularly tropical species. These chemicals hold promise in the development of anesthetics, painkillers, and other medicines that could greatly benefit humans.

REPTILES

Reptiles belong to the class Reptilia. Even though they may appear similar, reptiles differ from amphibians in that reptile skin is cornified (made of dead cells). All reptiles obtain oxygen from the air using lungs. Most reptiles lay shelled eggs, although some species, particularly lizards and snakes, give birth to live young. According to the IUCN, in *Red List of Threatened Species Version 2011.2*, 9,439 species of reptiles have been described.

Reptiles include crocodilians, lizards, snakes, and turtles. Birds are also technically reptiles (birds and crocodiles are actually close relatives), but have historically been treated separately.

There are four taxonomic orders of reptiles:

- Squamata—thousands of species of lizards, anoles, iguanas, Gila monsters, monitors, skinks, geckos, chameleons, snakes (including asps, boas, pythons, and vipers), racerunners, whiptails, and amphisbaenians (worm lizards)
- Testudines—hundreds of species of turtles, terrapins, and tortoises
- Crocodilia—around two dozen species, including alligators, caimans, crocodiles, and gavials
- Rhynchocephalia—two species of tuataras found only in New Zealand

Lizards and snakes represent the largest group of reptiles. Many reptiles are in serious decline. Several species are endangered due to habitat loss or degradation. In addition, humans hunt reptiles for their skins, shells, or meat. Global climate change has affected some reptile species, particularly turtles, in ominous ways—this is because in some reptiles ambient temperatures determine whether males or females are produced. Even a small increase in temperature can result in few or no males being born. Natural disasters, such as hurricanes, can also affect reptiles by killing the animals or damaging their habitat.

THREATENED AND ENDANGERED REPTILES

As of November 2011, there were 36 U.S. reptile species listed as threatened or endangered under the ESA. (See Table 6.4.) Some species have dual status because they have separate populations in the United States. In addition, a few reptiles are listed as SAT, which means threatened due to similarity of appearance. This listing is applied to animals, such as the American alligator, that closely resemble imperiled species—in this case the American crocodile.

Except for sea turtles, most of the imperiled U.S. reptiles are geographically clustered in either the Southeast, California, or Puerto Rico and the Virgin Islands. Sea turtles spend most of their life at sea and only come onto land to nest and lay young. Because there are many potential nesting sites along the U.S. coasts, sea turtles are listed in many states.

The following is a breakdown of imperiled U.S. reptiles by taxonomic order:

- Squamata—20 species (11 snakes and nine lizards)
- Testudines—14 species (six sea turtles, two [land-dwelling] tortoises, and six other turtle species)
- Crocodilia—two species

The 10 reptile species with the highest expenditures under the ESA during FY 2010 are shown in Table 6.5. The list is dominated by tortoise and sea turtle species.

Imperiled Tortoises in the United States

Terrestrial (land-dwelling) turtles inhabit inland areas and waterways; some species are generically called tortoises. As shown in Table 6.4, there were eight terrestrial turtle species listed under the ESA as of November 2011.

DESERT TORTOISES. The desert tortoise has dual listings under the ESA. (See Figure 6.4.) The population in Arizona, California, Nevada, and Utah was listed in 1980 as threatened except for those tortoises in Arizona south and east of the Colorado River. The latter were listed in 1990 as SAT. The population with a threatened listing is commonly known as the Mojave Desert population, whereas the population with an SAT listing is commonly known as the Sonoran Desert population. In "Genetic Analysis Splits Desert Tortoise into Two Species" (June 28, 2011, http://www.usgs.gov/newsroom/article.asp?ID=2842), the USGS reports that in 2011 researchers determined that the two populations are actually two different species. The so-called Sonoran Desert population has been officially named the Morafka's desert tortoise (*Gopherus morafkai*).

The decline of the desert tortoise has resulted from collection by humans, predation of young turtles by ravens, off-road vehicles, invasive plant species, and habitat destruction due to development for agriculture, mining, and livestock grazing.

Desert tortoise populations are constrained by the fact that females do not reproduce until they are 15 to 20 years of age (individuals can live 80 to 100 years) and by small clutch sizes (i.e., small numbers of eggs per laying). Juvenile mortality is also extremely high, in large

TABLE 6.4

Endangered and threatened reptile species, November 2011

Common name	Scientific name	Listing status[a]	U.S. or U.S./foreign listed	Recovery plan date	Recovery plan stage[b]
Alabama red-belly turtle	Pseudemysalabamensis	E	US	1/8/1990	F
Alameda whipsnake (=striped racer)	Masticophis lateralis euryxanthus	T	US	4/7/2003	D
American alligator	Alligator mississippiensis	SAT	US	None	—
American crocodile	Crocodylusacutus	T	US/foreign	5/18/1999	F
Atlantic salt marsh snake	Nerodia clarkii taeniata	T	US	12/15/1993	F
Bluetail mole skink	Eumeces egregiuslividus	T	US	5/18/1999	F
Blunt-nosed leopard lizard	Gambelia silus	E	US	9/30/1998	F
Bog (=Muhlenberg) turtle	Clemmysmuhlenbergii	T (1), SAT (1)	US	5/15/2001	F
Coachella Valley fringe-toed lizard	Uma inornata	T	US	9/11/1985	F
Concho water snake	Nerodia paucimaculata	T	US	9/27/1993	F
Copperbelly water snake	Nerodia erythrogaster neglecta	T	US	12/23/2008	F
Culebra Island giant anole	Anolis roosevelti	E	US	1/28/1983	F
Desert tortoise	Gopherus agassizii	T (1), SAT (1)	US/foreign	5/6/2011	RF(1)
Eastern indigo snake	Drymarchoncorais couperi	T	US	4/22/1982	F
Flattened musk turtle	Sternotherus depressus	T	US	2/26/1990	F
Giant garter snake	Thamnophisgigas	T	US	7/2/1999	D
Gopher tortoise	Gopherus polyphemus	T	US	12/26/1990	F
Green sea turtle	Chelonia mydas	E (1), T (1)	US/foreign	10/29/1991 and 1/12/1998	RF(1) and RF(1)
Hawksbill sea turtle	Eretmochelys imbricata	E	US/foreign	12/15/1993 and 1/12/1998	RF(1) and RF(1)
Island night lizard	Xantusia riversiana	T	US	1/26/1984	F
Kemp's ridley sea turtle	Lepidochelyskempii	E	US/foreign	3/16/2010	RD(2)
Leatherback sea turtle	Dermochelyscoriacea	E	US/foreign	4/6/1992 and 1/12/1998	RF(1) and RF(1)
Loggerhead sea turtle	Caretta caretta	T (2)	US/foreign	1/12/1998	RF(1)
Mona boa	Epicrates monensismonensis	T	US	4/19/1984	F
Mona ground Iguana	Cyclura cornuta stejnegeri	T	US	4/19/1984	F
Monito gecko	Sphaerodactylusmicropithecus	E	US	3/27/1986	F
New Mexican ridge-nosed rattlesnake	Crotalus willardi obscurus	T	US/foreign	3/22/1985	F
Olive ridley sea turtle	Lepidochelysolivacea	T	US/foreign	1/12/1998 and 1/16/2009	RF(1) and RF(2)
Plymouth Red-Bellied turtle	Pseudemys rubriventris bangsi	E	US	5/6/1994	RF(2)
Puerto Rican boa	Epicrates inornatus	E	US	3/27/1986	F
Ringed map turtle	Graptemys oculifera	T	US	4/8/1988	F
Sand skink	Neoseps reynoldsi	T	US	5/18/1999	F
San Francisco garter snake	Thamnophissirtalis tetrataenia	E	US	9/11/1985	F
St. Croix ground lizard	Ameiva polops	E	US	3/29/1984	F
Virgin Islands tree boa	Epicrates monensisgranti	E	US/foreign	3/27/1986	F
Yellow-blotched map turtle	Graptemys flavimaculata	T	US	3/15/1993	F

[a]E = endangered; T = threatened; SAT = threatened due to similarity of appareance; numbers in parenthesis indicate separate populations.
[b]F = final; D = draft; RD = draft revision; RF = final revision.

SOURCE: Adapted from "Generate Species List," in *Species Reports*, U.S. Department of the Interior, U.S. Fish & Wildlife Service, November 2011, http://ecos.fws.gov/tess_public/pub/adHocSpeciesForm.jsp (accessed November 8, 2011), and "Listed FWS/Joint FWS and NMFS Species and Populations with Recovery Plans (Sorted by Listed Entity)," in *Recovery Plans Search*, U.S. Department of the Interior, U.S. Fish & Wildlife Service, November 2011, http://ecos.fws.gov/tess_public/pub/speciesRecovery.jsp?sort=1 (accessed November 8, 2011)

part due to predation by ravens, whose populations in the desert tortoise's habitat have increased with increasing urbanization of desert areas—human garbage provides food for ravens and power lines provide perches.

Approximately $16.3 million was spent under the ESA during FY 2010 on conservation measures for the Mojave Desert population of the desert tortoise. (See Table 6.5.) In September 2010 the USFWS published *Mojave Population of the Desert Tortoise (*Gopherus agassizii*): 5-Year Review—Summary and Evaluation* (http://ecos.fws.gov/docs/five_year_review/doc3572.DT%205Year%20Review_FINAL.pdf), in which the agency decided to maintain the population's threatened listing. In May 2011 the agency published *Revised Recovery Plan for the Mojave Population of the Desert Tortoise (*Gopherus agassizii*)* (http://ecos.fws.gov/docs/recovery_plan/RRP%20for%20the%20Mojave%20Desert%20Tortoise%20-%20May%202011_1.pdf). The original recovery plan was finalized in 1994. The 2011 plan estimates that the species can be recovered by 2025 at a cost of at least $159 million.

As shown in Table 2.5 in Chapter 2, the Sonoran Desert tortoise DPS is a candidate species with a listing priority number (LPN) of 6. LPNs range from 1 to 12, with lower numbers indicating greater priority compared with other candidates. (See Table 2.4 in Chapter 2.)

As explained in Chapter 2, in late 2011 the agency settled long-standing lawsuits involving candidate species by agreeing to make final listing decisions for all

TABLE 6.5

Reptile species with the highest expenditures under the Endangered Species Act, fiscal year 2010

Ranking	Species	Expenditure
1	Tortoise, desert (Gopherus agassizii)—U.S.A., except in Sonoran Desert	$16,343,561
2	Sea turtle, loggerhead (Caretta caretta)	$11,285,625
3	Sea turtle, leatherback (Dermochelys coriacea)	$7,057,501
4	Sea turtle, hawksbill (Eretmochelys imbricata)	$6,598,089
5	Sea turtle, green (Chelonia mydas)—except where endangered	$5,662,140
6	Sea turtle, Kemp's ridley (Lepidochelys kempii)	$5,130,707
7	Tortoise, gopher (Gopherus polyphemus)—west of Mobile/Tombigbee rivers	$3,588,732
8	Snake, giant garter (Thamnophis gigas)	$3,490,313
9	Snake, eastern indigo (Drymarchon corais couperi)	$2,460,953
10	Alligator, American (Alligator mississippiensis)	$1,446,653

SOURCE: Adapted from "Table 2. Species Ranked in Descending Order of Total FY 2010 Reported Expenditures, Not Including Land Acquisition Costs," in *Federal and State Endangered and Threatened Species Expenditures: Fiscal Year 2010*, U.S. Department of the Interior, U.S. Fish and Wildlife Service, 2010, http://www.fws.gov/endangered/esa-library/pdf/2010.EXP.FINAL.pdf (accessed October 25, 2011)

FIGURE 6.4

The desert tortoise is threatened due to habitat destruction, livestock grazing, invasion of nonnative plant species, collection, and predation by ravens. *(© U.S. Fish and Wildlife Service.)*

FIGURE 6.5

Range of the gopher tortoise

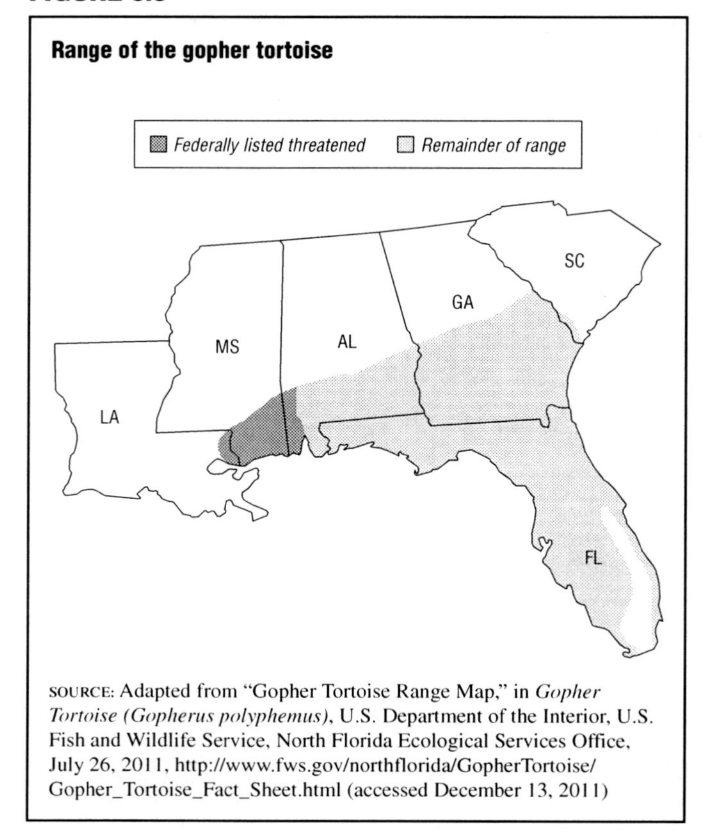

SOURCE: Adapted from "Gopher Tortoise Range Map," in *Gopher Tortoise (Gopherus polyphemus)*, U.S. Department of the Interior, U.S. Fish and Wildlife Service, North Florida Ecological Services Office, July 26, 2011, http://www.fws.gov/northflorida/GopherTortoise/Gopher_Tortoise_Fact_Sheet.html (accessed December 13, 2011)

of them by 2017. Regarding the Sonoran Desert tortoise, the USFWS (May 10, 2011, http://www.fws.gov/endangered/improving_ESA/exh_1_re_joint_motion_FILED.PDF) agreed to make such a decision by the end of FY 2015.

GOPHER TORTOISES. In 1987 the western population of the gopher tortoise was listed as threatened under the ESA. This population is found west of the Tombigbee and Mobile Rivers in Alabama and across Mississippi and into Louisiana. (See Figure 6.5.) Gopher tortoises are land-dwelling turtles that prefer habitat in longleaf pine ecosystems with sandy soils. The tortoises spend much of their time in sandy burrows that often provide shelter for other animals, such as snakes and frogs. A recovery plan for the gopher tortoise was completed in 1990. At that time the primary threats to the species were habitat degradation and illegal taking.

Coastal longleaf pine forests in the Southeast have been highly degraded by centuries of land development. As this habitat has become more fragmented, isolated pockets of the species have resulted in poor reproduction rates. Scientists fear that genetic drift and interbreeding are already occurring within the population. During the Great Depression (1929–1939) gopher tortoises were a highly prized meat source. Since that time their numbers have been further reduced by road strikes and collection for the pet trade. Natural predators include raccoons, foxes, snakes, and fire ants that prey on eggs or hatchlings.

In *Gopher Tortoise (*Gopherus polvphemus*) Recovery Plan* (December 26, 1990, http://ecos.fws.gov/docs/recovery_plan/901226.pdf), the USFWS notes that better management of government-owned forests in the region could help the survival status of the gopher tortoise. However, most lands with suitable habitat are privately owned and have already undergone, or are likely to undergo, development for agricultural, residential, or commercial purposes.

In 2001 the USFWS, in conjunction with the Mobile Area Water and Sewer System (MAWSS) and the conservation organizations Environmental Defense Fund and Southeastern Natural Resources, created a 222-acre (89.8-ha) gopher tortoise conservation bank near Big Creek Lake in Mobile, Alabama. Property owners can relocate gopher tortoises from their land to the conservation bank, which will remain undeveloped. As of 2011, the MAWSS (http://www.mawss.com/faq.html) charged $3,500 per gopher tortoise admitted to the bank. This money is used to manage the habitat. Each bank tortoise is also equipped with a radio collar so that the USFWS can monitor its movement.

As shown in Table 6.5, nearly $3.6 million was spent under the ESA in FY 2010 on the western population of the gopher tortoise. In April 2010 the USFWS (http://www.gpo.gov/fdsys/pkg/FR-2010-04-09/pdf/2010-8103.pdf#page=1) published notice of its intention to conduct a five-year review for 10 southeastern species, including the gopher tortoise. As of January 2012, the results of that review had not been published.

In 2006 the USFWS (September 9, 2009, http://www.gpo.gov/fdsys/pkg/FR-2009-09-09/pdf/E9-21481.pdf#page=1) received a petition submitted by two conservation groups—Save Our Big Scrub Inc. and Wild South—requesting that the eastern population of gopher tortoises be listed as threatened under the ESA. As shown in Figure 6.5, this population is found in Alabama (east of the Tombigbee and Mobile Rivers), Florida, Georgia, and into South Carolina. In July 2011 the agency (http://www.gpo.gov/fdsys/pkg/FR-2011-07-27/pdf/2011-18856.pdf) published a review finding the listing warranted, but was precluded by higher priorities. As shown in Table 2.5 in Chapter 2, the USFWS has assigned the candidate species an LPN of 8, which indicates a moderate-to-low priority for listing compared with other candidates.

FIGURE 6.6

Sea turtles that nest on U.S. coasts

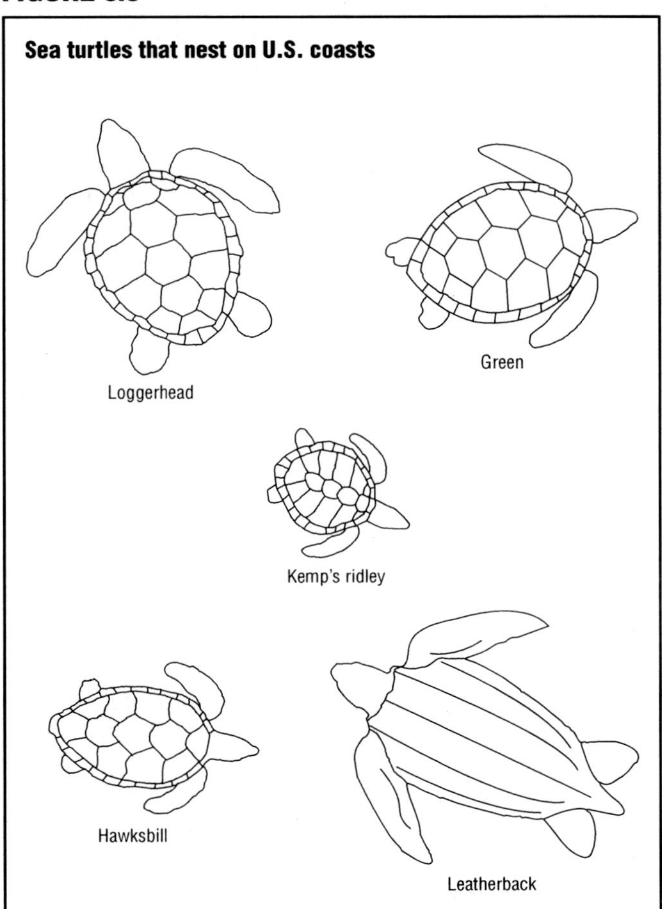

SOURCE: Adapted from "Untitled," in *You Can Help Protect Sea Turtles*, U.S. Department of the Interior, U.S. Fish & Wildlife Service, North Florida Ecological Services Office, July 2009, http://www.fws.gov/northflorida/SeaTurtles/20090700_You_Can_Help_ST.pdf (accessed November 9, 2011)

Imperiled Sea Turtles in the United States

Sea (or marine) turtles are excellent swimmers and spend nearly their entire life in water. They feed on a wide array of food items, including mollusks, vegetation, and crustaceans. Some sea turtles are migratory, swimming thousands of miles between feeding and nesting areas. Individuals are exposed to a variety of both natural and human threats. As a result, only an estimated one out of 10,000 sea turtles survives to adulthood.

There are seven species of sea turtles that exist worldwide. One species, the flatback turtle, occurs near Australia. The other six species spend part or all their life in U.S. territorial waters and, as of November 2011, were listed as endangered or threatened under the ESA. (See Table 6.4.) Imperiled sea turtles fall under the jurisdiction of the USFWS while they are on U.S. land and under the jurisdiction of the National Marine Fisheries Service (NMFS) while they are at sea. The green sea turtle, hawksbill sea turtle, Kemp's ridley sea turtle, leatherback sea turtle, and loggerhead sea turtle are shown in Figure 6.6.

The following provides information on the distribution, listing status, and recovery plan of each imperiled sea turtle species:

- Green sea turtles—found in U.S. waters around Hawaii, the U.S. Virgin Islands, Puerto Rico, and along the mainland coast from Texas to Massachusetts and from Southern California to Alaska. Key feeding grounds are in Florida coastal waters. Primary nesting sites are the Florida east coast, the U.S. Virgin Islands, Puerto Rico, and a remote atoll in Hawaii. As shown in Table 6.4, the green sea turtle has a dual listing under the ESA. The species is listed as endangered in Florida and as threatened throughout the rest of its U.S. range. Its recovery plan was published in 1991. In August 2007 the USFWS and the NMFS published *Green Sea Turtle (*Chelonia mydas*): 5-Year Review—Summary and Evaluation* (http://ecos.fws.gov/docs/five_year_review/doc1078.pdf). The agencies decided to maintain the dual listing that had been given to the species.

- Hawksbill sea turtles—found in U.S. waters primarily around Hawaii, the U.S. Virgin Islands, Puerto Rico, and along the Gulf and southeast Florida coasts. Key nesting sites are in Puerto Rico, the U.S. Virgin Islands, Hawaii, and the southeast coast and keys of Florida. The Hawksbill sea turtle has a listing of endangered. (See Table 6.4.) A recovery plan for the species was published in 1993. In August 2007 the USFWS and the NMFS published *Hawksbill Sea Turtle (*Eretmochelys imbricata*): 5-Year Review—Summary and Evaluation* (http://ecos.fws.gov/docs/five_year_review/doc1079.pdf), which indicated their decision to maintain its endangered listing under the ESA.
- Kemp's ridley sea turtles—found in U.S. waters along the Gulf and New England coasts. Primary nesting sites are in Mexico and Texas along the Gulf coast. As shown in Table 6.4, the species is listed as endangered under the ESA. In March 2010 the USFWS and the NMFS published *Draft Bi-national Recovery Plan for the Kemp's Ridley Sea Turtle (*Lepidochelys kempii*)* (http://ecos.fws.gov/docs/recovery_plan/100316.pdf), a revised recovery plan for the Kemp's ridley sea turtle that replaced a version published in 1992. In August 2007 the USFWS and the NMFS issued *Kemp's Ridley Sea Turtle (*Lepidochelys kempii*): 5-Year Review—Summary and Evaluation* (http://ecos.fws.gov/docs/five_year_review/doc1074.pdf). The agencies concluded that an endangered listing was still appropriate.
- Leatherback sea turtles—found in U.S. waters around Hawaii, the U.S. Virgin Islands, Puerto Rico, and along the entire Atlantic coast. Major nesting locations are in the U.S. Virgin Islands and Georgia. The leatherback sea turtle has a listing of endangered. (See Table 6.4.) Critical habitat was designated in 1979 and a recovery plan was published in 1992. In August 2011 the USFWS (http://www.gpo.gov/fdsys/pkg/FR-2011-08-04/pdf/2011-19676.pdf) published its 90-day finding on a petition submitted by the Sierra Club to expand the critical habitat to include coastal and marine areas off of Puerto Rico. The agency concluded that the request may be warranted and announced its intention to address the critical habitat designation in a "future planned status review" for the species. In August 2007 the USFWS and the NMFS issued *Leatherback Sea Turtle (*Dermochelys coriacea*): 5-Year Review—Summary and Evaluation* (http://ecos.fws.gov/docs/five_year_review/doc1076.pdf) and decided to maintain a listing of endangered.
- Loggerhead sea turtles—found in U.S. waters along the entire Atlantic and Pacific coasts. Primary nesting sites occur on the Gulf and east coast of Florida and in Georgia, South Carolina, and North Carolina. As shown in Table 6.4, two populations of the species are listed as threatened. These are the North Pacific Ocean and Northwest Atlantic Ocean DPSs. In September 2011 the USFWS and the NMFS (http://www.gpo.gov/fdsys/pkg/FR-2011-09-22/pdf/2011-23960.pdf) published a finding that the species actually includes nine DPSs: the North Pacific Ocean and Northwest Atlantic Ocean DPSs, plus DPSs in the Mediterranean Sea, North Indian Ocean, Northeast Atlantic Ocean, South Atlantic Ocean, South Pacific Ocean, Southeast Indo-Pacific Ocean, and Southwest Indian Ocean. The agencies also revised the listing for the North Pacific Ocean DPS to endangered. In 1998 a recovery plan was published for the U.S. Pacific populations of the species. In December 2008 the NMFS and the USFWS issued *Recovery Plan for the Northwest Atlantic Population of the Loggerhead Sea Turtle (*Caretta caretta*): Second Revision* (http://www.nmfs.noaa.gov/pr/pdfs/recovery/turtle_loggerhead_atlantic.pdf).
- Olive ridley sea turtles—found occasionally in southwestern U.S. waters. Major nesting sites are in Mexico along the Pacific coast and in other tropical locations. This species has a threatened listing under the ESA. (See Table 6.4.) Populations in Mexico are listed as endangered. In August 2007 the USFWS and the NMFS published *Olive Ridley Sea Turtle (*Lepidochelys olivacea*): 5-Year Review—Summary and Evaluation* (http://ecos.fws.gov/docs/five_year_review/doc1077.pdf) and indicated their decision to maintain the existing listing under the ESA.

THREATS TO NESTING TURTLES. Sea turtles bury their eggs in nests on sandy beaches. The building of beachfront resorts and homes has destroyed a large proportion of nesting habitat. Artificial lighting associated with coastal development also poses a problem—lights discourage females from nesting and cause hatchlings to become disoriented and wander inland instead of out to sea. Finally, beach nourishment (the human practice of rebuilding eroded beach soil) creates unusually compacted sand on which turtles are unable to nest.

SHRIMP NET CASUALTIES. Shrimp trawling is recognized as one of the most deadly human activities for sea turtles in the Gulf of Mexico and the Caribbean. During the late 1970s the NMFS began developing turtle excluder devices (TEDs), which allow sea turtles to escape from shrimp nets. In "Turtle Excluder Device (TED) Chronology" (2012, http://www.dnr.sc.gov/seaturtle/teds.htm), the South Carolina Department of Natural Resources indicates that by the early 1980s the NMFS had developed a TED that was estimated to exclude 97% of sea turtles from shrimp nets, while allowing no shrimp to escape. (See Figure 6.7.) At that time the NMFS estimated that shrimp trawling killed more than 12,000 sea turtles annually.

FIGURE 6.7

A bycatch reduction device designed to protect sea turtles

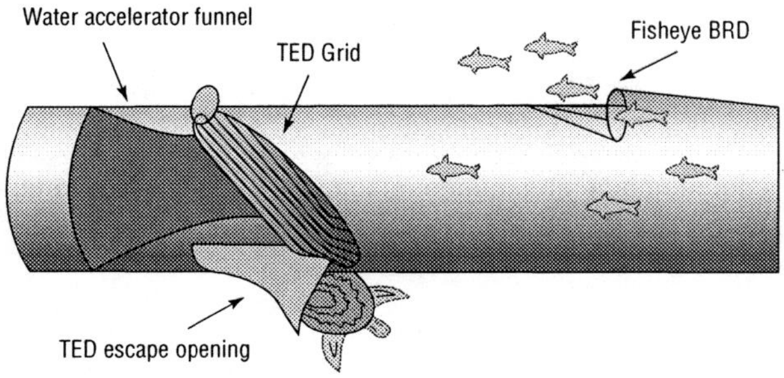

Notes: TED = turtle excluder device. BRD = bycatch reduction device.

SOURCE: Richard K. Wallace and Kristen M. Fletcher, "Figure 9," in *Understanding Fisheries Management: A Manual for Understanding the Federal Fisheries Management Process, Including Analysis of the 1996 Sustainable Fisheries Act*, 2nd ed., U.S. Department of Commerce, National Oceanic and Atmospheric Administration, National Sea Grant Office, Mississippi Alabama Sea Grant Consortium, 2001, http://nsgl.gso.uri.edu/masgc/masgch00001.pdf (accessed November 9, 2011)

In November 1989 Public Law 101-162, Section 609, was enacted in the United States banning the import of shrimp from countries that use harvesting methods deemed harmful to sea turtles. The law was challenged by India, Malaysia, Pakistan, and Thailand as violating commerce agreements under the World Trade Organization (WTO). In 1998 a WTO commission found that the United States was not implementing the law consistently with all countries. The United States responded by agreeing to change its implementation procedures and offer technical assistance to those countries that requested it.

Each year by May 1 the U.S. Department of State issues a list of nations that have been certified to import shrimp into the United States. Certification is based, in part, on the results of inspections conducted by the Department of State and the NMFS. In the press release "Sea Turtle Conservation and Shrimp Imports: Section 609 Certifications" (May 27, 2011, http://www.state.gov/e/oes/rls/fs/2011/164685.htm), the Department of State lists 38 nations and one "economy" that were certified in May 2011 for shrimp imports under Section 609. Certification means that the shrimp were obtained using TEDs or in some other manner that does not endanger sea turtles. Shrimp imports are allowed from noncertified countries on a shipment-by-shipment basis if the respective governments can show that the shrimp were harvested in a manner not harmful to sea turtles.

KEMP'S RIDLEY TURTLES. The Kemp's ridley turtle is the smallest sea turtle, with individuals measuring about 3 feet (0.9 m) in length and weighing less than 100 pounds (45 kg). Kemp's ridley is also the most endangered of the sea turtle species. It has two major nesting sites: Rancho Nuevo, Mexico (the primary nesting location), and the Texas Gulf coast.

The decline of the Kemp's ridley sea turtle is due primarily to human activities such as egg collecting, fishing for juveniles and adults, and killing of adults for meat or other products. In addition, the turtles have historically been subject to high levels of incidental take by shrimp trawlers. They are also affected by pollution from oil wells and by floating debris in the Gulf of Mexico, which can choke or entangle them. Now under strict protection, the population appears to be in the earliest stages of recovery. In 2001 the Texas Parks and Wildlife Department enacted restrictions on shrimp trawling within Gulf waters near nesting sea turtle populations. The National Park Service (NPS) reports in "Current Sea Turtle Nesting Season" (December 12, 2011, http://www.nps.gov/pais/naturescience/current-season.htm) that a record 199 Kemp's ridley nests along the Texas coast were found during the 2011 nesting season (April to July). According to the NPS, in "Sea Turtle Recovery Project" (January 17, 2012, http://www.nps.gov/pais/naturescience/strp.htm), laid eggs are collected and incubated by the NPS and the hatchlings are released into the wild.

Imperiled Snakes and Lizards

As of November 2011, there were 11 snakes (including whipsnakes and boas) and nine lizards (including skinks, anoles, iguanas, and geckos) listed as endangered or threatened under the ESA. (See Table 6.4.) The following sections discuss three of the species: giant garter snake, eastern indigo snake, and Monito gecko.

GIANT GARTER SNAKES. As shown in Table 6.5, almost $3.5 million was spent under the ESA during FY 2010 on the giant garter snake. This species is found only in California. It prefers agricultural wetlands (such as rice fields), canals, ponds, streams, and other small water bodies. Extensive land development in the Central Valley of the state has severely depleted the snake's habitat. Other threats to its survival include invasive predatory fish, water pollution, and flood control activities. In 1993 the giant garter snake was listed as threatened under the ESA. In 1999 the USFWS published *Draft Recovery Plan for the Giant Garter Snake (*Thamnopsis gigas*)* (http://www.fws.gov/pacific/news/1999/garter.pdf), and in September 2006 the agency published *Giant Garter Snake (*Thamnopsis gigas*): 5-Year Review—Summary and Evaluation* (http://www.fws.gov/cno/es/giant%20garter%20snake%205-year%20review.FINAL.pdf), in which it determined that a threatened listing was still appropriate. In May 2011 the USFWS (http://www.gpo.gov/fdsys/pkg/FR-2011-05-25/pdf/2011-12861.pdf) published notice that it was initiating a five-year review of dozens of West Coast species, including the giant garter snake. As of

FIGURE 6.8

Eastern indigo snake. *(© fivespots/Shutterstock.com.)*

January 2012, the results of the review had not been published.

EASTERN INDIGO SNAKES. As shown in Table 6.5, nearly $2.5 million was spent under the ESA during FY 2010 on the eastern indigo snake. This is a large thick-bodied black snake found only in Florida and Georgia. (See Figure 6.8.) In 1978 it was listed as threatened under the ESA. Its recovery plan was finalized in 1982. In May 2008 the USFWS published *Eastern Indigo Snake (*Drymarchon couperi*): 5-Year Review—Summary and Evaluation* (http://ecos.fws.gov/docs/five_year_review/doc1910.pdf). The agency decided to maintain the snake's threatened listing due to continued threats to its habitat.

MONITO GECKOS. The endangered Monito gecko is a small lizard less than 2 inches (5 cm) long. (See Figure 6.9.) This species exists only on the 38-acre (15.4-ha) Monito Island off the Puerto Rican coast. Endangerment of the Monito gecko has resulted from human activity and habitat destruction. After World War II (1939–1945) the U.S. military used Monito Island as a site for bombing exercises, causing large-scale habitat destruction. In 1982 the USFWS observed only 24 Monito geckos on the island. In 1985 Monito Island was designated as critical habitat for the species. The Commonwealth of Puerto Rico is now managing the island for the gecko and as a refuge for seabirds; unauthorized human visitation is prohibited. In 2007 the USFWS initiated a five-year review of 18 Caribbean wildlife and plant species, including the Monito gecko. As of January 2012, the results of that review had not been published.

FIGURE 6.9

Monito gecko

SOURCE: "Monito Gecko," in *Endangered Species Coloring Book: Save Our Species*, U.S. Environmental Protection Agency, May 2008, http://www.epa.gov/espp/coloring/cbook.pdf (accessed November 9, 2011)

Imperiled Crocodilians

Crocodilians play a crucial role in their habitat. They control fish populations and dig water holes, which are important to many species in times of drought. The disappearance of alligators and crocodiles has a profound effect on the biological communities these animals occupy. There are two imperiled crocodilian species in the United States: the American alligator and the American crocodile. (See Figure 6.10.) They are similar in appearance with only slight differences. The crocodile has a narrower, more pointed snout and an indentation in its upper jaw that allows a tooth to be seen when its mouth is closed.

The American alligator has a unique history under the ESA. It was on the first list of endangered species published in 1967. During the 1970s and 1980s populations of the species in many states rebounded in abundance and could have been delisted. Instead, it was reclassified as threatened. This measure was taken, in part, because federal officials acknowledged a certain amount of "public hostility" toward the creatures and were concerned that delisting would open the population to excessive hunting. Also, it was feared that the American alligator was so similar in appearance to the highly endangered American crocodile that delisting the alligator might lead to accidental taking of the crocodile species. By 1987 the alligator was considered fully recovered in the United States. As of November 2011, the alligator was listed as SAT. (See Table 6.4.)

The American crocodile is considered a success story of the ESA. When the species was originally listed as endangered in 1975, less than 300 individuals existed. Over the next three decades the species thrived and

FIGURE 6.10

American crocodile. *(© U.S. Fish and Wildlife Service.)*

expanded its nesting range to new locations on the east and west coasts of Florida. In 2005 the USFWS initiated a five-year status review of the species. In 2007 the western DPS of the species in Florida was downlisted from endangered to threatened. At that time an estimated 2,000 American crocodiles lived in the state, not including hatchlings. Figure 6.11 shows a map of the critical habitat for the American crocodile and areas for which the USFWS provides consultation services regarding the species.

Threatened and Endangered Foreign Reptile Species

The IUCN reports in *Red List of Threatened Species Version 2011.2* that 772 reptile species were threatened in 2011. A total of 3,336 reptile species were evaluated out of the 9,439 known species.

As of November 2011, there were 83 totally foreign reptile species listed under the ESA. (See Table 6.6.) Most are found in the tropical regions of Africa, Mexico, South America, and Southeast Asia.

FIGURE 6.11

Critical habitat of the American crocodile

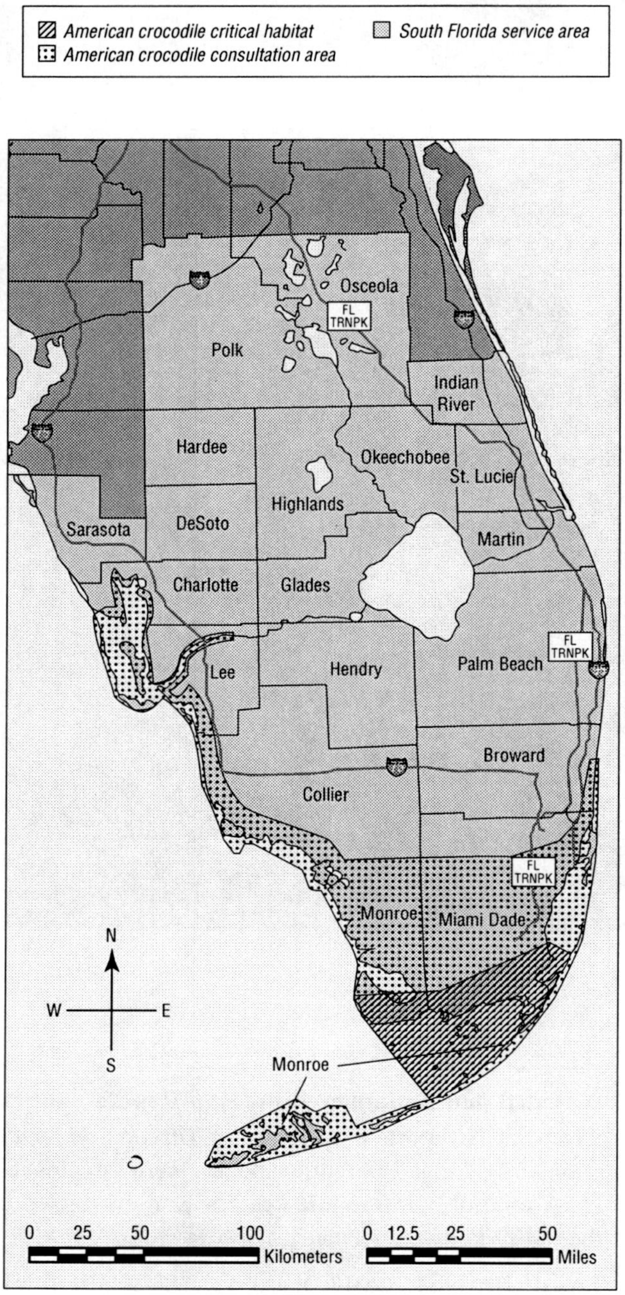

American crocodile critical habitat
American crocodile consultation area
South Florida service area

SOURCE: Adapted from "American Crocodile: Consultation Area Map," in *American Crocodile: Consultation Area*, U.S. Department of the Interior, U.S. Fish & Wildlife Service, South Florida Ecological Services Office, September 19, 2003, http://www.fws.gov/verobeach/ReptilesPDFs/AmericanCrocodileConsultationArea.pdf (accessed November 9, 2011)

TABLE 6.6

Foreign endangered and threatened reptile species, November 2011

Common name	Scientific name	Listing status*	Historic range
African dwarf crocodile	Osteolaemus tetraspis tetraspis	E	West Africa
African slender-snouted crocodile	Crocodylus cataphractus	E	Western and central Africa
Allen's Cay iguana	Cyclura cychlura inornata	T	West Indies_Bahamas
American crocodile	Crocodylus acutus	E	Mexico, Caribbean, Central and South America
Andros Island ground iguana	Cyclura cychlura cychlura	T	West Indies_Bahamas
Anegada ground iguana	Cyclura pinguis	E	West Indies_British Virgin Islands (Anegada Island)
Angulated tortoise	Geochelone yniphora	E	Malagasy Republic (=Madagascar)
Apaporis River caiman	Caiman crocodilus apaporiensis	E	Colombia
Aquatic box turtle	Terrapene coahuila	E	Mexico
Aruba Island rattlesnake	Crotalus unicolor	T	Aruba Island (Netherland Antilles)
Barrington land iguana	Conolophus pallidus	E	Ecuador (Galapagos Islands)
Black caiman	Melanosuchus niger	E	Amazon basin
Black softshell turtle	Trionyx nigricans	E	Bangladesh
Bolson tortoise	Gopherus flavomarginatus	E	Mexico
Brazilian sideneck turtle	Phrynops hogei	E	Brazil
Broad-snouted caiman	Caiman latirostris	E	Brazil, Argentina, Paraguay, Uruguay
Brother's Island tuatara	Sphenodon guntheri	E	New Zealand (N. Brother's Island)
Brown caiman	Caiman crocodilus fuscus (includes Caiman crocodilus chiapasius)	SAT	Mexico, Central America, Colombia, Ecuador, Venezuela, Peru
Burmese peacock turtle	Morenia ocellata	E	Burma
Cat Island turtle	Trachemys terrapen	E	West Indies_Jamaica, Bahamas
Cayman Brac ground iguana	Cyclura nubila caymanensis	T	West Indies_Cayman Islands
Central American river turtle	Dermatemys mawii	E	Mexico, Belize, Guatemala
Ceylon mugger crocodile	Crocodylus palustris kimbula	E	Sri Lanka
Chinese alligator	Alligator sinensis	E	China
Common caiman	Caiman crocodilus crocodilus	SAT	Brazil, Colombia, Ecuador, French Guiana, Guyana, Suriname, Venezuela, Bolivia, Peru
Congo dwarf crocodile	Osteolaemus tetraspis osborni	E	Congo R. drainage
Cuatro Cienegas softshell turtle	Trionyx ater	E	Mexico
Cuban crocodile	Crocodylus rhombifer	E	Cuba
Cuban ground iguana	Cyclura nubila nubila	T	Cuba
Day gecko	Phelsuma edwardnewtoni	E	Indian Ocean_Mauritius
Desert monitor	Varanus griseus	E	North Africa to Aral Sea, through Central Asia to Pakistan, Northwest India
Exuma Island iguana	Cyclura cychlura figginsi	T	West Indies_Bahamas
Fiji banded iguana	Brachylophus fasciatus	E	Pacific_Fiji, Tonga
Fiji crested iguana	Brachylophus vitiensis	E	Pacific_Fiji
Galapagos tortoise	Geochelone nigra (=elephantopus)	E	Ecuador (Galapagos Islands)
Gavial	Gavialis gangeticus	E	Pakistan, Burma, Bangladesh, India, Nepal
Geometric turtle	Psammobates geometricus	E	South Africa
Grand Cayman ground iguana	Cyclura nubila lewisi	E	West Indies_Cayman Islands
Hierro giant lizard	Gallotia simonyi simonyi	E	Spain (Canary Islands)
Ibiza wall lizard	Podarcis pityusensis	T	Spain (Balearic Islands)
Inagua Island turtle	Trachemys stejnegeri malonei	E	West Indies_Bahamas (Great Inagua Island)
Indian (=Bengal) monitor	Varanus bengalensis	E	Iran, Iraq, India, Sri Lanka, Malaysia, Afghanistan, Burma, Vietnam, Thailand
Indian python	Python molurus molurus	E	Sri Lanka and India
Indian sawback turtle	Kachuga tecta tecta	E	India
Indian softshell turtle	Trionyx gangeticus	E	Pakistan, India
Jamaican boa	Epicrates subflavus	E	Jamaica
Jamaican iguana	Cyclura collei	E	West Indies_Jamaica
Komodo Island monitor	Varanus komodoensis	E	Indonesia (Komodo, Rintja, Padar, and western Flores Island)
Lar Valley viper	Vipera latifii	E	Iran
Madagascar radiated tortoise	Geochelone radiata	E	Malagasy Republic (= Madagascar)
Maria Island ground lizard	Cnemidophorus vanzoi	E	West Indies_St. Lucia (Maria Islands)
Maria Island snake	Liophus ornatus	E	West Indies_St. Lucia (Maria Islands)
Mayaguana iguana	Cyclura carinata bartschi	T	West Indies_Bahamas
Morelet's crocodile	Crocodylus moreletii	E	Mexico, Belize, Guatemala
Mugger crocodile	Crocodylus palustris palustris	E	India, Pakistan, Iran, Bangladesh
Nile crocodile	Crocodylus niloticus	T	Africa, Middle East
Olive ridley sea turtle	Lepidochelys olivacea	E	Circumglobal in tropical and temperate seas

TABLE 6.6

Foreign endangered and threatened reptile species, November 2011 [CONTINUED]

Common name	Scientific name	Listing status*	Historic range
Orinoco crocodile	Crocodylus intermedius	E	South America_Orinoco River basin
Peacock softshell turtle	Trionyx hurum	E	India, Bangladesh
Philippine crocodile	Crocodylus novaeguineae mindorensis	E	Philippine Islands
River terrapin	Batagur baska	E	Malaysia, Bangladesh, Burma, India, Indonesia
Round Island bolyeria boa	Bolyeria multocarinata	E	Indian Ocean_Mauritius
Round Island casarea boa	Casarea dussumieri	E	Indian Ocean_Mauritius
Round Island day gecko	Phelsuma guentheri	E	Indian Ocean_Mauritius
Round Island skink	Leiolopisma telfairi	T	Indian Ocean_Mauritius
Saltwater crocodile	Crocodylus porosus	E	Southeast Asia, Australia, Papua New Guinea, Islands of the West Pacific Ocean
Saltwater crocodile	Crocodylus porosus	T	Southeast Asia, Australia, Papua New Guinea, Islands of the West Pacific Ocean
San Esteban Island chuckwalla	Sauromalus varius	E	Mexico
Serpent Island gecko	Cyrtodactylus serpensinsula	T	Indian Ocean_Mauritius
Short-necked or western swamp turtle	Pseudemydura umbrina	E	Australia
Siamese crocodile	Crocodylus siamensis	E	Southeast Asia, Malay Peninsula
South American red-lined turtle	Trachemys scripta callirostris	E	Colombia, Venezuela
Spotted pond turtle	Geoclemys hamiltonii	E	North India, Pakistan
Tartaruga	Podocnemis expansa	E	South America_Orinoco River and Amazon River basins
Three-keeled Asian turtle	Melanochelys tricarinata	E	Central India to Bangladesh and Burma
Tomistoma	Tomistoma schlegelii	E	Malaysia, Indonesia
Tracaja	Podocnemis unifilis	E	South America_Orinoco River and Amazon River basins
Tuatara	Sphenodon punctatus	E	New Zealand
Turks and Caicos iguana	Cyclura carinata carinata	T	West Indies_Turks and Caicos islands
Watling Island ground iguana	Cyclura rileyi rileyi	E	West Indies_Bahamas
White Cay ground iguana	Cyclura rileyi cristata	T	West Indies_Bahamas
Yacare caiman	Caiman yacare	T	Bolivia, Argentina, Peru, Brazil
Yellow monitor	Varanus flavescens	E	West Pakistan through India to Bangladesh

*E = endangered; T = threatened; SAT = threatened due to similarity of appearance.

SOURCE: Adapted from "Generate Species List," in *Species Reports*, U.S. Department of the Interior, U.S. Fish & Wildlife Service, November 2011, http://ecos.fws.gov/tess_public/pub/adHocSpeciesForm.jsp (accessed November 8, 2011)

CHAPTER 7
TERRESTRIAL MAMMALS

Terrestrial animals are animals that inhabit the land. Mammals are warm-blooded, breathe air, have hair at some point during their life, give birth to live young (as opposed to laying eggs), and nourish their young by secreting milk.

The biggest cause of terrestrial mammalian decline and extinction is habitat loss and degradation. As humans convert forests, grasslands, rivers, and wetlands for various uses, they relegate many species to precarious existences in small, fragmented habitat patches. In addition, some terrestrial mammals have been purposely eliminated by humans. For example, bison (buffalo), elk, and beaver stocks were severely depleted in the United States following colonization by European settlers. All three species were nearly hunted to extinction by the end of the 1800s. The disappearance of native large game had consequences on other species. Wolves and other predators began preying on livestock and became the target of massive kill-offs by humans.

Some terrestrial mammal species have been imperiled, in part, because they are considered dangerous to human life. This has been the case for many bears, wolves, and mountain lions. Changing attitudes have led to interest in preserving all species, and conservation measures have allowed several terrestrial mammals to recover.

ENDANGERED AND THREATENED U.S. SPECIES

As of November 2011, there were 68 species of terrestrial mammals in the United States listed under the Endangered Species Act (ESA) as endangered or threatened. (See Table 7.1.) Nearly all had an endangered listing, meaning that they are at risk of extinction, and most had recovery plans in place.

The imperiled species fall into nine broad categories:

- Bats
- Bears
- Canines—foxes and wolves
- Deer, caribou, pronghorns, bighorn sheep, and bison
- Felines—jaguars, jaguarundis, lynx, ocelots, panthers, and pumas
- Ferrets
- Rabbits
- Rodents—beavers, mice, prairie dogs, rats, squirrels, and voles
- Shrews

Table 7.2 shows the 10 terrestrial mammal species with the highest expenditures under the ESA during fiscal year (FY) 2010. The grizzly bear was the most expensive ($8.8 million), followed by the gray wolf ($6.9 million) and the Indiana bat ($6.7 million).

Bats

Bats belong to the taxonomic order Chiroptera, which means "hand-wing." They are the only true flying mammals. They typically weigh less than 2 ounces (57 g) and have wingspans of less than 20 inches (51 cm). Most are insectivores, meaning that insects are their primary food source. Bats prefer to sleep during the day and feed after dusk. Biologists believe that bats are vastly underappreciated for their role in controlling nighttime insect populations.

According to the U.S. Fish and Wildlife Service (USFWS), in "Introduction to Bats" (October 22, 2009, http://www.fws.gov/filedownloads/ftp_DJCase/endangered/bats/bats.htm), there are 45 species of bats in the United States. As of November 2011, nine of these species were listed by the USFWS as endangered or threatened under the ESA (see Table 7.1):

- Gray bat
- Hawaiian hoary bat

TABLE 7.1

Endangered and threatened terrestrial mammal species, November 2011

Common name	Scientific name	Listing status[a]	U.S. or U.S./foreign listed	Recovery plan date	Recovery plan stage[b]
Alabama beach mouse	Peromyscus polionotus ammobates	E	US	8/12/1987	F
Amargosa vole	Microtus californicus scirpensis	E	US	9/15/1997	F
American black bear	Ursus americanus	SAT	US	None	—
Anastasia Island beach mouse	Peromyscus polionotus phasma	E	US	9/23/1993	F
Black-footed ferret	Mustela nigripes	E (1), EXPN (1)	US/foreign	8/8/1988	RF(1)
Buena Vista Lake ornate shrew	Sorex ornatus relictus	E	US	9/30/1998	F
Canada Lynx	Lynx canadensis	T	US	9/14/2005	O
Carolina northern flying squirrel	Glaucomys sabrinus coloratus	E	US	9/24/1990	F
Choctawhatchee beach mouse	Peromyscus polionotus allophrys	E	US	8/12/1987	F
Columbian white-tailed deer	Odocoileus virginianus leucurus	E	US	6/14/1983	RF(1)
Delmarva peninsula fox squirrel	Sciurus niger cinereus	E (1), EXPN (1)	US	6/8/1993	RF(2)
Eastern puma (=cougar)	Puma (=Felis) concolor couguar	E	US/foreign	8/2/1982	F
Florida panther	Puma (=Felis) concolor coryi	E	US	12/18/2008	RF(3)
Florida salt marsh vole	Microtus pennsylvanicus dukecampbelli	E	US	9/30/1997	F
Fresno kangaroo rat	Dipodomys nitratoides exilis	E	US	9/30/1998	F
Giant kangaroo rat	Dipodomys ingens	E	US	9/30/1998	F
Gray bat	Myotis grisescens	E	US	7/8/1982	F
Gray wolf	Canis lupus	E (1), T (1), EXPN (2)	US/foreign	1/31/1992	RF(1)
Grizzly bear	Ursus arctos horribilis	T (1), EXPN (1)	US	3/13/2007	RF(1)
Gulf Coast jaguarundi	Herpailurus (=Felis) yagouaroundi cacomitli	E	US/foreign	8/22/1990	F
Hawaiian hoary bat	Lasiurus cinereus semotus	E	US	5/11/1998	F
Hualapai Mexican vole	Microtus mexicanus hualpaiensis	E	US	8/19/1991	F
Indiana bat	Myotis sodalis	E	US	4/16/2007	RD(1)
Jaguar	Panthera onca	E	US/foreign	None	—
Key deer	Odocoileus virginianus clavium	E	US	5/18/1999	F
Key Largo cotton mouse	Peromyscus gossypinus allapaticola	E	US	5/18/1999	F
Key Largo woodrat	Neotoma floridana smalli	E	US	5/18/1999	F
Lesser long-nosed bat	Leptonycteris curasoae yerbabuenae	E	US/foreign	3/4/1997	F
Little Mariana fruit bat	Pteropus tokudae	E	US	11/2/1990	F
Louisiana black bear	Ursus americanus luteolus	T	US	9/27/1995	F
Lower Keys marsh rabbit	Sylvilagus palustris hefneri	E	US	5/18/1999	F
Mariana fruit bat (=Mariana flying fox)	Pteropus mariannus mariannus	T	US	3/30/2010	RD(1)
Mexican long-nosed bat	Leptonycteris nivalis	E	US/foreign	9/8/1994	F
Morro Bay kangaroo rat	Dipodomys heermanni morroensis	E	US	1/25/2000	RD(1)
Mount Graham red squirrel	Tamiasciurus hudsonicus grahamensis	E	US	5/27/2011	RD(1)
Northern Idaho ground squirrel	Spermophilus brunneus brunneus	T	US	9/16/2003	F
Ocelot	Leopardus (=Felis) pardalis	E	US/foreign	8/26/2010	RD(1)
Ozark big-eared bat	Corynorhinus (=Plecotus) townsendii ingens	E	US	3/28/1995	RF(1)
Pacific pocket mouse	Perognathus longimembris pacificus	E	US	9/28/1998	F
Peninsular bighorn sheep	Ovis canadensis nelsoni	E	US/foreign	10/25/2000	F
Perdido Key beach mouse	Peromyscus polionotus trissyllepsis	E	US	8/12/1987	F
Point Arena mountain beaver	Aplodontia rufa nigra	E	US	6/2/1998	F
Preble's meadow jumping mouse	Zapus hudsonius preblei	T	US	None	—
Puma (=mountainlion)	Puma (=Felis) concolor (all subsp. except coryi)	SAT	US	None	—
Pygmy rabbit	Brachylagus idahoensis	E	US	6/29/2011	D
Red wolf	Canis rufus	E (1), EXPN (1)	US	10/26/1990	RF(2)
Rice rat	Oryzomys palustris natator	E	US	5/18/1999	F
Riparian brush rabbit	Sylvilagus bachmani riparius	E	US	9/30/1998	F
Riparian woodrat (=San Joaquin Valley)	Neotoma fuscipes riparia	E	US	9/30/1998	F
Salt marsh harvest mouse	Reithrodontomys raviventris	E	US	2/10/2010	D
San Bernardino Merriam's kangaroo rat	Dipodomys merriami parvus	E	US	None	—
San Joaquin kit fox	Vulpes macrotis mutica	E	US	9/30/1998	F
San Miguel Island fox	Urocyon littoralis littoralis	E	US	None	—
Santa Catalina Island fox	Urocyon littoralis catalinae	E	US	None	—
Santa Cruz Island fox	Urocyon littoralis santacruzae	E	US	None	—
Santa Rosa Island fox	Urocyon littoralis santarosae	E	US	None	—
Sierra Nevada bighorn sheep	Ovis canadensis sierrae	E	US	2/13/2008	F
Sinaloan Jaguarundi	Herpailurus (=Felis) yagouaroundi tolteca	E	US/foreign	None	—
Sonoran pronghorn	Antilocapra americana sonoriensis	E (1), EXPN (1)	US/foreign	12/3/1998	RF(1)
Southeastern beach mouse	Peromyscus polionotus niveiventris	T	US	9/23/1993	F
St. Andrew beach mouse	Peromyscus polionotus peninsularis	E	US	12/28/2010	F
Stephens' kangaroo rat	Dipodomys stephensi (incl. D. cascus)	E	US	6/23/1997	D
Tipton kangaroo rat	Dipodomys nitratoides nitratoides	E	US	9/30/1998	F
Utah prairie dog	Cynomys parvidens	T	US	9/17/2010	RD(1)
Virginia big-eared bat	Corynorhinus (=Plecotus) townsendii virginianus	E	US	5/8/1984	F

TABLE 7.1

Endangered and threatened terrestrial mammal species, November 2011 [CONTINUED]

Common name	Scientific name	Listing status[a]	U.S. or U.S./foreign listed	Recovery plan date	Recovery plan stage[b]
Virginia northern flying squirrel	Glaucomys sabrinus fuscus	E	US	9/24/1990	F
Wood bison	Bison bison athabascae	E	US/foreign	None	—
Woodland caribou	Rangifer tarandus caribou	E	US/foreign	3/4/1994	RF(2)

[a]E = endangered; T = threatened; SAT = similarity of appearance to a threatened taxon; EXPN = experimental population, non-essential; numbers in parentheses indicate separate populations.
[b]F = final; D = draft; RD = draft revision; RF = final revision; O = other.

SOURCE: Adapted from "Generate Species List," in *Species Reports*, U.S. Department of the Interior, U.S. Fish & Wildlife Service, November 2011, http://ecos.fws.gov/tess_public/pub/adHocSpeciesForm.jsp (accessed November 8, 2011), and "Listed FWS/Joint FWS and NMFS Species and Populations with Recovery Plans (Sorted by Listed Entity)," in *Recovery Plans Search*, U.S. Department of the Interior, U.S. Fish & Wildlife Service, November 2011, http://ecos.fws.gov/tess_public/pub/speciesRecovery.jsp?sort=1 (accessed November 8, 2011)

TABLE 7.2

Terrestrial mammal species with the highest expenditures under the Endangered Species Act, fiscal year 2010

Ranking	Species	Expenditure
1	Bear, grizzly (Ursus arctos horribilis)—lower 48 States, except where listed as an experimental population or delisted	$8,781,280
2	Wolf, gray (Canis lupus)—Lower 48 States, except MN, MT, ID, portions of eastern OR, eastern WA, north-central UT, and where EXPN. Mexico.	$6,954,633
3	Bat, Indiana (Myotis sodalis)	$6,742,846
4	Pronghorn, Sonoran (Antilocapra americana sonoriensis)	$5,827,743
5	Lynx, Canada (Lynx canadensis)—Contiguous U.S. DPS	$4,191,798
6	Bear, Louisiana black (Ursus americanus luteolus)	$1,915,352
7	Bear, grizzly (Ursus arctos horribilis)—EXPN, non-essential portions of ID and MT	$1,899,104
8	Wolf, gray (Canis lupus)—WY, ID, MT	$1,829,149
9	Prairie dog, Utah (Cynomys parvidens)	$1,720,072
10	Wolf, Mexican gray (Canis lupus baileyi)—portions of AZ, NM and TX	$1,628,931

EXPN = non-essential experimental population.
DPS = distinct population segment.

SOURCE: Adapted from "Table 2. Species Ranked in Descending Order of Total FY 2010 Reported Expenditures, Not Including Land Acquisition Costs," in *Federal and State Endangered and Threatened Species Expenditures: Fiscal Year 2010*, U.S. Department of the Interior, U.S. Fish and Wildlife Service, 2010, http://www.fws.gov/endangered/esa-library/pdf/2010.EXP.FINAL.pdf (accessed October 25, 2011)

- Indiana bat
- Lesser long-nosed bat
- Little Mariana fruit bat
- Mariana fruit bat
- Mexican long-nosed bat
- Ozark big-eared bat
- Virginia big-eared bat

Bats are imperiled for a variety of reasons, including habitat degradation, disturbance of hibernating and maternity colonies, direct extermination by humans, and the indirect effects of pesticide use on insects. During the first decade of the 21st century a new threat to bat survival emerged called white-nose syndrome (WNS). The U.S. Geological Survey (USGS) reports in the fact sheet "Investigating White-Nose Syndrome in Bats" (July 2009, http://www.nwhc.usgs.gov/publications/fact_sheets/pdfs/2009-3058_investigating_wns.pdf) that the disease was first documented during the winter of 2005–06 among bats hibernating in a cave in New York. It is evidenced by the appearance of a white fungus on the muzzle, ears, and wings of afflicted bats. USGS researchers believe WNS is caused by a previously unknown fungus called *Geomyces destructans*.

The USFWS states in the press release "Fish and Wildlife Service Unveils National Plan to Combat Deadly White-Nose Syndrome in Bats" (http://www.fws.gov/WhiteNoseSyndrome/pdf/051711_FinalReleaseNationalPlan.pdf) that as of May 2011 WNS had spread to 18 states and four Canadian provinces and had killed more than 1 million hibernating bats. The agency notes that $10.8 million had already been spent by U.S. Department of the Interior agencies fighting WNS.

In May 2011 the USFWS published *A National Plan for Assisting States, Federal Agencies, and Tribes in Managing White-Nose Syndrome in Bats* (http://www.fws.gov/WhiteNoseSyndrome/pdf/WNSnationalplanMay2011.pdf), which provides information about WNS and lays out a national plan for collaboration between federal, state, tribal, and local agencies for managing bats that are imperiled by WNS. The USFWS indicates that cave bats (i.e., species that hibernate in caves or mines during the winter months) are most susceptible to WNS. Species that have been afflicted included the little brown bat, the big brown bat, the Indiana bat, the northern long-eared bat, the eastern small-footed bat, and the tricolored bat. In 2010 three additional species—the gray bat, the cave myotis, and the southeastern myotis—were found to be infected with the fungus, but not yet showing WNS symptoms. As shown in Table 7.1, the gray bat and the Indiana bat have endangered listings under the ESA.

In June 2011 the USFWS (http://www.gpo.gov/fdsys/pkg/FR-2011-06-29/pdf/2011-16344.pdf) published notice that it received a petition submitted by the Center for Biological Diversity, a nonprofit conservation group, requesting that the eastern small-footed bat and the northern long-eared bat be listed as endangered under the ESA due to population declines caused by WNS. The agency found the request to be warranted and initiated a 12-month status review to make a final decision about the listings. As of January 2012, the results of the review had not been published.

INDIANA BATS. As shown in Table 7.2, more than $6.7 million was spent under the ESA during FY 2010 on the Indiana bat. It is a medium-sized, brown-colored bat found throughout a region encompassing the mid-Atlantic states and into the Midwest. The bats spend their winters in hibernation spots (or hibernacula) consisting primarily of large caves and abandoned mines. The bats are extremely sensitive to any disturbances during hibernation. In "Indiana Bat Fact Sheet" (2012, http://www.dec.ny.gov/animals/6972.html), the New York State Department of Environmental Conservation indicates that awakened bats become agitated and waste precious energy flying around frantically. This can leave them too weak and malnourished to survive the remainder of the winter. During the springtime adult females move to wooded areas and form maternity colonies. The loss of suitable habitat due to deforestation has disrupted this natural process. In addition, the bats have a low reproductive rate, producing only one baby per year. This makes it difficult for their populations to grow.

The Indiana bat was on the first list of endangered species, which was issued in 1967. (See Table 2.1 in Chapter 2.) In April 2007 the USFWS published *Indiana Bat (*Myotis sodalis*) Draft Recovery Plan: First Revision* (http://ecos.fws.gov/docs/recovery_plan/070416.pdf). The USFWS estimated that there were 883,300 Indiana bats in the United States in 1965. This number had dropped to 381,156 bats in 2001 before rising to 457,374 bats in 2005. As a result, the USFWS expressed optimism that the species would be fully recovered by the 2020s. However, this was before WNS emerged. As noted earlier, WNS poses a new and serious threat to the survival of the Indiana bat and other cave-hibernating bat species.

In September 2009 the USFWS published *Indiana Bat (*Myotis sodalis*): 5-Year Review—Summary and Evaluation* (http://ecos.fws.gov/docs/five_year_review/doc2627.pdf). The agency decided to maintain the endangered listing of the species. In July 2011 the USFWS (http://www.gpo.gov/fdsys/pkg/FR-2011-07-26/pdf/2011-18893.pdf) initiated another five-year review for the Indiana bat. As of January 2012, the results of the review had not been published.

Bears

Bears belong to the Ursidae family. Their furry bodies are large and heavy, with powerful arms and legs and short tails. For the most part, they feed on fruits and insects, but they also eat meat.

As of November 2011, there were three bear species listed under the ESA as endangered or threatened in the United States: the American black bear, the Louisiana black bear, and the grizzly bear. (See Table 7.1; note that the polar bear is not included in this table because it is considered a marine mammal and is discussed in Chapter 3.) The American black bear has a listing of SAT, which means threatened due to similarity of appearance. American black bears look similar to the imperiled Louisiana black bears. The listing is designed to prevent harmful actions by humans who might mistake Louisiana black bears for American black bears. Both species are found in Louisiana, Mississippi, and Texas. Grizzly bears inhabit Alaska and the northwestern states, primarily Idaho, Montana, Washington, and Wyoming.

In general, bears are endangered due to habitat loss. Some bears have been hunted because they are considered predatory or threatening, whereas others have been hunted for sport.

GRIZZLY BEARS. As shown in Table 7.2, nearly $8.8 million was spent under the ESA during FY 2010 to conserve grizzly bears. They are large animals, standing 4 feet (1.2 m) high at the shoulder when on four paws and as tall as 7 feet (2.1 m) when upright. Males weigh 500 pounds (227 kg) on average but are sometimes as large as 900 pounds (408 kg). Females weigh 350 pounds (159 kg) on average. Grizzlies have a distinctive shoulder hump, which actually represents a massive digging muscle. Their claws are 2 to 4 inches (5.1 to 10.2 cm) long.

The grizzly bear was originally found throughout the continental United States, but was eventually eliminated from all but Alaska and a handful of western habitats. It was first listed as endangered in 1967. (See Table 2.1 in Chapter 2.) As of December 2011, the grizzly bear population in Alaska was thriving and not listed under the ESA. Two populations in the lower 48 states were listed:

- Portions of Idaho and Montana—experimental population, nonessential (as explained in Chapter 2, this listing means that the survival of this population is not believed essential to the survival of the species as a whole; thus, the population receives less protection under the ESA)
- Remainder of the lower 48 states—threatened

The USFWS indicates in "Grizzly Bear (*Ursus arctos horribilis*)" (January 2012, http://ecos.fws.gov/speciesProfile/profile/speciesProfile.action?spcode=A001) that over the decades it has published numerous recovery plans and supplements for specific populations of grizzly bears. Figure 7.1 shows the recovery zones in place as of August 2011:

- NCASC—North Cascades ecosystem in north-central Washington
- SE—Selkirk ecosystem in northeastern Washington and northern Idaho
- CYE—Cabinet-Yaak ecosystem in northern Idaho and northwestern Montana
- NCDE—North Continental Divide ecosystem in northwestern Montana
- BE—Bitterroot ecosystem in eastern Idaho and western Montana
- GYA—Greater Yellowstone area ecosystem in eastern Idaho, southwestern Montana, and northwestern Wyoming

Recovery efforts for the grizzly bear are coordinated by the Interagency Grizzly Bear Committee (IGBC), which was created in 1983. The IGBC (2012, http://www.igbconline.org/html/about.html) is made up by representatives from the USFWS, the USGS, the U.S. Forest Service, the National Park Service, the Bureau of Land Management, the Canadian Wildlife Service, and the

FIGURE 7.1

Grizzly bear recovery zones

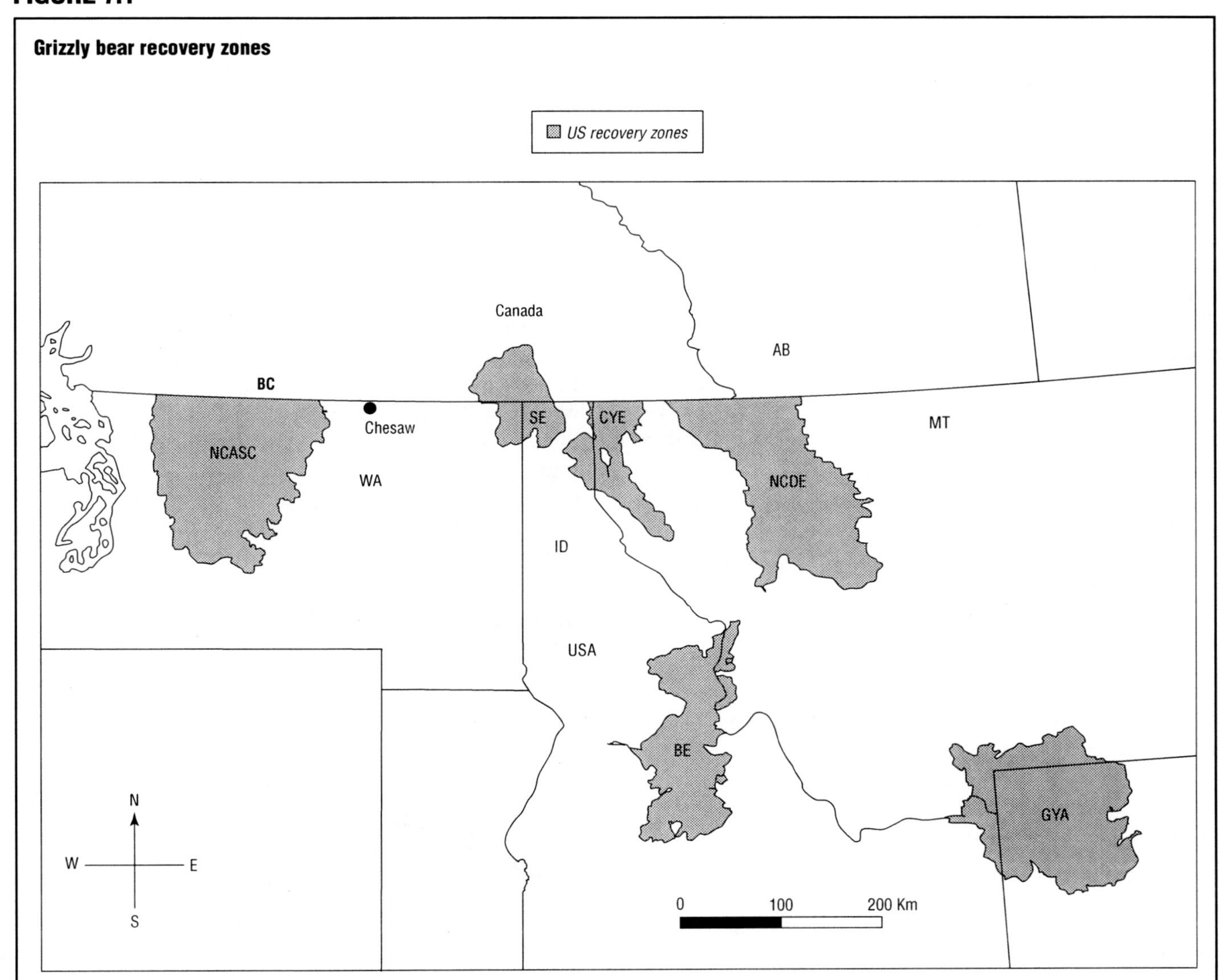

NCASC = North Cascades ecosystem. SE = Selkirk ecosystem. CYE = Cabinet-Yaak ecosystem.
NCDE = North Continental Divide ecosystem. BE = Bitterroot ecosystem. GYA = Greater Yellowstone Area ecosystem.

SOURCE: Adapted from "Figure 1. Current Grizzly Bear Recovery Ecosystems," in *Grizzly Bear (Ursus arctos horribilis) 5-Year Review: Summary and Evaluation*, U.S. Department of the Interior, U.S. Fish and Wildlife Service, Grizzly Bear Recovery Office, August 2011, http://ecos.fws.gov/docs/five_year_review/doc3847.review_August%202011.pdf (accessed November 10, 2011)

wildlife agencies of Idaho, Montana, Washington, and Wyoming.

In March 2007 the USFWS (http://frwebgate.access.gpo.gov/cgi-bin/getdoc.cgi?dbname=2007_register&docid=fr29mr07-14) established the GYA distinct population segment (DPS) of grizzly bears. As noted in Chapter 2, a DPS is a distinct population of a species that is capable of interbreeding and lives in a specific geographic area. In the rulemaking that created the DPS, the USFWS simultaneously delisted the DPS (i.e., removed its listing under the ESA) due to population gains. The delisting decision was highly unpopular with conservation organizations. In September 2009 Judge Donald William Molloy (1946–) of the U.S. District Court for the District of Montana overturned the delisting, ruling that the USFWS did not adequately consider the impacts of global warming and other factors on whitebark pine nuts, a grizzly food source, and had not ensured that adequate regulatory mechanisms are in place to protect the grizzly bear once it is delisted.

In August 2011 the USFWS published *Grizzly Bear (*Ursus arctos horribilis*): 5-Year Review—Summary and Evaluation* (http://ecos.fws.gov/docs/five_year_review/doc3847.review_August%202011.pdf), which summarized the lengthy and complicated legal history of the species under the ESA. The agency noted that between 1986 and 2007 it reviewed 10 petitions requesting listing status changes for individual grizzly bear populations. The USFWS concluded that the CYE, NCDE, and SE populations warrant listing as endangered under the ESA. However, as of January 2012 the agency had not finalized the listings due to higher priorities.

Canines

Canine is the common term used to describe a member of the Canidae family of carnivorous (meat-eating) animals. This family includes wolves, foxes, coyotes, jackals, and domestic dogs.

As of November 2011, there were five fox species and two wolf species listed under the ESA in the United States (see Table 7.1):

- San Joaquin kit fox
- San Miguel Island fox
- Santa Catalina Island fox
- Santa Cruz Island fox
- Santa Rosa Island fox
- Gray wolf
- Red wolf

All the species were listed as endangered except for specific wolf DPSs.

WOLVES. Wolves were once among the most widely distributed mammals on the earth. Before European settlement, wolves ranged over most of North America, from central Mexico to the Arctic Ocean. Their decline was mostly due to hunting. In 1914 Congress authorized funding for the removal of all large predators, including wolves, from federal lands. By the 1940s wolves had been eliminated from most of the contiguous United States. In 1967 the gray wolf (which was then called the timber wolf) and the red wolf (which is shown in Figure 7.2) were on the first list of endangered species issued by the USFWS. By that time both species had all but disappeared.

In 1991 Congress instructed the USFWS to prepare an environmental impact report on the possibility of reintroducing wolves to habitats in the United States. Reintroductions began in 1995. Over a two-year period 66 gray wolves from southwestern Canada were introduced to Yellowstone National Park and central Idaho.

Wolf reintroductions were not greeted with universal enthusiasm. Ranchers in particular were concerned that wolves would attack livestock. They were also worried that their land would be open to government restrictions as a result of the wolves' presence. Several measures were adopted to address the ranchers' concerns. The most significant was that ranchers would be reimbursed for livestock losses from a compensation fund. Until 2010 the fund was maintained by the Defenders of Wildlife, a private conservation group, and financed by private donors. In "Defenders of Wildlife Wolf Compensation Trust" (August 20, 2010, http://www.defenders.org/programs_and_policy/wildlife_conservation/solutions/wolf_compensation_trust/index.php), the Defenders of Wildlife states that between 1987 and October 2009 the fund paid out nearly $1.4 million to 893 ranchers, covering the losses

FIGURE 7.2

The red wolf is found in the eastern United States. *(© U.S. Fish and Wildlife Service.)*

of 3,832 livestock, mostly sheep and cattle. In 2010 the USFWS initiated its own compensation fund following passage of the Omnibus Public Lands Management Act of 2009, which established the Wolf Compensation and Prevention Program and provided $1 million in funding. In the press release "U.S. Fish and Wildlife Service Announces $1 Million to States for Wolf Livestock Compensation Project" (April 1, 2010, http://www.fws.gov/news/NewsReleases/showNews.cfm?newsId=CDF428BD-A0FF-2416-7FC0AA115458E74C), the USFWS states that the $1 million was to be divided among 10 states with gray wolf populations: Arizona, Idaho, Michigan, Minnesota, Montana, New Mexico, Oregon, Washington, Wisconsin, and Wyoming.

Wolf introductions were legally challenged in 1997, when the American Farm Bureau Federation initiated a lawsuit calling for the removal of wolves from Yellowstone. The farm coalition scored an initial victory, but in 2000 the 10th Circuit Court of Appeals in Denver, Colorado, overturned the decision on appeal by the Department of the Interior, the World Wildlife Fund, and other conservation groups.

As of November 2011, the listing status of the gray wolf was complicated because the USFWS recognized multiple populations of the species. In addition, efforts to delist gray wolf populations have been repeatedly challenged in court. The gray wolf was listed under the ESA as endangered throughout the lower 48 states, with the following exceptions or notes:

- Minnesota—threatened; this population was delisted in 2007 due to recovery, and then relisted in September 2009 following a court order
- Western Great Lakes DPS in Wisconsin, Michigan, North Dakota, South Dakota, Iowa, Illinois, Indiana, and Ohio—endangered; this population was delisted in 2007 due to recovery, and then relisted in September 2009 following a court order
- Northern Rocky Mountains DPS in Wyoming—experimental population, nonessential
- Portions of Arizona, New Mexico, and Texas—experimental population, nonessential of the Mexican gray wolf

The Northern Rocky Mountain DPS in the eastern one-third of Washington and Oregon, a small part of north-central Utah, and all of Montana, Idaho, and Wyoming (except the experimental population mentioned earlier) has been the subject of intense litigation and political controversy. The DPS was delisted in 2008, but that listing was vacated after 12 parties sued and prevailed in court. In April 2009 the USFWS again attempted to delist the DPS due to recovery. According to the agency (April 2, 2009, http://edocket.access.gpo.gov/2009/pdf/E9-5991.pdf), in 2008 this DPS consisted of 1,639 wolves in 95 breeding pairs. These values were approximately five times greater than the minimum population recovery goal and three times greater than the minimum breeding pair recovery goal.

In 2010 the delisting was overturned in court. As explained in Chapter 2, in April 2011 Congress passed a spending bill that included an amendment that essentially delisted the gray wolf Northern Rocky Mountain DPS. Actually, the rider (http://www.gpo.gov/fdsys/pkg/PLAW-112publ10/html/PLAW-112publ10.htm) reinstated the April 2009 delisting rule and forbade any further "judicial review" of the decision. The latter condition triggered multiple lawsuits by conservation and wildlife groups alleging that the prohibition against judicial review of a congressional decision is unconstitutional. As of January 2012, rulings on these lawsuits had not been issued.

The USFWS issues an annual report on the status of the Northern Rocky Mountain DPS. As of January 2012, the most recent report was *Rocky Mountain Wolf Recovery 2010 Interagency Annual Report* (March 9, 2011, http://www.fws.gov/mountain-prairie/species/mammals/wolf/annualrpt10/index.html). The agency placed the population of gray wolves at 1,614 individuals in 2010, up from only one or two individuals during the early 1980s. (See Table 7.3.)

Deer, Caribou, Pronghorns, Bighorn Sheep, and Bison

Deer and caribou are members of the Cervidae family, along with elk and moose. Pronghorn are the last surviving members of the Antilocapridae family and are often confused with antelopes. Bighorn sheep and bison belong to the large Bovidae family, which also contains antelopes, gazelles, and domesticated sheep, cattle, and goats. Even though these species are diverse in taxonomy, the wild populations share a common threat: they are popular big game animals for hunters.

As of November 2011, there were seven species of big game listed under the ESA:

- Columbian white-tailed deer
- Key deer
- Peninsular bighorn sheep
- Sierra Nevada bighorn sheep
- Sonoran pronghorn
- Wood bison
- Woodland caribou

As shown in Table 7.1, all these species had endangered listings except for one population of Sonoran pronghorn, which was designated as a nonessential experimental population.

TABLE 7.3

Minimum fall wolf population and number of breeding pairs in the northern Rocky Mountains, by Federal Recovery Area, 1980–2010

[Includes only Montana, Idaho, and Wyoming within the Northern Rocky Mountain Distinct Population Segment]

Year	80	81	82	83	84	85	86	87	88	89	90	91	92	93	94	95	96	97	98	99	00	01	02	03	04	05	06	07	08	09	10
Recovery area																															
NWMT	1	2	8	6	6	13	15	10	14	12	33	29	41	55	48	66	70	56	49	63	64	84	108	92	59	126	171	230	282	319	374
GYA																21	40	86	112	118	177	218	271	301	335	325	390	453	449	455	501
CID																14	42	71	114	156	196	261	284	368	452	565	739	830	924	913	739
Total	**1**	**2**	**8**	**6**	**6**	**13**	**15**	**10**	**14**	**12**	**33**	**29**	**41**	**55**	**48**	**101**	**152**	**213**	**275**	**337**	**437**	**563**	**663**	**761**	**846**	**1,016**	**1,300**	**1,513**	**1,655**	**1,687**	**1,614**

Breeding pairs by recovery area:

Year	80	81	82	83	84	85	86	87	88	89	90	91	92	93	94	95	96	97	98	99	00	01	02	03	04	05	06	07	08	09	10
Recovery area																															
NWMT							1	2	1	1	3	2	4	4	5	6	7	5	5	6	6	7	12	4	6	11	12	23	18	26	24
GYA																2	4	9	6	8	14	13	23	21	31	20	31	33	35	38	37
CID																	3	6	10	10	10	15	14	30	29	40	43	51	42	49	47
Total							**1**	**2**	**1**	**1**	**3**	**2**	**4**	**4**	**5**	**8**	**14**	**20**	**21**	**24**	**30**	**35**	**49**	**55**	**66**	**71**	**86**	**107**	**95**	**113**	**108**

NWMT = Northwest Montana Recovery Area; GYA = Greater Yellowstone Recovery Area; CID = Central Idaho Recovery Area.

Notes: By the standards of the Rocky Mountain Gray Wolf Recovery Plan and wolf reintroduction environmental impact statement, a breeding pair is defined as an adult male and an adult female wolf, accompanied by 2 pups that survived at least until Dec 31. Recovery goals call for 10 breeding pairs per area, or a total of 30 breeding pairs distributed through the 3 areas, for 3 years.

Each year, wolf packs discovered in the current year that contain ≥2 yearlings and ≥2 adults are added to the previous year's breeding pair and population totals; similarly, if evidence in the current year indicates that <2 pups or <2 adults survived on December 31 of the previous year, that wolf pack is deleted from the previous year's breeding pair counts and population totals. Therefore, breeding pair counts and population totals are updated in current annual reports.

SOURCE: "Table 4a. Northern Rocky Mountain Minimum Fall Wolf Population and Breeding Pairs 1980–2010, by Federal Recovery Area," in *Rocky Mountain Wolf Recovery 2010 Interagency Annual Report*, U.S. Department of the Interior, U.S. Fish and Wildlife Service, 2011, http://www.fws.gov/mountain-prairie/species/mammals/wolf/annualrpt10/tables/ FINAL_2010_%20BP_by_REC_%20AREA_Table_4a_03-02-11_csime.pdf (accessed November 10, 2011)

SONORAN PRONGHORNS. The Sonoran pronghorn is one of five subspecies of pronghorn, all of which were severely depleted in the United States by the end of the 19th century. It is an antelope-like creature that stands approximately 3 feet (0.9 m) tall when fully grown. In *Final Revised Sonoran Pronghorn Recovery Plan* (December 1998, http://ecos.fws.gov/docs/recovery_plan/981203.pdf), the USFWS indicates that it is the fastest mammal on land in North America. The species is found in the broad valleys of the Sonoran Desert in southern Arizona, and its diet consists primarily of cacti. As shown in Table 7.2, more than $5.8 million was spent under the ESA during FY 2010 to conserve the Sonoran pronghorn.

The Sonoran pronghorn was on the first list of endangered species issued by the USFWS in 1967. (See Table 2.1 in Chapter 2.) The first recovery plan for the species was published in 1982, and it was revised in 1998. At that time the USFWS estimated that less than 300 individuals of the species remained in the United States. Another 200 to 500 individuals were believed to be living in Mexico.

There are a variety of reasons for the imperiled status of the species. These include insufficient food and/or water, drought, predation, illegal hunting, and degradation and fragmentation of habitat due to development, primarily for livestock ranching. Pronghorn are not jumpers, so they are prevented from foraging by fencing. Historically, the species depended on the Gila River to provide a "greenbelt" of scrubby vegetation on which the animals survived when other food sources became scarce. Agricultural and residential development has dramatically reduced river flows and is believed to be a major factor in the endangered status of the species.

In May 2011 the USFWS (http://www.gpo.gov/fdsys/pkg/FR-2011-05-05/pdf/2011-10467.pdf) designated a nonessential experimental population of Sonoran pronghorn in southwestern Arizona. (See Figure 7.3.) According to the agency, the animals making up the population were raised in a captive-rearing pen at the Cabeza Prieta National Wildlife Refuge near Ajo, Arizona.

Felines

Feline is the common term used for a member of the Felidae family. This diverse family includes bobcats, cheetahs, cougars, jaguars, jaguarundis, leopards, lions, lynx, panthers, pumas, tigers, and domesticated cats. All the wild species are under threat as development has left them with less natural habitat in which to live.

As of November 2011, there were eight wild feline species listed as endangered or threatened under the ESA in the United States:

- Canada lynx
- Eastern puma (= cougar)
- Florida panther
- Gulf coast jaguarundi
- Jaguar
- Ocelot
- Puma (= mountain lion)
- Sinaloan jaguarundi

Almost all these species had endangered listings. (See Table 7.1.) The exceptions were the Canada lynx, which was threatened, and the puma (= mountain lion), which in Florida was threatened due to similarity of appearance to the Florida panther. (See Figure 7.4.)

CANADA LYNXES. The Canada lynx is a medium-sized feline. Adults average 30 to 35 inches (76 to 89 cm) in length and weigh about 20 pounds (9 kg). The animal has tufted ears, a short tail, long legs, and large flat paws that allow it to walk on top of the snow. The Canada lynx inhabits cold, moist northern forests that are dominated by coniferous trees. Its primary food source is the snowshoe hare. Habitat modification, chiefly forest fragmentation due to timber harvesting, forest fire suppression, and human development, is blamed for imperiling both the snowshoe hare and the Canada lynx. In FY 2010 nearly $4.2 million was spent under the ESA to conserve the Canada lynx. (See Table 7.2.)

In 1982 the USFWS designated the Canada lynx as a candidate species for listing. However, no action was taken on listing until 1994, when the agency proposed to list the species as threatened. The decision was challenged in court by a group of conservation organizations led by the Defenders of Wildlife. This began a protracted legal battle over the listing status and critical habitat designation for the species. In 2000 a DPS of the Canada lynx was officially listed as threatened under the ESA. The DPS did not include certain areas in the Northeast, the Great Lakes, and the southern Rockies. This decision sparked more legal action. In 2003 the USFWS clarified that the threatened listing applied to what it called a contiguous United States DPS, including the populations in Colorado, Idaho, Maine, Michigan, Minnesota, Montana, New Hampshire, New York, Oregon, Utah, Vermont, Washington, Wisconsin, and Wyoming.

In December 2009 the USFWS (http://frwebgate.access.gpo.gov/cgi-bin/getdoc.cgi?dbname=2009_register&docid=fr17de09-15) published a 12-month finding on a petition it had received in 2007 from seven conservation groups requesting that the DPS be expanded to include the mountains of north-central New Mexico. The agency concluded that the DPS change was warranted, but was "precluded by higher priority actions," and stated its intention to add the New Mexico population to the candidate species list with a listing priority number (LPN) of 12. As explained in Chapter 2, LPNs range from 1 to 12 with

FIGURE 7.3

Non-essential experimental population area for Sonoran pronghorn

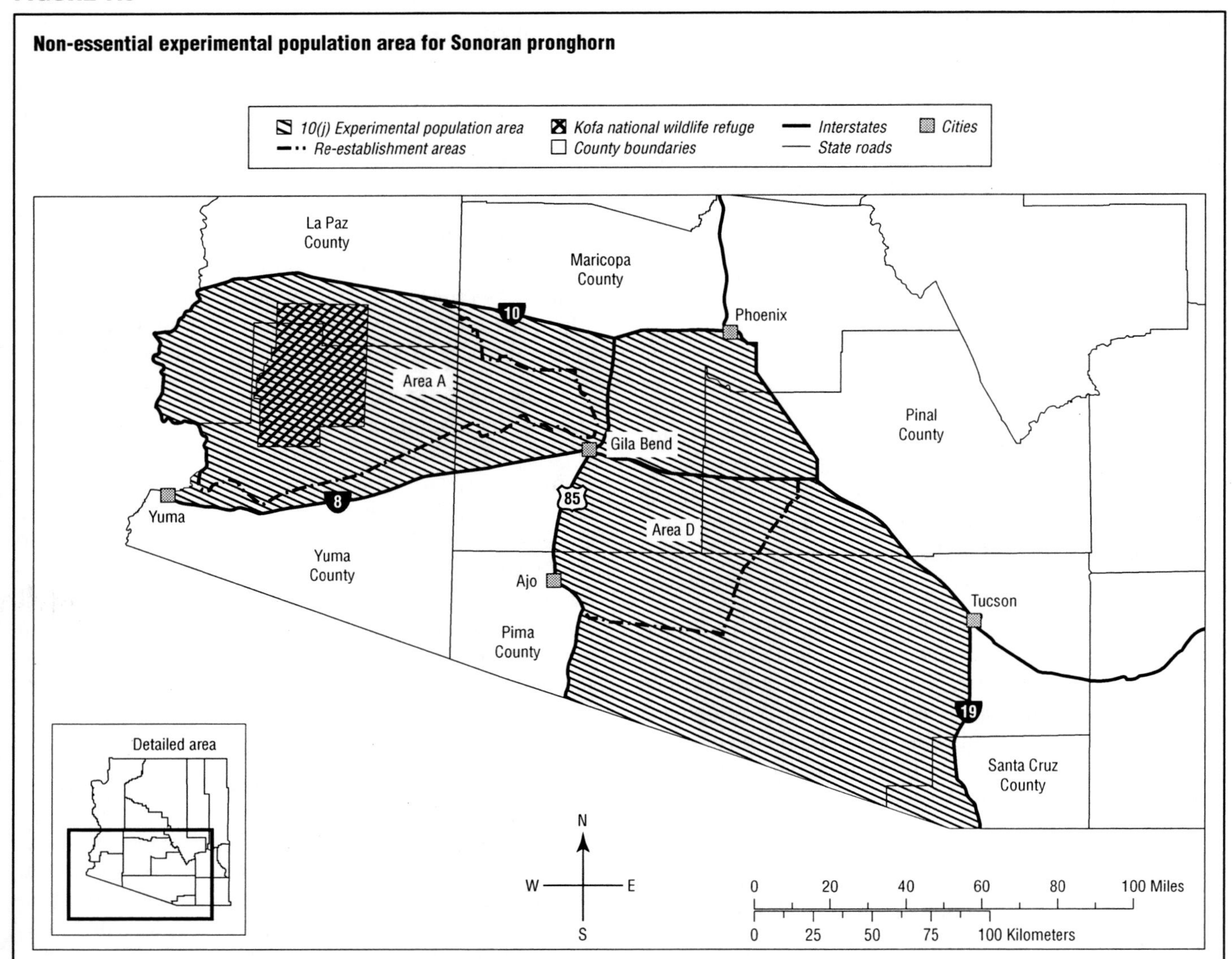

SOURCE: "10(j) Experimental Population Area for Sonoran Pronghorn," in "Endangered and Threatened Wildlife and Plants; Establishment of a Nonessential Experimental Population of Sonoran Pronghorn in Southwestern Arizona," *Federal Register*, vol. 76, no. 87, May 5, 2011, http://www.gpo.gov/fdsys/pkg/FR-2011-05-05/pdf/2011-10467.pdf (accessed December 13, 2011)

lower numbers indicating greater priority compared with other candidates. However, when the USFWS issued its annual Candidate Notice of Reviews (CNORs) for 2010 and 2011, the New Mexico lynx population was not among the candidate species. (The 2011 CNOR is shown in Table 2.5 in Chapter 2.) In 2011 the USFWS settled long-standing lawsuits involving candidate species by agreeing to make final listing decisions for all of them by 2017. In May 2011 the USFWS (http://www.fws.gov/endangered/improving_ESA/exh_1_re_joint_motion_FILED.PDF) also agreed to propose a rule before the end of FY 2013 to amend the DPS boundaries for the Canada lynx to include New Mexico.

A 2004 court order forced the USFWS to set critical habitat for the Canada lynx. Originally, 18,000 square miles (47,000 sq km) were proposed for this purpose. This area was substantially reduced to 1,800 square miles (4,700 sq km) when the final designation was made in 2006. In February 2009 the USFWS issued revised critical habitat totaling approximately 39,000 square miles (101,000 sq km) contained in five separate units in Idaho, Maine, Minnesota, Montana, Wyoming, and Washington. (See Figure 7.5.)

In 2007 the USFWS initiated a five-year status review to determine the appropriateness of the threatened listing for the Canada Lynx. As of January 2012, the results of that review had not been published.

Ferrets

The ferret is a member of the Mustelidae family, along with muskrats, badgers, otters, mink, skunks, and weasels. Ferrets are small, furry creatures with long,

FIGURE 7.4

Florida panther

SOURCE: Robert Savannah, artist, "Florida Panthers," in *Line Art (Drawings)*, U.S. Department of the Interior, U.S. Fish and Wildlife Service, undated, http://www.fws.gov/pictures/lineart/bobsavannah/floridapanthers.html (accessed November 11, 2011)

skinny bodies typically less than 2 feet (0.6 m) long. They have short legs and elongated necks with small heads. Ferrets are carnivores; in the wild they feed on rodents, rabbits, reptiles, and insects.

As shown in Table 7.1, as of November 2011 there was one U.S. species of ferret listed under the ESA: the black-footed ferret, which is depicted in Figure 7.6. The species was listed as endangered, except in nonessential experimental populations in portions of Arizona, Colorado, Montana, South Dakota, Utah, and Wyoming.

BLACK-FOOTED FERRETS. The black-footed ferret is a small, furrow-digging mammal. It is a nocturnal creature and helps control populations of snakes and rodents, including its primary prey: black-tailed prairie dogs. The black-footed ferret once ranged over 11 Rocky Mountain states and parts of Canada. Its population has declined because of the large-scale conversion of prairie habitats to farmland and because its main prey, the prairie dog, has been nearly exterminated by humans. Prairie dogs are considered pests because they dig holes and tunnels just beneath the ground surface. These holes can cause serious injury to horses or other large animals that step into them. Poisons that are used to kill prairie dogs may also kill some ferrets.

Black-footed ferret populations declined so drastically that the species was included on the first list of endangered species published in 1967. (See Table 2.1 in Chapter 2.) However, prairie dog poisonings continued, and by 1979 it was believed that the black-footed ferret was extinct. During the 1980s a small colony was discovered in Wyoming. These individuals were captured and entered into a captive breeding program.

The captive breeding of ferrets became markedly successful. The Black-Footed Ferret Recovery Implementation Team (BFFRIT) was created in 1996 to integrate the efforts of dozens of agencies and nonprofit organizations working to save the species. The BFFRIT (2011, http://www.blackfootedferret.org/reintroduction/) reports that every year approximately 150 to 220 of the animals are reintroduced into the wild. The ferrets are vaccinated against canine distemper and sylvatic plague before being released. The BFFRIT notes that in 2011 the wild population was estimated at approximately 1,000 individuals.

In November 2008 the USFWS published *Black-Footed Ferret (*Mustela nigripes*): 5-Year Status Review—Summary and Evaluation* (http://ecos.fws.gov/docs/five_year_review/doc2364.pdf). The agency acknowledged that the captive breeding program was successful, but noted that "recovery of the black-footed ferret cannot be achieved without more assertive restoration and management of sufficient prairie dog habitat." The USFWS also recommended that the 1988 recovery plan for the black-footed ferret be updated. As of January 2012, a revised recovery plan had not been published.

Rabbits

Rabbits are members of the Leporidae family, along with hares. Rabbits are generally smaller than hares and have somewhat shorter ears. Both species have tall slender ears and short bodies with long limbs and thick soft fur. Domesticated rabbits are all descended from European species.

As of November 2011, there were three rabbit species listed as endangered under the ESA in the United States. (See Table 7.1.) The species and their primary locations are:

- Lower Keys marsh rabbit—Florida
- Pygmy rabbit—Columbia Basin DPS in Washington
- Riparian brush rabbit—California

All the species had recovery plans in place. In September 2007 the USFWS published *Lower Keys Marsh*

FIGURE 7.5

Critical habitat for Canada lynx

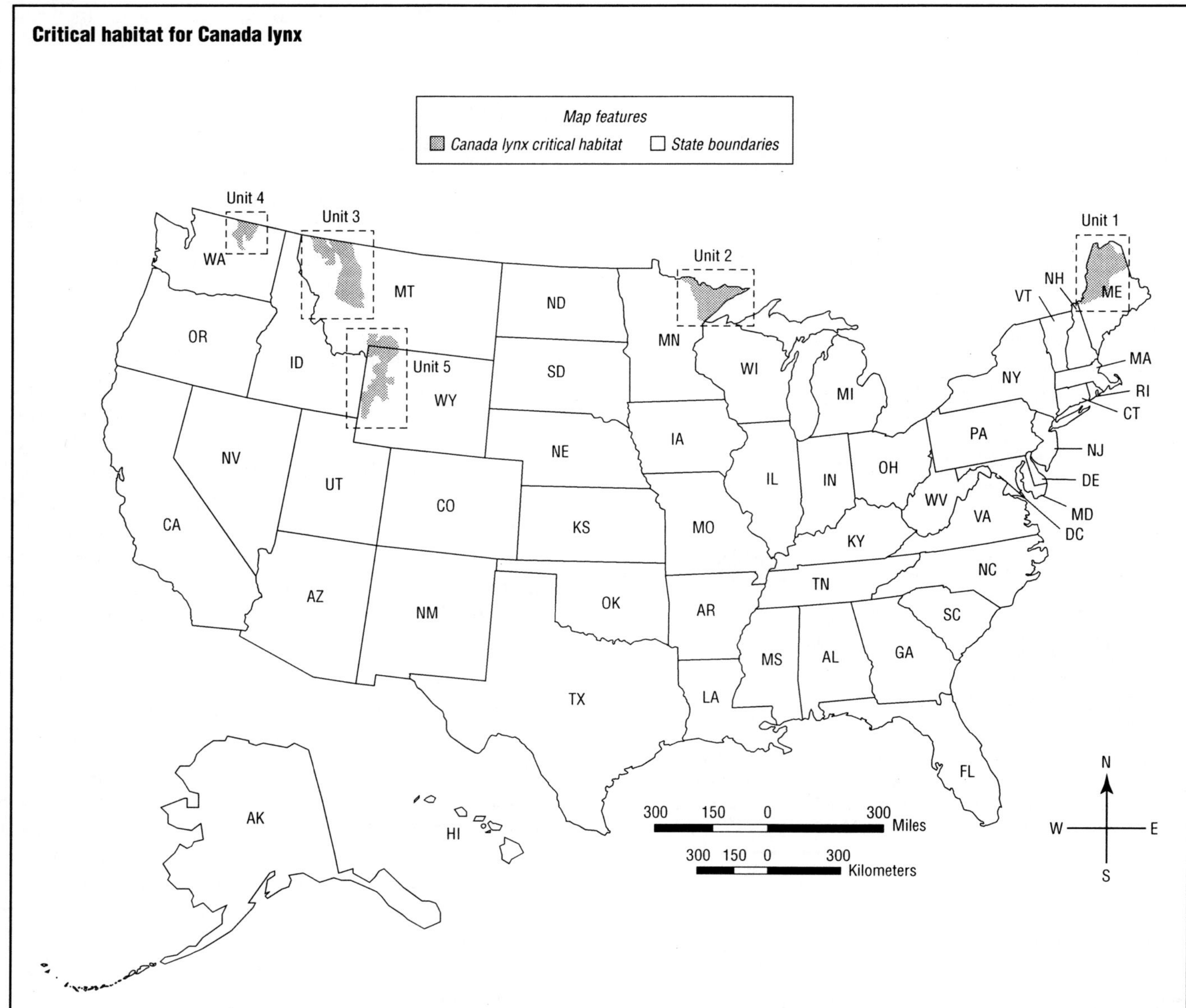

SOURCE: "Index Map: Critical Habitat for Lynx canadensis (Canada Lynx)," in "Endangered and Threatened Wildlife and Plants; Revised Designation of Critical Habitat for the Contiguous United States Distinct Population Segment of the Canada Lynx," *Federal Register*, vol. 74, no. 36, February 25, 2009, http://www.fws.gov/mountain-prairie/species/mammals/lynx/criticalhabitat_files/74FR8616.pdf (accessed November 10, 2011)

*Rabbit (*Sylvilagus palustris hefneri*): 5-Year Review—Summary and Evaluation* (http://ecos.fws.gov/docs/five_year_review/doc1110.pdf), in which the agency recommended maintaining the population's endangered listing.

The USFWS published notices in April 2010 (http://www.gpo.gov/fdsys/pkg/FR-2010-04-08/pdf/2010-7915.pdf#page=1) and May 2010 (http://www.gpo.gov/fdsys/pkg/FR-2010-05-21/pdf/2010-12170.pdf#page=1) indicating that it was initiating five-year reviews for dozens of western species, including the pygmy rabbit Columbia Basin DPS and the riparian brush rabbit, respectively. As of January 2012, the results of the reviews had not been issued.

Rodents

Rodents are members of the order Rodentia, the single largest group of mammals. This order includes mice, rats, beavers, chipmunks, squirrels, prairie dogs, voles, and many other species. Rodents are characterized by their distinctive teeth, particularly a pair of chisel-shaped incisors in each jaw. Even though most rodents are plant eaters, some species include insects in their diet.

As of November 2011, there were 29 rodent species listed under the ESA as endangered or threatened (see Table 7.1):

- Kangaroo rat—six species
- Mountain beaver—one species

FIGURE 7.6

Black-footed ferret. *(© Laura Romin & Larry Dalton/Alamy.)*

FIGURE 7.7

The prairie dog

SOURCE: Robert Savannah, artist, "Prairie Dogs," in *Line Art (Drawings)*, U.S. Department of the Interior, U.S. Fish and Wildlife Service, undated, http://www.fws.gov/pictures/lineart/bobsavannah/prairiedogs.html (accessed November 11, 2011)

- Mouse—10 species
- Prairie dog—one species
- Rat—three species
- Squirrel—five species
- Vole—three species

One imperiled rodent species of particular interest is the Utah prairie dog. The USFWS reports in *Federal and State Endangered and Threatened Species Expenditures, Fiscal Year 2010* (September 2011, http://www.fws.gov/endangered/esa-library/pdf/2010.EXP.FINAL.pdf) that in FY 2010 over $1.7 million was spent on this controversial species under the ESA.

UTAH PRAIRIE DOGS. Prairie dogs are members of the Sciuridae family, along with chipmunks and squirrels. (See Figure 7.7.) They are endemic (limited) to the United States and inhabit mostly arid grasslands. They are found from Montana and North Dakota south to Texas. The 19th-century explorers Meriwether Lewis (1774–1809) and William Clark (1770–1838) allegedly named the animals prairie dogs because of their barklike calls.

Before settlers moved into the West, it is believed that millions of prairie dogs inhabited the area. Prairie dogs are burrowing creatures and live in colonies. They produce holes, tunnels, and dirt mounds that can be damaging to land that is used for agriculture. The holes also pose a tripping hazard to horses. As a result, ranchers of the late 1800s and early 1900s tried to eradicate the prairie dog by using poison on a large scale. They were assisted in their efforts by the federal government.

According to the USFWS, in *Utah Prairie Dog (*Cynomys parvidens*) Draft Revised Recovery Plan* (August 2010, http://ecos.fws.gov/docs/recovery_plan/100917.pdf), the Utah prairie dog is a small furry creature that reaches 10 to 16 inches (25 to 41 cm) in length when fully grown. It is reddish brown in color with a short, white-tipped tail and a black spot above the eye. Before the control programs of the 1920s, 95,000 of the animals are estimated to have lived in Utah. By the mid-1970s the population had been reduced to about 3,300 individuals. Massive poisoning by humans, disease (a form of plague), and loss of suitable habitat are blamed for the population decline. In 1973 the USFWS listed the Utah prairie dog as endangered under the ESA. Over the next decade conservation efforts led to an increase in its population. However, angry farmers began reporting massive crop damage caused by the creatures, particularly to summer alfalfa crops, a favorite food source for the prairie dogs. The state of Utah petitioned the USFWS to downlist the species from endangered to threatened. In May 1984 the USFWS (http://ecos.fws.gov/docs/federal_register/fr838.pdf) reclassified the species.

The USFWS recognized that Utah farmers were not going to tolerate continuing threats to their crops from the prairie dogs. In a unique action, the agency established in 1984 a special regulation under section 4(d) of the ESA that allows the Utah Division of Wildlife Resources to issue permits to private landowners who wished to kill Utah prairie dogs on their property. The maximum allowable number killed was originally set at 5,000, but in 1991 was raised to 6,000. In addition, the USFWS began relocating Utah prairie dogs from private lands to lands under control of the federal government.

The USFWS (June 2, 2011, http://www.gpo.gov/fdsys/pkg/FR-2011-06-02/pdf/2011-13684.pdf) notes that its issuance of the 4(d) rule sparked years of litigation against the agency by Forest Guardians, a nonprofit conservation organization that later merged with other groups to form WildEarth Guardians. Forest Guardians petitioned the USFWS to reclassify the Utah prairie dog from threatened to endangered and challenged the 4(d) rule as biologically unsound. In June 2011 the USFWS proposed changes to the rule that would limit take (i.e., killing) to "agricultural lands and private property neighboring conservation properties." In addition, the total take would be based on a percentage of the colony population, rather than on a flat number. As of January 2012, the proposed changes had not been finalized.

VIRGINIA NORTHERN FLYING SQUIRRELS. The Virginia (or West Virginia) northern flying squirrel is a small furry animal that grows to be about 12 inches (30.5 cm) long and weighs less than 5 ounces (142 g). (See Figure 7.8.) According to the USFWS, in *West Virginia Northern Flying Squirrel* (June 16, 2011, http://www.fws.gov/northeast/newsroom/wvnfsq.html), the species is a relic from the last Ice Age and has existed for at least 10,000 years. It is found only in the highest mountains along the West Virginia and Virginia border. Its historical habitat (old-growth spruce and other hardwood forests) was virtually eliminated by industrial logging between the 1880s and 1940s. In 1985, when the squirrel was listed as endangered under the ESA, biologists had captured only 10 individuals throughout its range and feared that it was in imminent danger of extinction. However, the species managed to survive over the following two decades as the hardwood forests rebounded. In August 2008 the Virginia northern flying squirrel was delisted due to recovery. By 2009 its population was believed to be over 1,100 individuals. The USFWS notes that the squirrel's traditional forested habitat continues to spread due to protection efforts by federal and state government agencies and private conservation groups. However, in 2011 the delisting decision was overturned in court after being challenged by a coalition of nonprofit organizations. In June 2011 the USFWS (http://www.gpo.gov/fdsys/pkg/FR-2011-06-17/pdf/2011-15111.pdf) officially reinstated the endangered listing for the species.

FIGURE 7.8

West Virginia northern flying squirrel

SOURCE: Robert Savannah, artist, "Line Drawing of West Virginia Northern Flying Squirrel," in *Images in the News*, U.S. Department of the Interior, U.S. Fish and Wildlife Service, Northeast Region, September 18, 2008, http://www.fws.gov/northeast/images.html (accessed November 11, 2011)

Shrews

Shrews belong to the order Soricomorpha, along with moles and their relatives. They are small furry animals about the size of mice. Shrews are primarily insectivores (insect eaters) with long snouts and small eyes.

As of November 2011, only one shrew, the Buena Vista Lake ornate shrew, was listed under the ESA. (See Table 7.1.) It was designated as endangered. According to the USFWS, in "Buena Vista Lake Ornate Shrew (*Sorex ornatus* ssp. *relictus*)" (January 2012, http://ecos.fws.gov/speciesProfile/profile/speciesProfile.action?

spcode=A0DV), the species is found in California and is about 4 inches (10 cm) long when fully grown, including a 1.5-inch (3.8-cm) tail. In *Buena Vista Lake Ornate Shrew (*Sorex ornatus relictus*): 5-Year Review—Summary and Evaluation* (September 2011, http://ecos.fws.gov/docs/five_year_review/doc3889.pdf), the agency noted that the species has lost more than 95% of its historic wetland habitat due to channelization and agricultural and urban uses of water resources. The USFWS recommended maintaining an endangered listing for the shrew.

IMPERILED TERRESTRIAL MAMMALS AROUND THE WORLD

The International Union for Conservation of Nature (IUCN) indicates in *Red List of Threatened Species Version 2011.2* (2011, http://www.iucnredlist.org/documents/summarystatistics/2011_2_RL_Stats_Table1.pdf) that 1,138 mammal species were threatened as of 2011. The IUCN does not break down the number by marine and terrestrial species. However, as noted in Chapter 3 the IUCN has designated as threatened approximately three dozen marine mammal species. Thus, around 1,100 terrestrial mammal species were considered threatened by the IUCN in 2011.

As of November 2011, the USFWS listed 260 totally foreign species of terrestrial mammals as endangered or threatened under the ESA. (See Table 7.4.) Many of the species are from groups that are also imperiled in the United States, such as bats, bears, big game, canines, felines, rabbits, and rodents. In addition, there are exotic animals not native to this country, particularly elephants, pandas, primates, and rhinoceros.

Big Cats

The IUCN (2011, http://www.iucnredlist.org/apps/redlist/details/15955/0) recognizes five subspecies of tigers that are believed to be extant (still in existence) in the wild:

- Amur tiger (or Siberian tiger, see Figure 7.9)—eastern Russia and northeastern China
- Northern Indochinese tiger—Indochina north of Malaya
- Malayan tiger—Malayan peninsula
- Sumatran tiger—Sumatra
- Bengal tiger—Indian subcontinent

Wild tigers are found exclusively in Asia, from India to Siberia. Even though the world tiger population surpassed 100,000 during the 19th century, experts fear that fewer than 10,000 remained by the turn of the 21st century. Besides habitat loss, countless tigers fall victim to the illegal wildlife trade every year. Many tiger body parts are used as ingredients in traditional Chinese medicine, and the big cats are also prized in the exotic pet industry. In addition, ecologists warn that tigers, which hunt deer, wild pigs, cattle, antelope, and other large mammals, are seriously threatened by the loss of prey, much of which consists of nonprotected species that are being eliminated by hunters. Many tiger populations have also been weakened by inbreeding, which increases the possibility of reproductive problems and birth defects.

Elephants

Elephants are the largest land animals on the earth. They are frequently described as the "architects" of the savanna habitats in which they live. Elephants dig water holes, keep forest growth in check, and open up grasslands that support other species, including the livestock of African herders. Elephants are highly intelligent, emotional animals and form socially complex herds. There are two species of elephants: African elephants and Asian elephants, both of which are highly endangered. The African elephant, which can weigh as much as 6 tons (5.4 t), is the larger species. (See Figure 7.10.)

Elephants have huge protruding teeth (tusks) made of ivory. Ivory is valued by humans for several reasons, particularly for use in making jewelry and figurines. Piano keys were also once made almost exclusively of ivory; however, this practice has ceased. The market for ivory has had tragic consequences for African elephants. In "Africa's Environment in Crisis" (November 2002, http://diglib1.amnh.org/articles/Africa/Africa_environment.pdf), Gordy Slack of the American Museum of Natural History states that their numbers dropped from between 5 million and 10 million individuals in 1930 to only 600,000 in 1989. As a result of this decline, the Convention on International Trade in Endangered Species of Wild Fauna and Flora (CITES) banned worldwide commerce in ivory and other elephant products in 1990. However, like rhinoceros horns, elephant tusks continue to be illegally traded. The World Wildlife Fund claims in "Wildlife Trade" (2012, http://www.worldwildlife.org/what/globalmarkets/wildlifetrade/index.html) that up to 12,000 African elephants per year are illegally killed for their ivory.

Even though the ivory trade has always been the largest threat to elephants, conflicts between humans and elephants are an increasing issue. The ranges of many elephant herds now extend outside protected refuges, and elephants frequently come into contact with farmers, eating or otherwise destroying crops. Increasing human settlement in areas that are inhabited by elephants will likely result in more conflicts over time. According to the press release "African Elephant Fund Launched at

TABLE 7.4

Foreign endangered and threatened terrestrial mammal species, November 2011

Common name	Scientific name	Listing status*	Historic range
Addax	Addax nasomaculatus	E	North Africa
African elephant	Loxodonta africana	T	Africa
African wild ass	Equus asinus	E	Somalia, Sudan, Ethiopia
African wild dog	Lycaon pictus	E	Sub-Saharan Africa
Andean cat	Felis jacobita	E	Chile, Peru, Bolivia, Argentina
Apennine chamois	Rupicapra rupicapra ornata	E	Italy
Arabian gazelle	Gazella gazella	E	Arabian Peninsula, Palestine, Sinai
Arabian oryx	Oryx leucoryx	E	Arabian Peninsula
Arabian tahr	Hemitragus jayakari	E	Oman
Argali	Ovis ammon	E	Afganistan, China, India, Kazakhstan, Kyrgyzstan, Mongolia, Nepal, Pakistan, Russia, Tajikistan, Uzbekistan
Argali	Ovis ammon	T	Afganistan, China, India, Kazakhstan, Kyrgyzstan, Mongolia, Nepal, Pakistan, Russia, Tajikistan, Uzbekistan
Asian elephant	Elephas maximus	E	South-central and southeastern Asia
Asian golden (=Temmnick's) cat	Catopuma (=Felis) temminckii	E	Nepal, China, Southeast Asia, Indonesia (Sumatra)
Asian tapir	Tapirus indicus	E	Burma, Laos, Cambodia, Vietnam, Malaysia, Indonesia, Thailand
Asian wild ass	Equus hemionus	E	Southwestern and Central Asia
Asiatic lion	Panthera leo persica	E	Turkey to India
Australian native mouse	Notomys aquilo	E	Australia
Australian native mouse	Zyzomys pedunculatus	E	Australia
Avahi	Avahi laniger (entire genus)	E	Malagasy Republic (=Madagascar)
Aye-aye	Daubentonia madagascariensis	E	Malagasy Republic (=Madagascar)
Babirusa	Babyrousa babyrussa	E	Indonesia
Bactrian camel	Camelus bactrianus	E	Mongolia, China
Bactrian deer	Cervus elaphus bactrianus	E	Tajikistan, Uzbekistan, Afghanistan
Baluchistan bear	Ursus thibetanus gedrosianus	E	Iran, Pakistan
Banded hare wallaby	Lagostrophus fasciatus	E	Australia
Banteng	Bos javanicus	E	Southeast Asia
Barbary deer	Cervus elaphus barbarus	E	Morocco, Tunisia, Algeria
Barbary hyena	Hyaena hyaena barbara	E	Morocco, Algeria, Tunisia
Barbary serval	Leptailurus (=Felis) serval constantina	E	Algeria
Barbary stag	Cervus elaphus barbarus	E	Tunisia, Algeria
Barred bandicoot	Perameles bougainville	E	Australia
Beaver (Mongolian)	Castor fiber birulai	E	Mongolia
Black colobus monkey	Colobus satanas	E	Equatorial Guinea, People's Republic of Congo, Cameroon, Gabon
Black-faced impala	Aepyceros melampus petersi	E	Namibia, Angola
Black-footed cat	Felis nigripes	E	Southern Africa
Black howler monkey	Alouatta pigra	T	Mexico, Guatemala, Belize
Black rhinoceros	Diceros bicornis	E	Sub-Saharan Africa
Bontebok	Damaliscus pygarus (=dorcas) dorcas	E	South Africa
Brazilian three-toed sloth	Bradypus torquatus	E	Brazil
Brindled nail-tailed wallaby	Onychogalea fraenata	E	Australia
Brown bear	Ursus arctos arctos	E	Palearctic
Brown bear	Ursus arctos pruinosus	E	China (Tibet)
Brown hyena	Parahyaena (=Hyaena) brunnea	E	Southern Africa
Brush-tailed rat-kangaroo	Bettongia penicillata	E	Australia
Buff-headed marmoset	Callithrix flaviceps	E	Brazil
Bulmer's fruit bat (=flying fox)	Aproteles bulmerae	E	Papua New Guinea
Bumblebee bat	Craseonycteris thonglongyai	E	Thailand
Cabrera's hutia	Capromys angelcabrerai	E	Cuba
Calamianes (=Philippine) deer	Axis porcinus calamianensis	E	Philippines (Calamian Islands)
Capped langur	Trachypithecus (=Presbytis) pileatus	E	India, Burma, Bangladesh
Cedros Island mule deer	Odocoileus hemionus cerrosensis	E	Mexico (Cedros Island)
Central American tapir	Tapirus bairdii	E	Southern Mexico to Colombia and Ecuador
Cheetah	Acinonyx jubatus	E	Africa to India
Chiltan (=wild goat) markhor	Capra falconeri (=aegragrus) chiltanensis	E	Chiltan Range of west-central Pakistan
Chimpanzee	Pan troglodytes	E	Africa
Chimpanzee	Pan troglodytes	T	Africa
Chinchilla	Chinchilla brevicaudata boliviana	E	Bolivia
Clark's gazelle	Ammodorcas clarkei	E	Somalia, Ethiopia
Clouded leopard	Neofelis nebulosa	E	Southeastern and south-central Asia, Taiwan

CITES Meeting" (http://www.cites.org/eng/news/pr/2011/20110819_SC61.php), the CITES secretary-general John E. Scanlon announced in August 2011 the creation of the African Elephant Fund, a fund that will be used "to enhance law enforcement capacity and secure the long term survival of African elephant populations." He explained that the goal is to raise $100 million by 2014.

TABLE 7.4

Foreign endangered and threatened terrestrial mammal species, November 2011 [CONTINUED]

Common name	Scientific name	Listing status*	Historic range
Corsican red deer	Cervus elaphus corsicanus	E	Corsica, Sardinia
Costa Rican puma	Puma (=Felis) concolor costaricensis	E	Nicaragua, Panama, Costa Rica
Cotton-top marmoset	Saguinus oedipus	E	Costa Rica to Colombia
Crescent nail-tailed wallaby	Onychogalea lunata	E	Australia
Cuban solenodon	Solenodon cubanus	E	Cuba
Dama gazelle	Gazella dama	E	North Africa
Desert bandicoot	Perameles eremiana	E	Australia
Desert (=plain) rat-kangaroo	Caloprymnus campestris	E	Australia
Dhole	Cuon alpinus	E	C.I.S., Korea, China, India, Southeast Asia
Diana monkey	Cercopithecus diana	E	Coastal West Africa
Dibbler	Antechinus apicalis	E	Australia
Douc langur	Pygathrix nemaeus	E	Cambodia, Laos, Vietnam
Drill	Mandrillus (=Papio) leucophaeus	E	Equatorial West Africa
Dwarf hutia	Capromys nana	E	Cuba
Eastern jerboa marsupial	Antechinomys laniger	E	Australia
Eastern native-cat	Dasyurus viverrinus	E	Australia
Eld's brow-antlered deer	Cervus eldi	E	India to Southeast Asia
False water rat	Xeromys myoides	E	Australia
Fea's muntjac	Muntiacus feae	E	Northern Thailand, Burma
Field's mouse	Pseudomys fieldi	E	Australia
Flat-headed cat	Prionailurus (=Felis) planiceps	E	Malaysia, Indonesia
Formosan rock macaque	Macaca cyclopis	T	Taiwan
Formosan sika deer	Cervus nippon taiouanus	E	Taiwan
Formosan yellow-throated marten	Martes flavigula chrysospila	E	Taiwan
Francois' langur	Trachypithecus (=Presbytis) francoisi	E	China (Kwangsi), Indochina
Gaimard's rat-kangaroo	Bettongia gaimardi	E	Australia
Gelada baboon	Theropithecus gelada	T	Ethiopia
Giant armadillo	Priodontes maximus	E	Venezuela and Guyana to Argentina
Giant panda	Ailuropoda melanoleuca	E	China
Giant sable antelope	Hippotragus niger variani	E	Angola
Gibbons	Hylobates spp. (including Nomascus)	E	China, India, Southeast Asia
Goeldi's marmoset	Callimico goeldii	E	Brazil, Colombia, Ecuador, Peru, Bolivia
Golden langur	Trachypithecus (=Presbytis) geei	E	India (Assam), Bhutan
Golden-rumped tamarin	Leontopithecus spp.	E	Brazil
Goral	Naemorhedus goral	E	East Asia
Gorilla	Gorilla gorilla	E	Central and western Africa
Gould's mouse	Pseudomys gouldii	E	Australia
Gray (=entellus) langur	Semnopithecus (=Presbytis) entellus	E	China (Tibet), India, Pakistan, Kashmir, Sri Lanka, Sikkim, Bangladesh
Great Indian rhinoceros	Rhinoceros unicornis	E	India, Nepal
Grevy's zebra	Equus grevyi	T	Kenya, Ethiopia, Somalia
Guatemalan jaguarundi	Herpailurus (=Felis) yagouaroundi fossata	E	Mexico, Nicaragua
Guizhou snub-nosed monkey	Rhinopithecus brelichi	E	China
Haitian solenodon	Solenodon paradoxus	E	Dominican Republic, Haiti
Hartmann's mountain zebra	Equus zebra hartmannae	T	Namibia, Angola
Hispid hare	Caprolagus hispidus	E	India, Nepal, Bhutan
Indochina hog deer	Axis porcinus annamiticus	E	Thailand, Indochina
Indri	Indri indri (entire genus)	E	Malagasy Republic (=Madagascar)
Iriomote cat	Prionailurus (=Felis) bengalensis iriomotensis	E	Japan (Iriomote Island, Ryukyu Islands)
Japanese macaque	Macaca fuscata	T	Japan (Shikoku, Kyushu and Honshu Islands)
Javan rhinoceros	Rhinoceros sondaicus	E	Indonesia, Indochina, Burma, Thailand, Sikkim, Bangladesh, Malaysia
Jentink's duiker	Cephalophus jentinki	E	Sierra Leone, Liberia, Ivory Coast
Kabul markhor	Capra falconeri megaceros	E	Afghanistan, Pakistan
Kashmir stag	Cervus elaphus hanglu	E	Kashmir
Koala	Phascolarctos cinereus	T	Australia
Kouprey	Bos sauveli	E	Vietnam, Laos, Cambodia, Thailand
Kuhl's (=Bawean) deer	Axis porcinus kuhli	E	Indonesia
Large desert marsupial-mouse	Sminthopsis psammophila	E	Australia
Large-eared hutia	Capromys auritus	E	Cuba
Leadbeater's possum	Gymnobelideus leadbeateri	E	Australia
Lemurs	Lemuridae (incl. genera Lemur, Phaner, Hapalemur, Lepilemur, Microcebus, Allocebus, Cheirogaleus, Varecia)	E	Malagasy Republic (=Madagascar)
Leopard	Panthera pardus	E	Africa, Asia
Leopard	Panthera pardus	T	Africa, Asia
Leopard cat	Prionailurus (=Felis) bengalensis bengalensis	E	India, Southeast Asia
Lesser rabbit bandicoot	Macrotis leucura	E	Australia
Lesser slow loris	Nycticebus pygmaeus	T	Indochina

TABLE 7.4

Foreign endangered and threatened terrestrial mammal species, November 2011 [CONTINUED]

Common name	Scientific name	Listing status*	Historic range
Lesuer's rat-kangaroo	Bettongia lesueur	E	Australia
L'hoest's monkey	Cercopithecus lhoesti	E	Upper eastern Congo R. Basin, Cameroon
Lion-tailed macaque	Macaca silenus	E	India
Little earth hutia	Capromys sanfelipensis	E	Cuba
Little planigale	Planigale ingrami subtilissima	E	Australia
Long-tailed langur	Presbytis potenziani	T	Indonesia
Long-tailed marsupial-mouse	Sminthopsis longicaudata	E	Australia
Lowland anoa	Bubalus depressicornis	E	Indonesia
Malabar large-spotted civet	Viverra civettina (=megaspila c.)	E	India
Mandrill	Mandrillus (=Papio) sphinx	E	Equatorial West Africa
Maned wolf	Chrysocyon brachyurus	E	Argentina, Bolivia, Brazil, Paraguay, Uruguay
Mantled howler monkey	Alouatta palliata	E	Mexico to South America
Marbled cat	Pardofelis (=Felis) marmorata	E	Nepal, Southeast Asia, Indonesia
Margay	Leopardus (=Felis) wiedii	E	U.S.A. (TX), Central and South America
Marsh deer	Blastocerus dichotomus	E	Argentina, Uruguay, Paraguay, Bolivia, Brazil
McNeill's deer	Cervus elaphus macneilii	E	China (Sinkiang, Tibet)
Mexican bobcat	Lynx (=Felis) rufus escuinapae	E	Central Mexico
Mexican grizzly bear	Ursus arctos	E	Holarctic
Mexican prairie dog	Cynomys mexicanus	E	Mexico
Mhorr gazelle	Gazella dama mhorr	E	Morocco
Mongolian saiga (antelope)	Saiga tatarica mongolica	E	Mongolia
Moroccan gazelle	Gazella dorcas massaesyla	E	Morocco, Algeria, Tunisia
Mountain anoa	Bubalus quarlesi	E	Indonesia
Mountain (=Cuvier's) gazelle	Gazella cuvieri	E	Morocco, Algeria, Tunisia
Mountain pygmy possum	Burramys parvus	E	Australia
Mountain tapir	Tapirus pinchaque	E	Colombia, Ecuador and possibly Peru and Venezuela
Mountain zebra	Equus zebra zebra	E	South Africa
Musk deer	Moschus spp. (all species)	E	Central and eastern Asia
New Holland mouse	Pseudomys novaehollandiae	E	Australia
North Andean huemul	Hippocamelus antisensis	E	Ecuador, Peru, Chile, Bolivia, Argentina
North China sika deer	Cervus nippon mandarinus	E	China (Shantung and Chihli Provinces)
Northern swift fox	Vulpes velox hebes	E	U.S.A. (northern plains), Canada
Northern white rhinoceros	Ceratotherium simum cottoni	E	Zaire, Sudan, Uganda, Central African Republic
Numbat	Myrmecobius fasciatus	E	Australia
Orangutan	Pongo pygmaeus	E	Borneo, Sumatra
Pagi Island langur	Nasalis concolor	E	Indonesia
Pakistan sand cat	Felis margarita scheffeli	E	Pakistan
Pampas deer	Ozotoceros bezoarticus	E	Brazil, Argentina, Uruguay, Bolivia, Paraguay
Panamanian jaguarundi	Herpailurus (=Felis) yagouaroundi panamensis	E	Nicaragua, Costa Rica, Panama
Parma wallaby	Macropus parma	E	Australia
Pelzeln's gazelle	Gazella dorcas pelzelni	E	Somalia
Peninsular pronghorn	Antilocapra americana peninsularis	E	Mexico (Baja California)
Persian fallow deer	Dama mesopotamica (=dama m.)	E	Iraq, Iran
Philippine tarsier	Tarsius syrichta	T	Philippines
Pied tamarin	Saguinus bicolor	E	Brazil
Pig-footed bandicoot	Chaeropus ecaudatus	E	Australia
Pink fairy armadillo	Chlamyphorus truncatus	E	Argentina
Preuss' red colobus monkey	Procolobus (=Colobus) preussi (=badius p.)	E	Cameroon
Proboscis monkey	Nasalis larvatus	E	Borneo
Przewalski's horse	Equus przewalskii	E	Mongolia, China
Pudu	Pudu pudu	E	Southern South America
Purple-faced langur	Presbytis senex	T	Sri Lanka
Pygmy chimpanzee	Pan paniscus	E	Zaire
Pygmy hog	Sus salvanius	E	India, Nepal, Bhutan, Sikkim
Pyrenean ibex	Capra pyrenaica pyrenaica	E	Spain
Queensland hairy-nosed wombat (incl. Barnard's)	Lasiorhinus krefftii (formerly L. barnardi and L. gillespiei)	E	Australia
Queensland rat-kangaroo	Bettongia tropica	E	Australia
Quokka	Setonix brachyurus	E	Australia
Rabbit bandicoot	Macrotis lagotis	E	Australia
Red-backed squirrel monkey	Saimiri oerstedii	E	Costa Rica, Panama
Red-bellied monkey	Cercopithecus erythrogaster	E	Western Nigeria
Red-eared nose-spotted monkey	Cercopithecus erythrotis	E	Nigeria, Cameroon, Fernando Po
Red lechwe	Kobus leche	T	Southern Africa
Rio de Oro Dama gazelle	Gazella dama lozanoi	E	Western Sahara
Rodrigues fruit bat (=flying fox)	Pteropus rodricensis	E	Indian Ocean_Rodrigues Island
Ryukyu rabbit	Pentalagus furnessi	E	Japan (Ryukyu Islands)
Ryukyu sika deer	Cervus nippon keramae	E	Japan (Ryukyu Islands)
Sand gazelle	Gazella subgutturosa marica	E	Jordan, Arabian Peninsula
Saudi Arabian gazelle	Gazella dorcas saudiya	E	Israel, Iraq, Jordan, Syria, Arabian Peninsula
Scaly-tailed possum	Wyulda squamicaudata	E	Australia

TABLE 7.4

Foreign endangered and threatened terrestrial mammal species, November 2011 [CONTINUED]

Common name	Scientific name	Listing status*	Historic range
Scimitar-horned oryx	Oryx dammah	E	North Africa
Seledang	Bos gaurus	E	Bangladesh, Southeast Asia, India
Serow	Naemorhedus (=Capricornis) sumatraensis	E	East Asia, Sumatra
Shansi sika deer	Cervus nippon grassianus	E	China (Shansi Province)
Shapo	Ovis vignei vignei	E	Kashmir
Shark Bay mouse	Pseudomys praeconis	E	Australia
Shortridge's mouse	Pseudomys shortridgei	E	Australia
Shou	Cervus elaphus wallichi	E	Tibet, Bhutan
Siamang	Symphalangus syndactylus	E	Malaysia, Indonesia
Sichuan snub-nosed monkey	Rhinopithecus roxellana	E	China
Sifakas	Propithecus spp.	E	Malagasy Republic (=Madagascar)
Simien fox	Canis simensis	E	Ethiopia
Singapore roundleaf horseshoe bat	Hipposideros ridleyi	E	Malaysia
Slender-horned gazelle	Gazella leptoceros	E	Sudan, Egypt, Algeria, Libya
Smoky mouse	Pseudomys fumeus	E	Australia
Snow leopard	Uncia (=Panthera) uncia	E	Central Asia
South American (=Brazilian) tapir	Tapirus terrestris	E	Colombia and Venezuela south to Paraguay and Argentina
South Andean huemul	Hippocamelus bisulcus	E	Chile, Argentina
South China sika deer	Cervus nippon kopschi	E	Southern China
Southern bearded saki	Chiropotes satanas satanas	E	Brazil
Southern planigale	Planigale tenuirostris	E	Australia
Spanish lynx	Felis pardina	E	Spain, Portugal
Spider monkey	Ateles geoffroyi frontatus	E	Costa Rica, Nicaragua
Spider monkey	Ateles geoffroyl panamensis	E	Costa Rica, Panama
Spotted linsang	Prionodon pardicolor	E	Nepal, Assam, Vietnam, Cambodia, Laos, Burma
Stick-nest rat	Leporillus conditor	E	Australia
Straight-horned markhor	Capra falconeri jerdoni	E	Afghanistan, Pakistan
Stump-tailed macaque	Macaca arctoides	T	India (Assam) to southern China
Sumatran rhinoceros	Dicerorhinus sumatrensis	E	Bangladesh to Vietnam to Indonesia (Borneo)
Swamp deer	Cervus duvauceli	E	India, Nepal
Swayne's hartebeest	Alcelaphus buselaphus swaynei	E	Ethiopia, Somalia
Tamaraw	Bubalus mindorensis	E	Philippines
Tana River mangabey	Cercocebus galeritus galeritus	E	Kenya
Tana River red colobus monkey	Procolobus (=Colobus) rufomitratus (=badius r.)	E	Kenya
Tasmanian forester kangaroo	Macropus giganteus tasmaniensis	E	Australia (Tasmania)
Tasmanian tiger	Thylacinus cynocephalus	E	Australia
Temnick's ground pangolin	Manis temminckii	E	Africa
Thin-spined porcupine	Chaetomys subspinosus	E	Brazil
Tibetan antelope	Pantholops hodgsonii	E	China, India, Nepal
Tiger	Panthera tigris	E	Temperate and tropical Asia
Tiger cat	Leopardus (=Felis) tigrinus	E	Costa Rica to northern Argentina
Tonkin snub-nosed monkey	Rhinopithecus avunculus	E	Vietnam
Toque macaque	Macaca sinica	T	Sri Lanka
Tora hartebeest	Alcelaphus buselaphus tora	E	Ethiopia, Sudan, Egypt
Uakari (all species)	Cacajao spp.	E	Peru, Brazil, Ecuador, Colombia, Venezuela
Urial	Ovis musimon ophion	E	Cyprus
Vancouver Island marmot	Marmota vancouverensis	E	Canada (Vancouver Island)
Vicuna	Vicugna vicugna	E	South America (Andes)
Visayan deer	Cervus alfredi	E	Philippines
Volcano rabbit	Romerolagus diazi	E	Mexico
Walia ibex	Capra walie	E	Ethiopia
Western giant eland	Taurotragus derbianus derbianus	E	Senegal to Ivory Coast
Western hare wallaby	Lagorchestes hirsutus	E	Australia
Western mouse	Pseudomys occidentalis	E	Australia
White-collared mangabey	Cercocebus torquatus	E	Senegal to Ghana; Nigeria to Gabon
White-eared (=buffy tufted-ear) marmoset	Callithrix aurita (=jacchus a.)	E	Brazil
White-footed tamarin	Saguinus leucopus	T	Colombia
White-nosed saki	Chiropotes albinasus	E	Brazil
Wild yak	Bos mutus (=grunniens m.)	E	China (Tibet), India
Woolly spider monkey	Brachyteles arachnoides	E	Brazil
Yarkand deer	Cervus elaphus yarkandensis	E	China (Sinkiang)
Yellow-footed rock wallaby	Petrogale xanthopus	E	Australia
Yellow-tailed woolly monkey	Lagothrix flavicauda	E	Andes of northern Peru
Yunnan snub-nosed monkey	Rhinopithecus bieti	E	China
Zanzibar red colobus monkey	Procolobus (=Colobus) pennantii (=kirki) kirki	E	Tanzania
Zanzibar suni	Neotragus moschatus moschatus	E	Zanzibar (and nearby islands)

*E = endangered; T = threatened.

SOURCE: Adapted from "Generate Species List," in *Species Reports*, U.S. Department of the Interior, U.S. Fish & Wildlife Service, November 2011, http://ecos.fws.gov/tess_public/pub/adHocSpeciesForm.jsp (accessed November 8, 2011)

FIGURE 7.9

The Siberian tiger is one of the most endangered species in the world. It now occupies forest habitats in the Amur-Ussuri region of Siberia. *(© H. Damke/Shutterstock.com.)*

FIGURE 7.10

Elephants are highly intelligent and social animals. Once on the verge of extinction, elephants have recovered somewhat after a worldwide ban on the ivory trade. *(© Four Oaks/Shutterstock.com.)*

Pandas

Few creatures have engendered more human affection than the giant panda, with its roly-poly character, small ears, and black eye patches on a snow-white face. Giant pandas are highly endangered. According to the IUCN (http://www.iucnredlist.org/apps/redlist/details/712/0), in 2011 there were less than 2,500 adult giant pandas in the wild. Pandas are endemic to portions of southwestern China, where they inhabit a few fragmentary areas of high-altitude bamboo forest. Unlike other bear species, to which they are closely related, pandas have a vegetarian diet that consists entirely of bamboo.

Pandas have become star attractions at many zoos, where they draw scores of visitors. Zoos typically pay China millions of dollars for the loan of adult pandas. These funds are used to support panda conservation efforts in China, including the purchase of land for refuges as well as the development of habitat corridors to link protected areas.

Big Game

Most big game species are members of the Artiodactyla order. This order contains a variety of ungulates (hoofed animals), including antelopes, bison, buffalo, camels, deer, gazelles, goats, hartebeests, hippos, impalas, and sheep. Many of the wild species have been overhunted for their meat, bones, or horns. Horns are used in traditional Chinese medicine and are popular trophies for big game hunters. Big game species also face threats from domesticated livestock due to competition for habitat and food resources.

Primates

The endangerment of primate species is mostly due to loss of habitat and overhunting. Primates are dependent on large expanses of tropical forests, a habitat that is under siege worldwide. Countries with large numbers of primate species include Brazil, the Democratic Republic of Congo, Indonesia, and Madagascar. Many of the most endangered primate species are found on Madagascar, which has a diverse and unique primate fauna. Most of Madagascar's primate species are endemic.

Habitat loss, especially the fragmentation and conversion of tropical forests for road building and agriculture, contributes to the decline of nearly all imperiled primates. For example, in Indonesia and Borneo, which are home to most of the world's orangutans, deforestation has dramatically shrunk orangutan habitat. (See Figure 7.11.) Logging and extensive burning have caused many orangutans to flee the forests for villages, where they have been killed or captured by humans.

Some threatened primates face pressures from excessive hunting and poaching. They are also used in medical research because of their close biological relationship with humans. As of 2012, almost all countries had either banned or strictly regulated the trade of primates, but these laws were often hard to enforce.

Rhinoceros

Rhinoceros are among the largest land mammals. They weigh up to 4 tons (3.6 t) and are herbivorous (plant-eating) grazers. The name *rhinoceros* consists of two Greek words meaning "nose" and "horn," and rhinos are in fact the only animals on the earth that have a horn on their nose. Rhinoceros are found in the wild only

FIGURE 7.11

The orangutan is highly endangered, along with the majority of the world's primate species. *(© Eric Gevaert/Shutterstock.com.)*

in Africa and Asia. Figure 7.12 shows an African white rhinoceros with two horns. The female may be identified by her longer, more slender primary horn.

Rhinoceros have roamed the earth for more than 40 million years, but in less than a century humans—their only predators—have reduced populations to dangerously low levels. Hunting has been the primary cause of rhinoceros decline. Rhinoceros horn is highly prized as an aphrodisiac (even though its potency has never been shown), as well as an ingredient in traditional Chinese medicine. Rhinoceros were first listed by CITES in 1976. This banned international trade in the species and its products. In 1992 CITES also started requiring the destruction of horn caches that were confiscated from poachers. Nonetheless, people continue to buy and consume rhinoceros horn, and many poachers are willing to risk death to acquire it.

FIGURE 7.12

The white rhinoceros is native to Africa and can weigh up to 8,000 pounds. *(© U.S. Fish and Wildlife Service.)*

CHAPTER 8
BIRDS

Birds belong to the class Aves, which contains dozens of orders. Birds are warm-blooded vertebrates with wings, feathers, and light hollow bones. The vast majority of birds are capable of flight. In *Red List of Threatened Species Version 2011.2* (2011, http://www.iucnredlist.org/documents/summarystatistics/2011_2_RL_Stats_Table1.pdf), the International Union for Conservation of Nature (IUCN) states that 10,052 species of birds have been identified around the world. An estimated 600 to 900 of these species spend all or part of their life in the United States.

Besides taxonomy, birds are broadly classified by their physical characteristics (such as feet or beak structure), eating habits, primary habitats, or migratory habits. For example, raptors (birds of prey) have curved beaks and talons that are well suited for catching prey. This category includes buzzards, eagles, hawks, owls, and vultures. Perching birds have a unique foot structure with three toes in front and one large flexible toe to the rear. Ducks and geese are known as open-water or swimming birds and have webbed feet. Habitat categories include seabirds, shore birds, and arboreal (tree-dwelling) birds. Some birds migrate over long distances and others, such as turkeys and quail, do not migrate at all.

ENDANGERED AND THREATENED U.S. SPECIES

As of November 2011, there were 90 bird species listed under the Endangered Species Act (ESA) as endangered or threatened in the United States. (See Table 8.1.) The vast majority had an endangered listing, meaning that they are at risk of extinction. Nearly all had a recovery plan in place.

The imperiled birds come from many different genera (plural of genus) and represent a variety of habitats. Most are perching birds, seabirds, or shore birds. There are also a handful of other bird types, including woodpeckers and raptors, such as the northern spotted owl and the Puerto Rican broad-winged hawk.

Table 8.2 shows the 10 bird species with the highest expenditures under the ESA during fiscal year (FY) 2010. The three most expensive species were the red-cockaded woodpecker ($25.2 million), northern spotted owl ($14.1 million), and piping plover ($11.2 million).

The following sections describe the categories of birds that are found on the list of endangered and threatened species.

Woodpeckers

Woodpeckers belong to the order Piciformes and the family Picidae. They are characterized by their physiology. They have hard chisel-like beaks and a unique foot structure with two toes pointing forward and two toes pointing backward. This allows them to take a firm grip on tree trunks and extend horizontally from vertical surfaces. Woodpeckers prefer arboreal habitats, primarily dead trees in old-growth forests. The birds hammer away at the bark on the trees to dig out insects living there. They often form deep cavities in the tree to use as roosting and nesting holes.

RED-COCKADED WOODPECKERS. The red-cockaded woodpecker is named for the red patches, or cockades, of feathers on the head of the male. (See Figure 8.1.) This species is found in old pine forests in the southeastern United States, where family groups—consisting of a breeding male and female as well as several helpers—nest within self-dug cavities in pine trees. Tree cavities serve as nesting sites and provide protection from predators. Because red-cockaded woodpeckers rarely nest in trees less than 80 years old, heavy logging has destroyed much of their former habitat.

The red-cockaded woodpecker was first listed under the ESA in 1970. It is currently found in fragmented

TABLE 8.1

Endangered and threatened bird species, November 2011

Common name	Scientific name	Listing status[a]	U.S. or U.S./foreign listed	Recovery plan date	Recovery plan stage[b]
Akiapola`au (honeycreeper)	Hemignathus munroi	E	US	9/22/2006	RF(1)
Akikiki	Oreomystis bairdi	E	US	9/22/2006	RF(1)
Attwater's greater prairie-chicken	Tympanuchus cupido attwateri	E	US	4/26/2010	RF(2)
Audubon's crested caracara	Polyborus plancus audubonii	T	US	5/18/1999	F
Bachman's warbler (=wood)	Vermivora bachmanii	E	US/foreign	None	—
Black-capped vireo	Vireo atricapilla	E	US/foreign	9/30/1991	F
Bridled white-eye	Zosterops conspicillatus conspicillatus	E	US	9/28/1990	F
California clapper rail	Rallus longirostris obsoletus	E	US	2/10/2010	D
California condor	Gymnogyps californianus	E (1), EXPN (1)	US	4/25/1996	RF(3)
California least tern	Sterna antillarum browni	E	US/foreign	9/27/1985	RF(1)
Cape Sable seaside sparrow	Ammodramus maritimus mirabilis	E	US	5/18/1999	F
Coastal California gnatcatcher	Polioptila californica californica	T	US/foreign	None	—
Crested honeycreeper	Palmeria dolei	E	US	9/22/2006	RF(1)
Eskimo curlew	Numenius borealis	E	US/foreign	None	—
Everglade snail kite	Rostrhamus sociabilis plumbeus	E	US	5/18/1999	F
Florida grasshopper sparrow	Ammodramus savannarum floridanus	E	US	5/18/1999	F
Florida scrub-jay	Aphelocoma coerulescens	T	US	5/9/1990	F
Golden-cheeked warbler (=wood)	Dendroica chrysoparia	E	US/foreign	9/30/1992	F
Guam Micronesian kingfisher	Halcyon cinnamomina cinnamomina	E	US	11/14/2008	RF(1)
Guam rail	Rallus owstoni	E (1), EXPN (1)	US	9/28/1990	F
Hawaii akepa (honeycreeper)	Loxops coccineus coccineus	E	US	9/22/2006	RF(1)
Hawaiian (=`alala) crow	Corvus hawaiiensis	E	US	4/17/2009	RF(1)
Hawaiian common moorhen	Gallinula chloropus sandvicensis	E	US	8/24/2005	RD(2)
Hawaiian coot	Fulica americana alai	E	US	8/24/2005	RD(2)
Hawaiian dark-rumped petrel	Pterodroma phaeopygia sandwichensis	E	US	4/25/1983	F
Hawaiian goose	Branta (=Nesochen) sandvicensis	E	US	9/24/2004	RD(1)
Hawaiian (=koloa) duck	Anas wyvilliana	E	US	8/24/2005	RD(2)
Hawaiian (=`lo) hawk	Buteo solitarius	E	US	5/9/1984	F
Hawaiian stilt	Himantopus mexicanus knudseni	E	US	8/24/2005	RD(2)
Hawaii creeper	Oreomystis mana	E	US	9/22/2006	RF(1)
Heinroth's shearwater	Puffinus heinrothi	T	US	None	—
Inyo California towhee	Pipilo crissalis eremophilus	T	US	4/10/1998	F
Ivory-billed woodpecker	Campephilus principalis	E	US/foreign	7/19/2010	F
Kauai akialoa (honeycreeper)	Hemignathus procerus	E	US	9/22/2006	RF(1)
Kauai `o`o (honeyeater)	Moho braccatus	E	US	9/22/2006	RF(1)
Kirtland's warbler	Dendroica kirtlandii	E	US/foreign	9/30/1985	RF(1)
Large Kauai (=kamao) thrush	Myadestes myadestinus	E	US	9/22/2006	RF(1)
Laysan duck	Anas laysanensis	E	US	9/22/2009	RF(1)
Laysan finch (honeycreeper)	Telespyza cantans	E	US	10/4/1984	F
Least Bell's vireo	Vireo bellii pusillus	E	US/foreign	5/6/1998	D
Least tern	Sterna antillarum	E	US	9/19/1990	F
Light-footed clapper rail	Rallus longirostris levipes	E	US	6/24/1985	RF(1)
Marbled murrelet	Brachyramphus marmoratus	T	US	9/24/1997	F
Mariana (=aga) crow	Corvus kubaryi	E	US	1/11/2006	RD(1)
Mariana common moorhen	Gallinula chloropus guami	E	US	9/30/1991	F
Mariana gray swiftlet	Aerodramus vanikorensis bartschi	E	US	9/30/1991	F
Masked bobwhite (quail)	Colinus virginianus ridgwayi	E	US/foreign	4/21/1995	RF(2)
Maui akepa (honeycreeper)	Loxops coccineus ochraceus	E	US	9/22/2006	RF(1)
Maui parrotbill (honeycreeper)	Pseudonestor xanthophrys	E	US	9/22/2006	RF(1)
Mexican spotted owl	Strix occidentalis lucida	T	US/foreign	6/28/2011	RD(1)
Micronesian megapode	Megapodius laperouse	E	US/foreign	4/10/1998	F
Mississippi sandhill crane	Grus canadensis pulla	E	US	9/6/1991	RF(3)
Molokai creeper	Paroreomyza flammea	E	US	9/22/2006	RF(1)
Molokai thrush	Myadestes lanaiensis rutha	E	US	9/22/2006	RF(1)
Newell's Townsend's shearwater	Puffinus auricularis newelli	T	US	4/25/1983	F
Nightingale reed warbler (old world warbler)	Acrocephalus luscinia	E	US	4/10/1998	F
Nihoa finch (honeycreeper)	Telespyza ultima	E	US	10/4/1984	F
Nihoa millerbird (old world warbler)	Acrocephalus familiaris kingi	E	US	10/4/1984	F
Northern aplomado falcon	Falco femoralis septentrionalis	E	US/foreign	6/8/1990	F
Northern aplomado falcon	Falco femoralis septentrionalis	EXPN	US/foreign	6/8/1990	F
Northern spotted owl	Strix occidentalis caurina	T	US/foreign	7/1/2011	RF(1)
Nukupu`u (honeycreeper)	Hemignathus lucidus	E	US	9/22/2006	RF(1)
Oahu creeper	Paroreomyza maculata	E	US	9/22/2006	RF(1)
Oahu elepaio	Chasiempis sandwichensis ibidis	E	US	9/22/2006	RF(1)
`O`u (honeycreeper)	Psittirostra psittacea	E	US	9/22/2006	RF(1)
Palila (honeycreeper)	Loxioides bailleui	E	US	9/22/2006	RF(1)
Piping plover	Charadrius melodus	E	US/foreign	5/2/1996	RF(1)
Piping plover	Charadrius melodus	T	US/foreign	5/2/1996	RF(1)
Po`ouli (honeycreeper)	Melamprosops phaeosoma	E	US	9/22/2006	RF(1)

TABLE 8.1

Endangered and threatened bird species, November 2011 [CONTINUED]

Common name	Scientific name	Listing status[a]	U.S. or U.S./foreign listed	Recovery plan date	Recovery plan stage[b]
Puerto Rican broad-winged hawk	Buteo platypterus brunnescens	E	US	9/8/1997	F
Puerto Rican nightjar	Caprimulgus noctitherus	E	US	4/19/1984	F
Puerto Rican parrot	Amazona vittata	E	US	6/17/2009	RF(1)
Puerto Rican plain pigeon	Columba inornata wetmorei	E	US	10/14/1982	F
Puerto Rican sharp-shinned hawk	Accipiter striatus venator	E	US	9/8/1997	F
Red-cockaded woodpecker	Picoides borealis	E	US	3/20/2003	RF(2)
Roseate tern	Sterna dougallii dougallii	E (1), T (1)	US/foreign	11/5/1998	RF(1)
Rota bridled white-eye	Zosterops rotensis	E	US	10/19/2007	F
San Clemente loggerhead shrike	Lanius ludovicianus mearnsi	E	US	1/26/1984	F
San Clemente sage sparrow	Amphispiza belli clementeae	T	US	1/26/1984	F
Short-tailed albatross	Phoebastria (=Diomedea) albatrus	E	US/foreign	5/20/2009	F
Small Kauai (=puaiohi) thrush	Myadestes palmeri	E	US	9/22/2006	RF(1)
Southwestern willow flycatcher	Empidonax traillii extimus	E	US/foreign	8/30/2002	F
Spectacled eider	Somateria fischeri	T	US/foreign	8/12/1996	F
Steller's eider	Polysticta stelleri	T	US	9/30/2002	F
Western snowy plover	Charadrius alexandrinus nivosus	T	US/foreign	9/24/2007	F
White-necked crow	Corvus leucognaphalus	E	US	None	—
Whooping crane	Grus americana	E (1), EXPN (3)	US/foreign	5/29/2007	RF(3)
Wood stork	Mycteria americana	E	US	1/27/1997	RF(1)
Yellow-shouldered blackbird	Agelaius xanthomus	E	US	11/12/1996	RF(1)
Yuma clapper rail	Rallus longirostris yumanensis	E	US	2/10/2010	RD(1)

[a]E = endangered; T = threatened; EXPN = experimental population, non-essential; numbers in parentheses indicate separate populations.
[b]F = final; D = draft; RD = draft revision; RF = final revision.

SOURCE: Adapted from "Generate Species List," in *Species Reports*, U.S. Department of the Interior, U.S. Fish & Wildlife Service, November 2011, http://ecos.fws.gov/tess_public/pub/adHocSpeciesForm.jsp (accessed November 8, 2011), and "Listed FWS/Joint FWS and NMFS Species and Populations with Recovery Plans (Sorted by Listed Entity)," in *Recovery Plans Search*, U.S. Department of the Interior, U.S. Fish & Wildlife Service, November 2011, http://ecos.fws.gov/tess_public/pub/speciesRecovery.jsp?sort=1 (accessed November 8, 2011)

TABLE 8.2

Bird species with the highest expenditures under the Endangered Species Act, fiscal year 2010

Ranking	Species	Expenditure
1	Woodpecker, red-cockaded (Picoides borealis)	$25,194,067
2	Owl, northern spotted (Strix occidentalis caurina)	$14,086,571
3	Plover, piping (Charadrius melodus)—except Great Lakes watershed	$11,235,785
4	Flycatcher, southwestern willow (Empidonax traillii extimus)	$10,579,072
5	Tern, least (Sterna antillarum)—interior population	$9,614,921
6	Tern, California least (Sterna antillarum browni)	$7,463,329
7	Vireo, least Bell's (Vireo bellii pusillus)	$6,925,553
8	Murrelet, marbled (Brachyramphus marmoratus)—CA, OR, WA	$6,173,301
9	Plover, western snowy (Charadrius alexandrinus nivosus)—Pacific coastal population	$4,697,911
10	Scrub-jay, Florida (Aphelocoma coerulescens)	$3,997,085

SOURCE: Adapted from "Table 2. Species Ranked in Descending Order of Total FY 2010 Reported Expenditures, Not Including Land Acquisition Costs," in *Federal and State Endangered and Threatened Species Expenditures: Fiscal Year 2010*, U.S. Department of the Interior, U.S. Fish and Wildlife Service, 2010, http://www.fws.gov/endangered/esa-library/pdf/2010.EXP.FINAL.pdf (accessed October 25, 2011)

populations in the southeastern seaboard westward into Texas. In January 2003 the USFWS published *Recovery Plan for the Red-Cockaded Woodpecker (*Picoides borealis*): Second Revision* (http://www.fws.gov/rcwrecovery/finalrecoveryplan.pdf) and estimated that approximately 14,000 of the birds still existed in the wild.

In *Red-Cockaded Woodpecker (*Picoides borealis*): Five-Year Review—Summary and Evaluation* (October 2006, http://ecos.fws.gov/docs/five_year_review/doc787.pdf), the USFWS concludes that the endangered status listing is still appropriate for the species. The agency reports that over 6,000 active clusters (occupied territories) of the bird have been documented, up from around 4,700 active clusters reported during the early 1990s. However, the USFWS believes the bird still faces significant threats from the loss, degradation, and fragmentation of nesting and foraging habitat.

The USFWS indicates in "Red-Cockaded Woodpecker (*Picoides borealis*)" (January 2012, http://ecos.fws.gov/speciesProfile/profile/speciesProfile.action?spcode=B04F) that critical habitat has not been designated for the red-cockaded woodpecker. However, eight states (Alabama, Florida, Georgia, Louisiana, North Carolina, South Carolina, Texas, and Virginia) have established statewide Safe Harbor Agreements with the USFWS. These are voluntary agreements in which nonfederal landowners agree to conserve and manage listed species on their property. In exchange, the federal government provides assurances that it will not impose certain restrictions on the conservation activities. For example, if a landowner improves or expands red-cockaded woodpecker habitat and more of the birds move into the habitat, the landowner will not face increased regulatory requirements.

FIGURE 8.1

The red cockaded woodpecker

SOURCE: Robert Savannah, artist, "Red Cockaded Woodpeckers," in *Line Art (Drawings)*, U.S. Department of the Interior, U.S. Fish and Wildlife Service, undated, http://www.fws.gov/pictures/lineart/bobsavannah/redcockadedwoodpeckers.html (accessed November 11, 2011)

Passerines

Just over half of all bird species belong to the order Passeriformes and are called passerines. They are informally known as perching birds or songbirds, although not all passerines are truly songbirds. According to the Integrated Taxonomic Information System (2012, http://www.itis.gov/index.html), a taxonomic tracking system that is operated by various North American government agencies, there are 86 families in this order, and they include many well-known species, such as blackbirds, cardinals, crows, finches, larks, mockingbirds, sparrows, starlings, swallows, and wrens. Over one-third of the U.S. species of endangered and threatened birds listed in Table 8.1 are passerine (perching) birds. The following sections describe some species of note.

SOUTHWESTERN WILLOW FLYCATCHERS. As shown in Table 8.2, in FY 2010 nearly $10.6 million was spent under the ESA to conserve the southwestern willow flycatcher. This small bird has a grayish-green back and wings with a pale yellow belly and a white-colored throat. It was first listed as endangered in 1995. In August 2002 the USFWS published *Final Recovery Plan Southwestern Willow Flycatcher (*Empidonax traillii extimus*)* (http://ecos.fws.gov/docs/recovery_plans/2002/020830c.pdf). At that time the agency reported that approximately 900 to 1,100 pairs of the bird were believed to exist. The species migrates from the southwestern United States to Mexico and Central and South America for the winter. The bird feeds on insects and prefers riparian areas (dense vegetation near rivers or streams) for its habitat. It is endangered primarily due to the loss of riparian vegetation. In ranching areas this vegetation is often stripped by grazing livestock. Another factor in its decline is harm from brood parasites—bird species that lay their eggs in the nests of other species. Brown-headed cowbirds are brood parasites that threaten southwestern willow flycatchers. They lay their eggs in the flycatchers' nests, and the unsuspecting flycatchers raise the cowbirds' young as their own.

Since 1997 the USFWS has repeatedly designated and revised critical habitat for the southwestern willow flycatcher in response to litigation. In August 2011 the agency (http://www.gpo.gov/fdsys/pkg/FR-2011-08-15/pdf/2011-19713.pdf) proposed 2,090 stream miles (3,360 stream km) in Arizona, California, Colorado, Nevada, New Mexico, and Utah. As of January 2012, the proposal had not been finalized.

In 2008 the USFWS launched a five-year status review of more than two dozen southwestern species, including the southwestern willow flycatcher. As of January 2012, the results of that review had not been published.

HAWAIIAN HONEYCREEPERS. Hawaiian honeycreepers are a group of songbirds endemic (limited) to Hawaii. The honeycreepers are named for the characteristic creeping behavior some species exhibit as they search for nectar. Hawaiian honeycreepers are extremely diverse in their diet—different species are insect, nectar, or seed eaters. Species also differ in the shapes of their beaks and in plumage coloration. The birds are found in forest habitats at high elevations. According to the article "Native and Endangered Species of the Hawaiian Islands" (2012, http://www.hawaiianencyclopedia.com/native-and-endangered-species.asp), there were originally at least 50 Hawaiian honeycreeper species and subspecies, but 21 of them were already extinct by the time Europeans arrived in 1778.

As of November 2011, 12 species of Hawaiian honeycreepers were listed under the ESA as endangered:

- Akiapola'au
- Crested honeycreeper
- Hawaii akepa
- Kauai akialoa
- Laysan finch
- Maui akepa
- Maui parrotbill

- Nihoa finch
- Nukupu‘u
- ‘O‘u
- Palila
- Po‘ouli

Some honeycreeper species are among the most endangered animals on the earth, with only a few individuals left. One of the primary factors involved in honeycreeper endangerment is loss of habitat. In addition, the introduction of predators that hunt birds or eat their eggs, such as cats, mongooses, and rats, have contributed to the decline of many species. The introduction of bird diseases, particularly those spread by introduced mosquitoes, has decimated honeycreeper populations. Finally, competition with introduced bird species for food and habitat has also been a significant cause of decline.

In *Revised Recovery Plan for Hawaiian Forest Birds* (September 22, 2006, http://ecos.fws.gov/docs/recovery_plan/060922a.pdf), the USFWS covers 19 endangered Hawaiian forest birds. The agency reports that 10 of these species have not been observed in at least a decade and may well be extinct. Most of these species are native to rain forests at elevations above 4,000 feet (1,200 m) on the islands of Hawaii (Big Island), Maui, and Kauai. Major threats to endangered forest species include habitat loss and modification, other human activity, disease, and predation. Of particular importance are nonnative plants, which have converted native plant communities to alien ecosystems that are unsuitable as habitat.

Between 2008 and 2011 the USFWS (http://ecos.fws.gov/ecos/indexPublic.do) published individual five-year reviews for all 12 Hawaiian honeycreeper species. The agency decided to maintain the endangered listings for all of them.

MIGRATORY SONGBIRDS. In the fact sheet "Neotropical Migratory Bird Basics" (2012, http://nationalzoo.si.edu/scbi/migratorybirds/fact_sheets/default.cfm?fxsht=9), Mary Deinlein of the Smithsonian Migratory Bird Center states that there are nearly 200 species of songbirds known as neotropical migrators. Every year these birds migrate between the United States and tropical areas in Mexico, the Caribbean, and Central and South America. Even though some songbirds are appreciated by humans for their beautiful songs and colorful plumage, migratory songbirds also play a vital role in many ecosystems. For example, during spring migration in the Ozarks dozens of migratory bird species arrive and feed on the insects that inhabit oak trees, thereby helping control insect populations.

Migratory species are particularly vulnerable because they are dependent on suitable habitat in both their winter and spring ranges. In North America, land development has eliminated many forest habitats. Migratory songbird habitats are also jeopardized in Central and South America, where farmers and ranchers have been burning and clearing tropical forests to plant crops and graze livestock. Some countries, including Belize, Costa Rica, Guatemala, and Mexico, have set up preserves for songbirds, but improved forest management is needed to save them.

Raptors

The term *raptor* is derived from the Latin word *raptores*, which was once the order on the taxonomy table to which birds of prey were assigned. Eventually, scientists split the birds into three orders:

- Accipitriformes—includes buzzards, eagles, and hawks
- Falconiformes—falcons
- Strigiformes—owls

As of November 2011, there were less than a dozen raptors listed as endangered or threatened in the United States. (See Table 8.1.) Species of note include the northern spotted owl and the California condor.

NORTHERN SPOTTED OWLS. The northern spotted owl occupies old-growth forests in the Pacific Northwest, where it nests in the cavities of trees 200 years old or older. (See Figure 8.2.) According to the Sierra Club, in "Species at Risk: Northern Spotted Owl" (2012, http://www.sierraclub.org/lewisandclark/species/owl.asp), owl pairs may forage across areas as large as 2,200 acres (890 ha).

Northern spotted owl populations have declined primarily due to habitat loss. Most of the private lands in its range have been heavily logged, leaving only public lands, such as national forests and national parks, for habitat. Because logging has also been permitted in many old-growth national forest areas, the Sierra Club indicates that the species has lost approximately 90% of its original habitat. In 1990 the northern spotted owl was listed under the ESA. Court battles began over continued logging in national forest habitats.

In 1992 the USFWS set aside 7 million acres (2.8 million ha) as critical habitat for the species and published a recovery plan. Two years later the Northwest Forest Plan (http://www.fs.fed.us/r5/nwfp/) was established. It reduced logging in 13 national forests by about 85% to protect northern spotted owl habitats. However, populations of the northern spotted owl continued to decline—this despite the unanticipated discovery of 50 pairs of nesting adults in California's Marin County, just north of the Golden Gate Bridge.

In November 2004 the USFWS published *Northern Spotted Owl: Five-Year Review—Summary and Evaluation*

FIGURE 8.2

The northern spotted owl, which inhabits old-growth forests in the Pacific Northwest, was the subject of a lengthy battle pitting environmentalists against logging interests. (© *U.S. Fish and Wildlife Service.*)

(http://ecos.fws.gov/docs/five_year_review/doc743.pdf). The review was conducted as part of a settlement agreement for litigation that was filed by the American Forest Resource Council, the Rough & Ready Lumber Company, Swanson Group Inc., and the Western Council of Industrial Workers. The agency concluded that the bird should continue to have a threatened listing under the ESA. The USFWS found that habitat loss on federal lands had been minimized after the species was listed. This success was attributed to the Northwest Forest Plan. However, the agency noted that the population of northern spotted owls in California, Oregon, and Washington continued to decline and that the species faced emerging threats from forest fires, West Nile virus, Sudden Oak Death (a plant disease that has killed hundreds of thousands of trees in California and Oregon), and competition for habitat from barred owls.

The settlement agreement also required the USFWS to review critical habitat based on economic impacts. In August 2008 the USFWS (http://frwebgate.access.gpo.gov/cgi-bin/getdoc.cgi?dbname=2008_register&docid=fr13au08-17) designated revised critical habitat that covered approximately 5.3 million acres (2.1 million ha) of federal lands in California, Oregon, and Washington. This represented nearly 1.6 million less acres (647,000 ha) than were originally designated in 1992. As of January 2012, the 2008 critical habitat designation remained in effect.

In November 2010 the USFWS (http://www.gpo.gov/fdsys/pkg/FR-2010-11-24/pdf/2010-29584.pdf#page=1) published notice that it was initiating a five-year review for 58 western species, including the northern spotted owl. As of January 2012, the results of the review had not been published.

In June 2011 the USFWS issued *Revised Recovery Plan for the Northern Spotted Owl (*Strix occidentalis caurina*)* (http://ecos.fws.gov/docs/recovery_plan/RevisedNSORecPlan2011_1.pdf). Figure 8.3 shows the bird's current range divided into 12 physiographic provinces based on physical and environmental features. The agency indicates that competition from the barred owl is the most important threat facing the spotted owl. Other threats include ongoing habitat losses due to logging, wildfires, and other disturbances and lost habitat due to past actions. The USFWS estimates that at least $127.1 million will be required to fund recovery efforts over the next 30 years.

CALIFORNIA CONDORS. The California condor has a wingspan of more than 9 feet (2.7 m) and is among the continent's most impressive birds. Ten thousand years ago this species soared over most of North America. However, its range contracted at the end of the Ice Age, and eventually it was found only along the Pacific coast. Like other vulture species, the California condor is a carrion eater and feeds on the carcasses of deer, sheep, and smaller species such as rodents. Random shooting, egg collection, poisoning (particularly from lead in bullets used by hunters to kill game), and loss of habitat devastated the condor population. The species was listed as endangered in 1967. (See Table 2.1 in Chapter 2.) In the press release "California Condor Chick Takes Flight in Southern California" (November 15, 2006, http://www.fws.gov/news/newsreleases/showNews.cfm?newsId=EEBC36AD-E3FC-17D3-F1EEEF6BCD8C7611), the USFWS states that by late 1984 only 15 condors remained in the wild. After seven of these birds died, the agency decided to capture the remaining population.

An intense captive breeding program for the California condor was initiated in 1987. (See Figure 8.4.) The breeding program was successful enough that California condors were released into the wild beginning in 1992. The introduced birds in parts of Arizona, Nevada, and Utah were designated a nonessential experimental

FIGURE 8.3

Range of the spotted owl

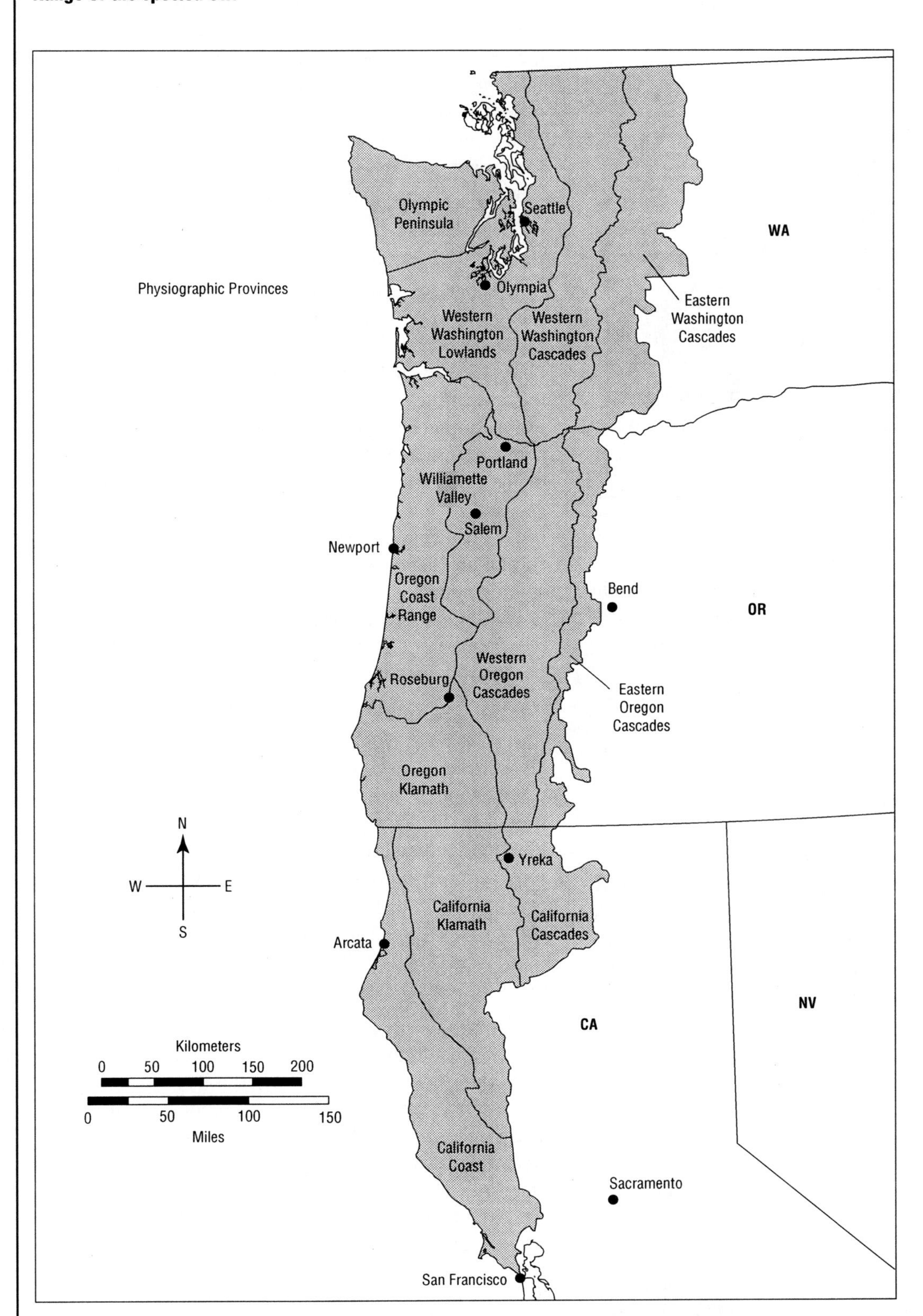

SOURCE: "Figure A-1. Physiographic Provinces within the Range of the Spotted Owl in the United States," in *Revised Recovery Plan for the Northern Spotted Owl (Strix occidentalis caurina)*, U.S. Department of the Interior, U.S. Fish and Wildlife Service, June 28, 2011, http://www.fws.gov/arcata/es/birds/NSO/documents/USFWS2011RevisedRecoveryPlanNorthernSpottedOwl.pdf (accessed November 11, 2011)

FIGURE 8.4

Condors bred in captivity

Zookeepers use hand puppets that look like adult condors to feed captively bred condor chicks.

SOURCE: "Captively-Bred Condors," in *California Condor: Gymnogyps Californianus*, U.S. Fish and Wildlife Service, August 1998, http://training.fws.gov/library/Pubs/condor.pdf (accessed November 11, 2011)

population. (Nonessential experimental populations are not believed essential to the survival of a species as whole. Thus, they receive less rigorous protections under the ESA.) In December 2008 the USFWS announced in the press release "California Condor Recovery Reaches Milestone: With Seven Wild Chicks Fledged, Wild Condor Population Now Exceeds Captive Population" (http://www.fws.gov/news/newsreleases/showNews.cfm?newsId=47108004-F5F1-6A82-E0E70AF5FF07D0E2) that a significant milestone was reached in the condor recovery program: there were more condors in the wild population than in the captive population. At that time there were 167 wild condors and 160 captive condors. According to Michael Martinez, in "Once Nearly Extinct, the California Condor Nears New Milestones" (CNN.com, April 26, 2011), the total condor population was expected to reach 400 during the spring of 2011, with approximately 200 birds living in the wild.

Water Birds

Water birds live in and around bodies of water. Some prefer marine (ocean) habitats and others are found only near freshwater. Many species inhabit swamps, marshes, and wetlands. These areas may be inland or intertidal (along the sea coast).

As of November 2011, there were more than two dozen water birds listed as endangered or threatened in the United States. (See Table 8.1.) They include a variety of species from many different taxonomic orders.

MIGRATORY SHORE BIRDS. Migratory shore birds are found most often in marshes, mudflats, estuaries, and other wetland areas where the sea meets freshwater. This category includes avocets, oystercatchers, plovers, sandpipers, shearwaters, snipes, and stilts. These birds vary greatly in size and color, but nearly all migrate over long distances. Most of them breed near the North Pole during the spring and spend their winters anywhere from the southern United States to South America. During their migrations the birds stop to rest and feed at specific locations, known as staging areas, in the United States. Major staging areas include Cheyenne Bottoms in Kansas, the Copper River delta in Alaska, Delaware Bay, the Great Salt Lake in Utah, and San Francisco Bay.

The piping plover is an imperiled migratory shorebird first listed under the ESA in 1985. The USFWS recognizes three distinct populations of the species. (See Figure 8.5.) As of November 2011, the Great Lakes population was listed as endangered. Piping plovers in the rest of the United States, including the northern Great Plains population and the Atlantic coast population, were listed as threatened. The birds breed and raise their young in the northern areas during the spring and summer and migrate to the south Atlantic coast, the Gulf coast, and the Caribbean and Mexican coasts for the winter. As shown in Table 8.2, approximately $11.2 million was spent under the ESA during FY 2010 on the northern Great Plains and Atlantic coast populations of this species. Recovery plans have been published for all three populations.

In September 2009 the USFWS published *Piping Plover (*Charadrius melodus*): 5-Year Review—Summary and Evaluation* (http://ecos.fws.gov/docs/five_year_review/doc3009.pdf) and concluded that no changes were warranted to its listings under the ESA. The agency noted that the endangered Great Lakes population contained only 63 breeding pairs, which was well below the recovery goal of 150 breeding pairs. The USFWS also indicated that all three populations face continuing threats to their survival from development of coastal lands in their breeding and wintering habitats.

Another imperiled migratory shorebird is the least tern. (See Figure 8.6.) The least tern is the smallest member of the gull and tern family. According to the Texas Parks and Wildlife Department, in "Interior Least Tern (*Sterna antillarum athalassos*)" (2012, http://www.tpwd.state.tx.us/huntwild/wild/species/leasttern/), there are three North American populations of the least tern. An Atlantic coast population is not listed under the ESA. As of November 2011, the California least tern and what is known as the interior population were listed as endangered. The latter is distributed throughout the nation's midsection from Montana to Texas and as far east as Tennessee. As shown in Table 8.2, in FY 2010 over $9.6 million was spent under the ESA on the interior population of the least tern. In April 2008 the USFWS launched a five-year status review of seven midwestern

FIGURE 8.5

Range of the piping plover

SOURCE: Adapted from "Piping Plover Range," in *All About Piping Plovers*, U.S. Department of the Interior, U.S. Fish and Wildlife Service, undated, http://www.fws.gov/plover/piplchmaps/NA_range.gif (accessed November 11, 2011)

species, including the interior population of the least tern. As of January 2012, the results of the review had not been published.

SEABIRDS. Seabirds spend most of their time out at sea, but nest on land. They are also known as pelagic birds, because pelagic means oceanic (associated with the open seas). Seabird species include albatrosses, auks and auklets, cormorants, gulls, kittiwakes, murres and murrelets, petrels, penguins, and puffins.

The marbled murrelet is one of a handful of seabirds listed under the ESA. The bird was first listed under the ESA in 1992, and as of November 2011 it was designated as threatened in California, Oregon, and Washington. The marbled murrelet is about 9 inches (23 cm) long and has a distinctive two-tone pattern of light and dark markings. (See Figure 8.7.) The species prefers to nest in the trees of old-growth forests along the northwest Pacific coastline. Logging and other types of habitat degradation have resulted in population declines.

In May 2008 the USFWS received a petition from the American Forest Resource Council; the Carpenters Industrial Council of Douglas County, Oregon; and Ron Stuntzner (the founder of Stuntzner Engineering and

FIGURE 8.6

Least tern

SOURCE: Laurel Ovitt, artist, "Untitled," in *Threatened and Endangered Species: Least Tern Sterna antillarum Fact Sheet*, U.S. Department of Agriculture Natural Resources Conservation Service, November 2005, http://efotg.nrcs.usda.gov/references/public/MT/LeastTern.pdf (accessed November 11, 2011)

FIGURE 8.7

Marbled murrelet. (© *Universal Images Group Limited/Alamy.*)

Forestry) requesting that the California/Oregon/Washington distinct population segment (a distinct population of a species that is capable of interbreeding and lives in a specific geographic area) be delisted. After finding the request to be warranted, the agency initiated a status review. In January 2010 the USFWS (http://www.gpo.gov/fdsys/pkg/FR-2010-01-21/pdf/2010-951.pdf#page=1) declined to delist the bird because it "continues to be subject to a broad range of threats, such as nesting habitat loss, habitat fragmentation, and predation." The agency noted that even though some threats have been reduced since the species was first listed, new threats have arisen, such as becoming entangled in abandoned fishing gear and being harmed by algal blooms and declining quality of marine prey.

In June 2009 the USFWS published *Marbled Murrelet (*Brachyramphus marmoratus*): 5-Year Review* (http://ecos.fws.gov/docs/five_year_review/doc2417.pdf). The agency concluded that the marbled murrelet should retain its listing as threatened. However, the USFWS noted deep concern about "the apparent substantial downward trend of the population and the species' continued vulnerability from a broad range of threats" and indicated that "a change in listing status to endangered may be warranted in the future."

WADING BIRDS. Wading birds are unusual birds characterized by long skinny legs and extended necks and beaks. They wade in shallow waters of swamps, wetlands, and bays, where they feed on aquatic life forms. Wading birds include species of crane, egret, ibis, and stork. As of November 2011, there were three wading birds of note listed under the ESA: the wood stork, the whooping crane, and the Mississippi sandhill crane.

The wood stork weighs about 5 pounds (2.3 kg) and stands up to 3 feet (0.9 m) tall with a 5-foot (1.5-m) wingspan. At one time tens of thousands of the birds inhabited the southeastern coastline. In 1984 the species was listed under the ESA as endangered in Alabama, Florida, Georgia, and South Carolina. A recovery plan for the bird was published in 1999. At that time about 5,000 breeding pairs lived in the wild. Throughout the first decade of the 21st century populations declined in the Everglades in southern Florida, but increased in coastal areas farther north. In September 2010 the USFWS (http://www.gpo.gov/fdsys/pkg/FR-2010-09-21/pdf/2010-23138.pdf#page=1) published a 90-day finding on a petition to reclassify the wood stork from endangered to threatened. The petition was submitted by the Pacific Legal Foundation on behalf of the Florida Homebuilders Association. The USFWS found that the downlisting may be warranted and initiated a status review. As of January 2012, the results of the review had not been published.

Standing 5 feet (1.5 m) tall, the whooping crane is North America's tallest bird and among the best-known endangered species in the United States. (See Figure 8.8.) Its name comes from its loud and distinctive call, which can be heard for miles. Historically, whooping cranes lived in the Great Plains and along the southeastern coast of the United States. The birds were once heavily hunted for their meat as well as for their beautiful, long white feathers. In addition, the heavy loss of wetland areas in the United States deprived whooping cranes of much of their original habitat. In 1937 it was discovered that fewer than 20 whooping cranes were left in the wild in two small populations: a migratory population that nested in Canada and wintered on the Texas coast and a nonmigratory population that lived in Louisiana.

Every fall the migratory whooping cranes fly 2,500 miles (4,000 km) from nesting grounds in Wood Buffalo, Canada, to Aransas, Texas, for the winter before returning north during the spring to breed. Whooping cranes

FIGURE 8.8

The whooping crane is highly endangered. Each year whooping cranes migrate from breeding grounds in Canada to wintering grounds in south Texas. (© *Al Mueller/Shutterstock.com.*)

FIGURE 8.9

Mississippi Sandhill Crane

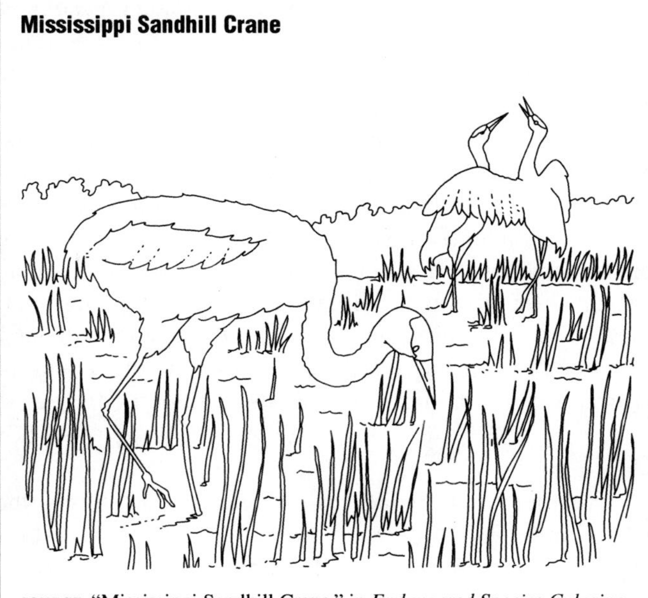

SOURCE: "Mississippi Sandhill Crane," in *Endangered Species Coloring Book: Save Our Species*, U.S. Environmental Protection Agency, May 2008, http://www.epa.gov/espp/coloring/cbook.pdf (accessed November 11, 2011)

return to the same nesting site each year with the same mate. In 1937 the Aransas National Wildlife Refuge was established in southern Texas to protect the species' wintering habitat. Conservation efforts for the whooping crane are coordinated with the Canadian government, which manages the birds' breeding areas.

As of November 2011, the whooping crane was listed under the ESA as endangered in Colorado, Kansas, Montana, Nebraska, North Dakota, Oklahoma, South Dakota, and Texas. Since 1993 three nonessential experimental populations have been designated in dozens of states from Wyoming to Florida. In 2001 the first introduced cranes in Wisconsin were led to their Florida wintering grounds along the migration route by ultralight aircraft. Ever since, the birds have been successfully migrating between their nesting and wintering grounds.

The Mississippi sandhill crane is a gray-colored bird that is about 4 feet (1.2 m) tall when standing upright. (See Figure 8.9.) It was first listed as endangered under the ESA in 1973. In September 1991 the USFWS (http://ecos.fws.gov/docs/recovery_plan/910906.pdf) published a recovery plan for the bird. At that time the species was in critical danger of becoming extinct with only one population existing in a small area of Jackson County, Mississippi. The bird was imperiled due to loss and degradation of its pine savanna habitat and reproductive isolation from other sandhill crane subspecies. The USFWS operates a captive breeding program at its Mississippi Sandhill Crane National Wildlife Refuge in Gautier, Mississippi. According to the agency (http://www.fws.gov/mississippisandhillcrane/), as of October 2011 the refuge was home to 110 of the birds, up from only 30 to 35 individuals reported during the 1970s. Refuge staff are working to increase the acreage of species habitat by burning or mechanically removing unsuitable shrubs and trees.

OTHER BIRDS. Other types of birds listed under the ESA include nonmigratory shore birds, such as the California clapper rail and the Guam rail (a flightless bird); swimming birds, including coots, ducks, eiders, and geese; and ground-dwelling birds, such as the prairie chicken.

GENERAL THREATS TO IMPERILED U.S. BIRD SPECIES

The driving force behind current declines in most imperiled bird species in the United States is the destruction, degradation, and fragmentation of habitat due to increasing human population size and consumption of resources. For example, natural habitats are lost due to agriculture, urban sprawl and other development, logging, mining, and road building. Other general threats include chemical contaminants, such as pesticides and oil, that are released to the environment and invasive species that prey on imperiled birds or compete with them for food and resources.

Pesticides

During the latter half of the 20th century pesticides and other toxic chemicals were recognized as a major cause of avian mortality and a primary factor in the endangerment of several species, including the bald eagle and the peregrine falcon. Many harmful pesticides, such as dichlorodiphenyltrichloroethane (DDT) and toxaphene, have been banned. The U.S. Environmental Protection Agency (EPA) evaluates the potential risks of pesticides to listed species and any designated critical habitat through a process called an "effects determination." The EPA (http://www.epa.gov/oppfead1/endanger/litstatus/effects/) has published the results of dozens of effects determinations on hundreds of imperiled species.

Oil

Another threat to imperiled birds is oil. During the 1970s and 1980s oil spills from seagoing tanker ships were a major environmental problem for coastal birds and seabirds. Birds coming into contact with spilled oil often died immediately, either from losing the insulation of their feathers or by ingesting lethal amounts of oil when they tried to clean themselves. (See Figure 8.10.) According to the *Exxon Valdez* Oil Spill Trustee Council (2012, http://www.evostc.state.ak.us/facts/qanda.cfm), the worst oil tanker spill in U.S. waters occurred on March 24, 1989, when the *Exxon Valdez* tanker released approximately 257,000 barrels of crude oil into Alaska's Prince William Sound. Hundreds of thousands of seabirds are estimated to have died as a result of the accident. Ever since, large-volume oil tanker spills have become much less common due to strict international regulation of oil tankers.

In April 2010 the largest oil spill in U.S. history occurred when an explosion aboard the *Deepwater Horizon*, an offshore drilling platform in the Gulf of Mexico, released nearly 5 million barrels of oil. The drilling platform was leased by the British Petroleum oil company and located approximately 41 miles (66 km) off the coast of Louisiana. The oil spill impacted aquatic wildlife and water quality across a huge area of the Gulf of Mexico.

The U.S. Department of the Interior has compiled a bird database as part of its Natural Resource Damage Assessment (NRDA) process, in which live and dead birds were collected following the spill. As of January 2012, the most recent NRDA database available was dated May 12, 2011. It should be noted that the data were

FIGURE 8.10

An oil-contaminated guillemot is cleaned after a spill. (© *corepics/Shutterstock.com.*)

provisional (i.e., not finalized) and that not all collected birds were necessarily killed or affected by the oil spill. The database (http://www.fws.gov/home/dhoilspill/pdfs/Bird%20Data%20Species%20Spreadsheet%2005122011.pdf) listed a total of 7,258 birds, including 6,381 dead birds. Of the latter, 2,121 were described as "visibly oiled." Nearly half of the visibly oiled dead birds were identified as laughing gulls. Other species in this category included the northern gannet (225), the brown pelican (152), and the royal tern (116). Comparison of the database against the ESA-listed bird species in Table 8.1 indicates that only two listed species were collected alive or dead: 106 least terns (of which 46 were visibly oiled dead birds) and one dead, but not visibly oiled, piping plover. It should be noted that dozens of birds included the database are not identified by species.

A much greater threat to bird species than oil spills is oil field waste pits. These pits hold oil-containing water that is pumped out of the ground during oil extraction. In "Contaminant Issues—Oil Field Waste Pits" (2012, http://www.fws.gov/mountain-prairie/contaminants/contaminants1a.html), the USFWS notes that an estimated 500,000 to 1 million birds per year die in these pits and similar oilfield wastewater facilities, such as lagoons. Birds mistake the pits and lagoons for freshwater and become oiled when they light on the water or dive into it after prey. The main avian victims are waterfowl; however, songbirds and raptors, such as hawks and owls, are also killed. It is unknown how many endangered and threatened species might be among the victims.

Invasive Species

Invasive species have damaged bird populations in some parts of the world, particularly those that occupy islands. Guam's unique bird fauna has been all but wiped out by the brown tree snake, an invasive species that probably arrived in ship cargo decades ago. According to the National Zoo, in the fact sheet "Where Have the Birds of Guam Gone?" (2012, http://nationalzoo.si.edu/Animals/Birds/Facts/FactSheets/fact-guambirds.cfm), there are believed to be as many as 14,000 snakes in a single square mile (2.6 sq km) in some Guam forest habitats. Nine out of 18 bird species have already gone extinct on Guam, including the Guam flycatcher, the Rufus fantail, the white-throated ground dove, and the cardinal honey-eater. Several other Guam bird species are close to extinction. Many of these birds are or were unique to Guam. Measures have been implemented to try to keep this destructive snake from invading other islands, including careful inspection of all cargo arriving from Guam. The removal of the brown tree snake in select habitat areas on Guam (which is a high-effort project, requiring the constant trapping of snakes) allowed the reintroduction of one bird, the flightless Guam rail, in 1998. The Guam rail had gone extinct in the wild, but a population is being maintained in captivity.

Other particularly destructive invasive species include several that are associated with humans, including cats, dogs, and rats, which often prey on birds and their eggs.

BACK FROM THE BRINK: SUCCESS STORIES

As shown in Table 2.13 in Chapter 2, nine bird species under ESA protection have recovered (i.e., been delisted). Four of the species—the Palua fantail flycatcher, the Palua ground dove, the Palua owl, and the Tinian (old world flycatcher) monarch—inhabit the small islands of Palua and Tinian, respectively, in the Pacific Ocean. All four species were listed in 1970. The Palua species were delisted in 1985 and the Tinian species in 2004.

Bald Eagles

The bald eagle is a raptor with special emphasis in the United States. (See Figure 8.11.) A symbol of honor, courage, nobility, and independence (eagles do not fly in flocks), the bald eagle is found only in North America, and its image is engraved on the official seal of the United States of America.

Bald eagles came dangerously close to extinction during the 20th century, largely due to DDT, which was introduced in 1947. Like other carnivorous (meat-eating) species, bald eagles ingested large amounts of DDT by eating prey that had been exposed to it. DDT either prevents birds from laying eggs or causes the eggshells to be so thin that they are unable to protect the eggs until they hatch. The Bald Eagle Protection Act of 1940, which made it a federal offense to kill bald eagles, helped protect the species. However, numbers continued to

FIGURE 8.11

The bald eagle was once endangered due to habitat destruction and pollution by pesticides, such as DDT. Its populations have recovered with protection and a ban on DDT. (© *nialat/Shutterstock.com.*)

dwindle. The bald eagle was listed as endangered in 1967. (See Table 2.1 in Chapter 2.)

Bald eagle populations started to recover with the banning of DDT in 1972. The species also benefited from habitat protection and attempts to clean up water pollution. In 1995 the bald eagle's status under the ESA was changed from endangered to threatened. In 1999 the species was proposed for delisting. A year later all delisting criteria contained in species recovery plans were achieved. However, the USFWS was slow to complete the delisting process. In June 2007 the bald eagle was finally delisted due to recovery. At that time the USFWS estimated there were 9,789 nesting pairs of the bird in the United States.

Peregrine Falcons

Many falcon species have declined with the spread of humans. Like other predatory species, falcons were often hunted, either for sport or because they were considered a threat to chickens or livestock. They also declined due to being exposed to DDT. In 1970 the American peregrine falcon (see Figure 8.12) and the arctic peregrine falcon were listed as endangered. Both subspecies ranged throughout the United States.

The recovery of the peregrine falcons was made possible by the banning of DDT and by the establishment of special captive breeding centers on several continents. The arctic peregrine falcon and the American peregrine falcon were delisted in 1994 and 1999, respectively.

Aleutian Canada Geese

Aleutian Canada geese inhabit the Aleutian Islands off the coast of Alaska and migrate back and forth to the northern California and Oregon coasts. (See Figure 8.13.) The species was first listed as endangered in 1967. (See Table 2.1 in Chapter 2.) It was officially delisted in 2001. In the delisting notice the USFWS (March 20, 2001, http://www.fws.gov/policy/library/66fr15643.html) indicates that the population increased from less than 1,000 individuals during the 1960s to around 37,000 individuals between 1999 and 2000. The conservation efforts that led to recovery included captive breeding, removal of foxes that preyed on the birds, and relocation and reintroduction of geese to unoccupied islands.

FIGURE 8.12

Peregrine falcon. (© *Cheryl Ann Quigley/Shutterstock.com.*)

FIGURE 8.13

Aleutian Canada geese. (© *U.S. Fish and Wildlife Service.*)

Brown Pelicans

The brown pelican is a coastal bird found in the United States along the Atlantic coast from Virginia southward, along the Gulf coast in Alabama, Louisiana, and Texas, and along the Pacific coast. (See Figure 8.14.) Its brownish gray feathers were highly sought after for women's hats during the late 1800s and early 1900s. As a result, the species underwent a dramatic decline. Its numbers were further decimated following World War I (1914–1918), when fishermen killed thousands of the birds, claiming they were competing for food fish. The use of DDT over the following decades took a huge toll on the birds by causing severe weakening of the shells of their eggs.

In 1970 the brown pelican was listed as endangered throughout its range. By the mid-1980s DDT restrictions and conservation efforts had allowed the Atlantic coast population of the species to rebound. In 1985 that population was delisted due to recovery. In 2009 the brown pelican throughout the remainder of the United States was also delisted.

FIGURE 8.14

Brown pelican

SOURCE: Robert Savannah, artist, "Brown Pelicans," in *Line Art (Drawings)*, U.S. Department of the Interior, U.S. Fish and Wildlife Service, undated, http://www.fws.gov/pictures/lineart/bobsavannah/brownpelicans.html (accessed November 11, 2011)

FOREIGN SPECIES OF ENDANGERED AND THREATENED BIRDS

The IUCN indicates in *Red List of Threatened Species Version 2011.2* that in 2011 a total of 1,253 bird species were considered threatened out of 10,052 evaluated and known species.

As of November 2011, there were 217 totally foreign species of birds listed under the ESA. (See Table 8.3.) Categories include various species, such as cranes, eagles, owls, parakeets, parrots, pheasants, pigeons, and warblers.

TABLE 8.3

Foreign endangered and threatened bird species, November 2011

Common name	Scientific name	Listing status*	Historic range
African penguin	Spheniscus demersus	E	
Alaotra grebe	Tachybaptus rufolavatus	E	Madagascar
Aldabra warbler (old world warbler)	Nesillas aldabranus	E	Indian Ocean_Seychelles (Aldabra Island)
Algerian nuthatch	Sitta ledanti	E	Algeria
Amsterdam albatross	Diomedia amsterdamensis	E	Indian Ocean—Amsterdam Island
Andean condor	Vultur gryphus	E	Colombia to Chile and Argentina
Andean flamingo	Phoenicoparrus andinus	E	(Current distribution in Argentina, Bolivia, Chile, Peru)
Andrew's frigatebird	Fregata andrewsi	E	East Indian Ocean
Anjouan Island sparrowhawk	Accipiter francesii pusillus	E	Indian Ocean_Comoro Islands
Anjouan scops owl	Otus rutilus capnodes	E	Indian Ocean_Comoro Island
Arabian ostrich	Struthio camelus syriacus	E	Jordan, Saudi Arabia
Atitlan grebe	Podilymbus gigas	E	Guatemala
Audouin's gull	Larus audouinii	E	Mediterranean Sea
Aukland Island rail	Rallus pectoralis muelleri	E	New Zealand
Azores wood pigeon	Columba palumbus azorica	E	East Atlantic Ocean_Azores
Bahaman or Cuban parrot	Amazona leucocephala	E	West Indies_Cuba, Bahamas, Caymans
Banded cotinga	Cotinga maculata	E	Brazil
Banded wattle-eye	Platysteira laticincta	E	Cameroon
Bannerman's turaco	Tauraco bannermani	E	Cameroon
Barbados yellow warbler (=wood)	Dendroica petechia petechia	E	West Indies_Barbados
Bar-tailed pheasant	Syrmaticus humaie	E	Burma, China
Bermuda petrel	Pterodroma cahow	E	Entire
black-breasted puffleg	Eriocnemis nigrivestis	E	(Current distribution in Ecuador)
Black-fronted piping-guan	Pipile jacutinga	E	Argentina
Black hooded (black-headed) antwren	Formicivora erythronotos	E	Brazil
Black-necked crane	Grus nigricollis	E	China (Tibet)
Black stilt	Himantopus novaezelandiae	E	New Zealand
Blue-throated (=ochre-marked) parakeet	Pyrrhura cruentata	E	Brazil
Blyth's tragopan pheasant	Tragopan blythii	E	Burma, China, India
Brazillian merganser	Mergus octosetaceus	E	(Current distribution in Brazil)
Brown eared pheasant	Crossoptilon mantchuricum	E	China
Cabot's tragopan pheasant	Tragopan caboti	E	China
Caerulean Paradise flycatcher	Eutrichomyias rowleyi	E	
Campbell Island flightless teal	Anas aucklandica nesiotis	E	New Zealand (Campbell Island)
Canarian black oystercatcher	Haematopus meadewaldoi	E	Atlantic Ocean_Canary Islands
Cantabrian capercaillie	Tetrao urogallus cantabricus	E	Spain and Portugal
Cebu black shama (thrush)	Copsychus niger cebuensis	E	Philippines
Chatham Island petrel	Pterodroma axillaris	E	Pacific Ocean—New Zealand (Chatham Island)
Chatham Island pigeon	Hemiphaga novaeseelandiae chathamensis	E	New Zealand
Chatham Island robin	Petroica traversi	E	New Zealand
Cheer pheasant	Catreus wallichii	E	India, Nepal, Pakistan
Cherry-throated tanager	Nemosia rourei	E	Brazil
Chilean woodstar	Eulidia yarrellii	E	Chile
Chinese egret	Egretta eulophotes	E	China, Korea
Chinese monal pheasant	Lophophorus lhuysii	E	China
Christmas Island goshawk	Accipiter fasciatus natalis	E	Indian Ocean_Christmas Island
Clarke's weaver	Ploceus golandi	E	Kenya
Cloven-feathered dove	Drepanoptila holosericea	E	Southwest Pacific Ocean_New Caledonia
Cuba hook-billed kite	Chondrohierax uncinatus wilsonii	E	West Indies_Cuba
Cuba sandhill crane	Grus canadensis nesiotes	E	West Indies_Cuba
Dappled mountain robin	Arcanator orostruthus	T	Mozambique, Tanzania
Djibouti francolin	Francolinus ochropectus	E	Djibouti
Edward's pheasant	Lophura edwardsi	E	Vietnam
Eiao Polynesian warbler	Acrocephalus percernis aquilonis	E	Eiao, marqueses Islands, Polynesia
Elliot's pheasant	Syrmaticus ellioti	E	China
Erect-crested penguin	Eudyptes sclateri	T	
Euler's flycatcher	Empidonax euleri johnstonei	E	West Indies_Grenada
Eurasian peregrine falcon	Falco peregrinus peregrinus	E	Europe, Eurasia south to Africa and Mideast
Eyrean grasswren (flycatcher)	Amytornis goyderi	E	Australia
Fiji petrel	Pterodroma macgillivrayi	E	Pacific Ocean, Fiji
fiordland crested penguin	Eudyptes pachyrhynchus	T	
Floreana tree-finch	Camarhynchus pauper	E	Ecuador
Forbes' parakeet	Cyanoramphus auriceps forbesi	E	New Zealand
Freira	Pterodroma madeira	E	Atlantic Ocean_Madeira Island
Fringe-backed fire-eye	Pyriglena atra	E	Brazil
Galapagos hawk	Buteo galapagoensis	E	Ecuador (Galapagos Islands)
Galapagos penguin	Spheniscus mendiculus	E	Ecuador (Galapagos Islands)
Galapagos petrel	Pterodroma phaeopygia	T	Pacific Ocean, Ecuador
Giant ibis	Pseudibis gigantea	E	
Giant scops owl	Mimizuku (=Otus) gurneyi	E	Philippines_Marinduque and Mindanao Island

TABLE 8.3

Foreign endangered and threatened bird species, November 2011 [CONTINUED]

Common name	Scientific name	Listing status*	Historic range
Glaucous macaw	Anodorhynchus glaucus	E	Paraguay, Uruguay, Brazil
Golden parakeet	Aratinga guarouba	E	Brazil
Golden-shouldered parakeet	Psephotus chrysopterygius	E	Australia
Greater adjutant	Leptoptilos dubius	E	
Great Indian bustard	Ardeotis (=Choriotis) nigriceps	E	India, Pakistan
Greenland white-tailed eagle	Haliaeetus albicilla groenlandicus	E	Greenland and adjacent Atlantic islands
Grenada gray-fronted dove	Leptotila rufaxilla wellsi	E	West Indies_Grenada
Grenada hook-billed kite	Chondrohierax uncinatus mirus	E	West Indies_Grenada
Grey-necked rockfowl	Picathartes oreas	E	Cameroon, Gabon
Ground parrot	Pezoporus wallicus	E	Australia
Guadeloupe house wren	Troglodytes aedon guadeloupensis	E	West Indies_Guadeloupe
Gurney's pitta	Pitta gurneyi	E	
Harpy eagle	Harpia harpyja	E	Mexico south to Argentina
Helmeted honeyeater	Lichenostomus melanops cassidix (=Meliphaga c.)	E	Australia
Helmeted hornbill	Buceros (=Rhinoplax) vigil	E	Thailand, Malaysia
Hooded crane	Grus monacha	E	Japan, Russia
Hook-billed hermit (hummingbird)	Ramphodon (=Glaucis) dohrnii	E	Brazil
Horned guan	Oreophasis derbianus	E	Guatemala, Mexico
Humboldt penguin	Spheniscus humboldti	T	
Ibadan malimbe	Malimbus ibadanensis	E	Nigeria
Imperial parrot	Amazona imperialis	E	West Indies_Dominica
Imperial pheasant	Lophura imperialis	E	Vietnam
Imperial woodpecker	Campephilus imperialis	E	Mexico
Indigo macaw	Anodorhynchus leari	E	Brazil
Japanese crane	Grus japonensis	E	China, Japan, Korea, Russia
Japanese crested ibis	Nipponia nippon	E	China, Japan, Russia, Korea
Jerdon's courser	Rhinoptilus bitorquatus	E	(Current distribution in India)
Kaempfer's tody-tyrant	Hemitriccus kaempferi	E	(Current distribution in Brazil)
Kagu	Rhynochetos jubatus	E	South Pacific Ocean_New Caledonia
Kakapo	Strigops habroptilus	E	New Zealand
Koch's pitta	Pitta kochi	E	Philippines
Kokako (wattlebird)	Callaeas cinerea	E	New Zealand
Lesser rhea (incl. Darwin's)	Rhea (=Pterocnemia) pennata	E	Argentina, Bolivia, Peru, Uruguay
Little blue macaw	Cyanopsitta spixii	E	Brazil
Long-legged warbler	Trichocichla rufa	E	Fiji, south pacific
Long-tailed ground roller	Uratelornis chimaera	E	Malagasy Republic (=Madagascar)
Lord Howe wood rail	Gallirallus (=Tricholimnas) sylvestris	E	Australia (Lord Howe Island)
Madagascar pochard	Aythya innotata	E	Madagascar
Madagascar red owl	Tyto soumagnei	E	Madagascar
Madagascar sea eagle	Haliaeetus vociferoides	E	Madagascar
Madagascar serpent eagle	Eutriorchis astur	E	Madagascar
Magenta petrel	Pterodroma magentae	E	Pacific Ocean, New Zealand
Maleo megapode	Macrocephalon maleo	E	Indonesia (Celebes)
Margaretta's hermit	Phaethornis malaris margarettae	E	Brazil
Marquesas pigeon	Ducula galeata	E	Polynesia,Marquesas Islands
Martinique trembler (thrasher)	Cinclocerthia ruficauda gutturalis	E	West Indies_Martinique
Marungu sunbird	Nectarinia prigoginei	E	Zaire
Mascarene black petrel	Pterodroma aterrima	E	Indian Ocean_Mauritius (Reunion Island)
Mauritius cuckoo-shrike	Coquus typicus	E	Indian Ocean_Mauritius
Mauritius fody	Foudia rubra	E	Indian Ocean_Mauritius
Mauritius kestrel	Falco punctatus	E	Indian Ocean_Mauritius
Mauritius olivaceous bulbul	Hypsipetes borbonicus olivaceus	E	Indian Ocean_Mauritius
Mauritius parakeet	Psittacula echo	E	Indian Ocean_Mauritius
Merriam's Montezuma quail	Cyrtonyx montezumae merriami	E	Mexico (Vera Cruz)
Mikado pheasant	Syrmaticus mikado	E	Taiwan
Mindoro imperial (=zone-tailed) pigeon	Ducula mindorensis	E	Philippines
Morden's owlet	Otus ireneae	E	Kenya
New Zealand bushwren	Xenicus longipes	E	New Zealand
New Zealand shore plover	Thinornis novaeseelandiae	E	New Zealand
New Zealand thrush (wattlebird)	Turnagra capensis	E	New Zealand
Night (=Australian) parrot	Geopsittacus occidentalis	E	Australia
Noisy scrub-bird	Atrichornis clamosus	E	Australia
Nordmann's greenshank	Tringa guttifer	E	Russia, Japan, south to Malaya, Borneo
Norfolk Island parakeet	Cyanoramphus cookii (=novaezelandiae c.)	E	Australia (Norfolk Island)
Norfolk Island white-eye	Zosterops albogularis	E	Indian Ocean_Norfolk Islands
Northern bald ibis	Geronticus eremita	E	Southern Europe, southwestern Asia, northern Africa
Orange-bellied parakeet	Neophema chrysogaster	E	Australia
Oriental white stork	Ciconia boyciana (=ciconia b.)	E	China, Japan, Korea, Russia
Palawan peacock pheasant	Polyplectron emphanum	E	Philippines
Paradise parakeet	Psephotus pulcherrimus	E	Australia
Philippine eagle	Pithecophaga jefferyi	E	Philippines

TABLE 8.3

Foreign endangered and threatened bird species, November 2011 [CONTINUED]

Common name	Scientific name	Listing status*	Historic range
Pink-headed duck	Rhodonessa caryophyllacea	E	India
Pink pigeon	Columba mayeri	E	Indian Ocean_Mauritius
Plain wanderer (=collared-hemipode)	Pedionomous torquatus	E	Australia
Pollen's vanga	Xenopirostris polleni	T	Madagascar
Ponape greater white-eye	Rukia longirostra	E	West Pacific Ocean_Federated States of Micronesia
Ponape mountain starling	Aplonis pelzelni	E	West Pacific Ocean_Federated States of Micronesia
Raso lark	Alauda razae	E	Atlantic Ocean_Raso Island (Cape Verde)
Razor-billed curassow	Mitu mitu mitu	E	Brazil (Eastern)
Red-billed curassow	Crax blumenbachii	E	Brazil
Red-browed parrot	Amazona rhodocorytha	E	Brazil
Red-capped parrot	Pionopsitta pileata	E	Brazil
Red-faced malkoha (cuckoo)	Phaenicophaeus pyrrhocephalus	E	Sri Lanka (=Ceylon)
Red-necked parrot	Amazona arausiaca	E	West Indies_Dominica
Red siskin	Carduelis cucullata	E	South America
Red-spectacled parrot	Amazona pretrei pretrei	E	Brazil, Argentina
Red-tailed parrot	Amazona brasiliensis	E	Brazil
Relict gull	Larus relictus	E	India, China
Resplendent quetzel	Pharomachrus mocinno	E	Mexico to Panama
Reunion cuckoo-shrike	Coquus newtoni	E	Indian Ocean_Reunion
Rodrigues fody	Foudia flavicans	E	Indian Ocean_Rodrigues Island (Mauritius)
Rodrigues warbler (old world warbler)	Bebrornis rodericanus	E	Mauritius (Rodrigues Islands)
Rothschild's starling (myna)	Leucopsar rothschildi	E	Indonesia (Bali)
Salmon-crested cockatoo	Cacatua moluccensis	T	(Current distribution in Indonesia)
Sao Miguel bullfinch (finch)	Pyrrhula pyrrhula murina	E	Eastern Atlantic Ocean_Azores
Scarlet-breasted robin (flycatcher)	Petroica multicolor multicolor	E	Australia (Norfolk Island)
Scarlet-chested parakeet	Neophema splendida	E	Australia
Sclater's monal pheasant	Lophophorus sclateri	E	Burma, China, India
Semper's warbler (=wood)	Leucopeza semperi	E	West Indies_St. Lucia
Seychelles fody (weaver-finch)	Foudia sechellarum	E	Indian Ocean_Seychelles
Seychelles kestrel	Falco araea	E	Indian Ocean_Seychelles Islands
Seychelles lesser vasa parrot	Coracopsis nigra barklyi	E	Indian Ocean_Seychelles (Praslin Island)
Seychelles magpie-robin (thrush)	Copsychus sechellarum	E	Indian Ocean_Seychelles Islands
Seychelles paradise flycatcher	Terpsiphone corvina	E	Indian Ocean_Seychelles
Seychelles scops owl	Otus magicus (=insularis) insularis	E	Indian Ocean_Seychelles Islands
Seychelles turtle dove	Streptopelia picturata rostrata	E	Indian Ocean_Seychelles
Seychelles warbler (old world warbler)	Bebrornis sechellensis	E	Indian Ocean_Seychelles Island
Seychelles white-eye	Zosterops modesta	E	Indian Ocean_Seychelles
Siberian white crane	Grus leucogeranus	E	C.I.S. (Siberia) to India, including Iran and China
Slender-billed curlew	Numenius tenuirostris	E	(Current distribution in Bulgaria, China, Algeria, Greece, Hungary, Italy, Kazakstan, Morocco, Romania, Russia, Slovenia, Turkmenistan, Tunisia, Turkey, Ukraine, Uzbekistan, Yugoslavia)
Slender-billed grackle	Quisicalus palustris	E	Mexico
Socorro mockingbird	Mimus graysoni	E	
Solitary tinamou	Tinamus solitarius	E	Brazil, Paraguay, Argentina
Southeastern rufous-vented ground cuckoo	Neomorphus geoffroyi dulcis	E	Brazil
southern rockhopper penguin	Eudyptes chrysocome	T	New Zealand—Campbell Plateau
Spanish imperial eagle	Aquila heliaca adalberti	E	Spain, Morocco, Algeria
St. Lucia forest thrush	Cichlherminia iherminieri santaeluciae	E	St. Lucia, WestIndies
St. Lucia house wren	Troglodytes aedon mesoleucus	E	West Indies_St. Lucia
St. Lucia parrot	Amazona versicolor	E	West Indies_St. Lucia
St Vincent parrot	Amazona guildingii	E	West Indies_St. Vincent
Swinhoe's pheasant	Lophura swinhoii	E	Taiwan
Tahiti flycatcher	Pomarea nigra	E	South Pacific Ocean_Tahiti
Taita thrush	Turdus olivaceus helleri	E	Kenya
Thick-billed parrot	Rhynchopsitta pachyrhyncha	E	Mexico, U.S.A. (AZ, NM)
Thyolo alethe	Alethe choloensis	E	Malawi, Mozambique
Trinidad white-headed curassow	Pipile pipile pipile	E	West Indies_Trinidad
Tristam's woodpecker	Dryocopus javensis richardsi	E	Korea
Turquoise parakeet	Neophema pulchella	E	Australia
Ulugura bush-shrike	Malaconotus alius	T	Tanzania
Van Dam's vanga	Xenopirostris damii	T	Madagascar
Vinaceous-breasted parrot	Amazona vinacea	E	Brazil
West African ostrich	Struthio camelus spatzi	E	Spanish Sahara
Western bristlebird	Dasyornis longirostris (=brachypterus l.)	E	Australia
Western rufous bristlebird	Dasyornis broadbenti littoralis	E	Australia
Western tragopan pheasant	Tragopan melanocephalus	E	India, Pakistan
Western whipbird	Psophodes nigrogularis	E	Australia

TABLE 8.3

Foreign endangered and threatened bird species, November 2011 [CONTINUED]

Common name	Scientific name	Listing status*	Historic range
White-breasted guineafowl	Agelastes meleagrides	T	West Africa
White-breasted thrasher	Ramphocinclus brachyurus	E	West Indies_St. Lucia, Martinique
White eared pheasant	Crossoptilon crossoptilon	E	China (Tibet), India
white-flippered penguin	Eudyptula albosignata	T	
White-naped crane	Grus vipio	E	Mongolia
White-necked rockfowl	Picathartes gymnocephalus	E	Africa_Togo to Sierra Leone
White-tailed laurel pigeon	Columba junoniae	T	Atlantic Ocean_Canary Islands
White-winged cotinga	Xipholena atropurpurea	E	Brazil
White-winged guan	Penelope albipennis	E	Peru
White-winged wood duck	Cairina scutulata	E	India, Malaysia, Indonesia, Thailand
yellow-eyed penguin	Megadyptes antipodes	T	

*E = endangered; T = threatened.

SOURCE: Adapted from "Generate Species List," in *Species Reports*, U.S. Department of the Interior, U.S. Fish & Wildlife Service, November 2011, http://ecos.fws.gov/tess_public/pub/adHocSpeciesForm.jsp (accessed November 8, 2011)

CHAPTER 9

INSECTS AND SPIDERS

Insects are members of the Animalia kingdom and belong to the phylum Arthropoda, along with crustaceans. There are many classes of arthropods, including insects and arachnids. Both are invertebrates, but insects have six legs, whereas arachnids have eight legs. The arachnids include spiders, mites, ticks, scorpions, and harvestmen.

Insects are the most diverse group in the animal kingdom. Scientists are not certain of the total number of insect species; estimates range as high as 30 million species. The International Union for Conservation of Nature (IUCN) indicates in *Red List of Threatened Species Version 2011.2* (2011, http://www.iucnredlist.org/documents/summarystatistics/2011_2_RL_Stats_Table1.pdf) that 1 million of the species have been described. Insects have not been as thoroughly studied as the vertebrate groups, so there are likely to be many endangered insects whose state is unknown.

Insects and arachnids, like many other species, suffer from diminished habitat as a result of encroaching development, industrialization, changing land use patterns, and invasive species.

THREATENED AND ENDANGERED INSECT SPECIES IN THE UNITED STATES

As of November 2011, there were 61 U.S. insect species listed under the Endangered Species Act (ESA). (See Table 9.1.) The predominant species types include 19 butterflies, 18 beetles, and 12 pomace flies. The remaining insects include three skippers, two moths, two flies, two damselflies, a dragonfly, a grasshopper, and a naucorid. (See Figure 9.1.) Most of the listed insects are endangered, and nearly all have recovery plans in place.

According to the U.S. Fish and Wildlife Service (USFWS), most of the imperiled insects are found exclusively in either California, Hawaii, or Texas.

Table 9.2 shows the 10 insect species with the highest expenditures under the ESA during fiscal year (FY) 2010. The list is dominated by butterflies, but topped by a beetle. Nearly $2.5 million was devoted to the valley elderberry longhorn beetle, a species found only in California. Over $1.7 million was spent on the Karner blue butterfly, which inhabits midwestern states.

Butterflies, Moths, and Skippers

Butterflies, moths, and skippers are flying insects that belong to the order Lepidoptera. Scientists believe there could be several hundred thousand species in this order. Skippers have stockier bodies than butterflies and are structurally different from moths. As such, they are considered to be an intermediate between butterflies and moths.

Like amphibians, many butterflies and moths are considered to be indicator species (meaning that their well-being gives scientists a good indication of the general health of their habitat) because they are particularly sensitive to environmental degradation. The decline of these species serves as a warning to humans about the condition of the environment. For example, certain moth species feed on lichen, a fungus-based organism that grows on trees, rocks, and other solid surfaces. Lichen are very susceptible to air pollutants. Reductions in the number of lichen-feeding moths can indicate the presence of harmful air pollution in an area. Part of the reason butterflies are sensitive to many aspects of the environment is that these species undergo a drastic metamorphosis, or change, from larva to adult as a natural part of their life cycle. Butterfly larvae are generally crawling, herbivorous (plant-eating) caterpillars, whereas butterfly adults fly and eat nectar. Butterflies can thrive only when intact habitats are available for both caterpillars and adults. Consequently, healthy butterfly populations tend to occur in areas with healthy ecosystems. Because many

TABLE 9.1

Endangered and threatened insect species, November 2011

Common name	Scientific name	Listing status[a]	U.S.or U.S./foreign listed	Recovery plan date	Recovery plan stage[b]
Beetle, American burying	Nicrophorus americanus	E	US/foreign	9/27/1991	F
Beetle, Caseys June	Dinacoma caseyi	E	US	None (listed in 2011)	—
Beetle, Coffin Cave mold	Batrisodes texanus	E	US	8/25/1994	F
Beetle, Comal Springs dryopid	Stygoparnus comalensis	E	US	2/14/1996	RF(1)
Beetle, Comal Springs riffle	Heterelmis comalensis	E	US	2/14/1996	RF(1)
Beetle, Delta green ground	Elaphrus viridis	T	US	12/15/2005	F
Beetle, Helotes mold	Batrisodes venyivi	E	US	10/4/2011	F
Beetle, Hungerford's crawling water	Brychius hungerfordi	E	US/foreign	9/28/2006	F
Beetle, Kretschmarr Cave mold	Texamaurops reddelli	E	US	8/25/1994	F
Beetle, Mount Hermon June	Polyphylla barbata	E	US	9/28/1998	F
Beetle, Northeastern beach tiger	Cicindela dorsalis dorsalis	T	US	9/29/1994	F
Beetle, Ohlone tiger	Cicindela ohlone	E	US	9/28/1998	F
Beetle, Puritan tiger	Cicindela puritana	T	US	9/29/1993	F
Beetle, Salt Creek tiger	Cicindela nevadica lincolniana	E	US	2/20/2009	O
Beetle, Tooth Cave ground	Rhadine persephone	E	US	8/25/1994	F
Beetle, [Unnamed] ground	Rhadine exilis	E	US	10/4/2011	F
Beetle, [Unnamed] ground	Rhadine infernalis	E	US	10/4/2011	F
Beetle, Valley elderberry longhorn	Desmocerus californicus dimorphus	T	US	6/28/1984	F
Butterfly, Bay checkerspot	Euphydryas editha bayensis	T	US	9/30/1998	F
Butterfly, Behren's silverspot	Speyeria zerene behrensii	E	US	1/20/2004	D
Butterfly, Callippe silverspot	Speyeria callippe callippe	E	US	None	—
Butterfly, El Segundo blue	Euphilotes battoides allyni	E	US	9/28/1998	F
Butterfly, Fender's blue	Icaricia icarioides fenderi	E	US	6/29/2010	F
Butterfly, Karner blue	Lycaeides melissa samuelis	E	US/foreign	9/19/2003	F
Butterfly, Lange's metalmark	Apodemia mormo langei	E	US	4/25/1984	RF(1)
Butterfly, Lotis blue	Lycaeides argyrognomon lotis	E	US	12/26/1985	F
Butterfly, Mission blue	Icaricia icarioides missionensis	E	US	10/10/1984	F
Butterfly, Mitchell's satyr	Neonympha mitchellii mitchellii	E	US	4/2/1998	F
Butterfly, Myrtle's silverspot	Speyeria zerene myrtleae	E	US	9/29/1998	F
Butterfly, Oregon silverspot	Speyeria zerene hippolyta	T	US	8/22/2001	RF(1)
Butterfly, Palos Verdes blue	Glaucopsyche lygdamus palosverdesensis	E	US	1/19/1984	F
Butterfly, Quino checkerspot	Euphydryas editha quino (=E. e. wrighti)	E	US/foreign	9/17/2003	F
Butterfly, Saint Francis' satyr	Neonympha mitchellii francisci	E	US	4/23/1996	F
Butterfly, San Bruno elfin	Callophrys mossii bayensis	E	US	10/10/1984	F
Butterfly, Schaus swallowtail	Heraclides aristodemus ponceanus	E	US	5/18/1999	F
Butterfly, Smith's blue	Euphilotes enoptes smithi	E	US	11/9/1984	F
Butterfly, Uncompahgre fritillary	Boloria acrocnema	E	US	3/17/1994	F
Damselfly, Flying earwig	Megalagrion nesiotes	E	US	None (listed in 2010)	—
Damselfly, Pacific Hawaiian	Megalagrion pacificum	E	US	None (listed in 2010)	—
Dragonfly, Hine's emerald	Somatochlora hineana	E	US	9/27/2001	F
Fly, Delhi Sands flower-loving	Rhaphiomidas terminatus abdominalis	E	US	9/14/1997	F
Fly, Hawaiian picture-wing	Drosophila sharpi	E	US	None (listed in 2010)	—
Grasshopper, Zayante band-winged	Trimerotropis infantilis	E	US	9/28/1998	F
Moth, Blackburn's sphinx	Manduca blackburni	E	US	9/28/2005	F
Moth, Kern primrose sphinx	Euproserpinus euterpe	T	US	2/8/1984	F
Naucorid, Ash Meadows	Ambrysus amargosus	T	US	9/28/1990	F
Pomace fly, [Unnamed]	Drosophila aglaia	E	US	None	—
Pomace fly, [Unnamed]	Drosophila differens	E	US	None	—
Pomace fly, [Unnamed]	Drosophila hemipeza	E	US	None	—
Pomace fly, [Unnamed]	Drosophila heteroneura	E	US	None	—
Pomace fly, [Unnamed]	Drosophila montgomeryi	E	US	None	—
Pomace fly, [Unnamed]	Drosophila mulli	T	US	None	—
Pomace fly, [Unnamed]	Drosophila musaphila	E	US	None	—
Pomace fly, [Unnamed]	Drosophila neoclavisetae	E	US	None	—
Pomace fly, [Unnamed]	Drosophila obatai	E	US	None	—
Pomace fly, [Unnamed]	Drosophila ochrobasis	E	US	None	—
Pomace fly, [Unnamed]	Drosophila substenoptera	E	US	None	—
Pomace fly, [Unnamed]	Drosophila tarphytrichia	E	US	None	—
Skipper, Carson wandering	Pseudocopaeodes eunus obscurus	E	US	9/13/2007	F
Skipper, Laguna Mountains	Pyrgus ruralis lagunae	E	US	None	—
Skipper, Pawnee montane	Hesperia leonardus montana	T	US	9/21/1998	F

[a]E = endangered; T = threatened.
[b]F = final; D=draft; RF = final revision; O = other.

SOURCE: Adapted from "Generate Species List," in *Species Reports*, U.S. Department of the Interior, U.S. Fish & Wildlife Service, November 2011, http://ecos.fws.gov/tess_public/pub/adHocSpeciesForm.jsp (accessed November 8, 2011), and "Listed FWS/Joint FWS and NMFS Species and Populations with Recovery Plans (Sorted by Listed Entity)," in *Recovery Plans Search*, U.S. Department of the Interior, U.S. Fish & Wildlife Service, November 2011, http://ecos.fws.gov/tess_public/pub/speciesRecovery.jsp?sort=1 (accessed November 8, 2011)

FIGURE 9.1

The ash meadows naucorid

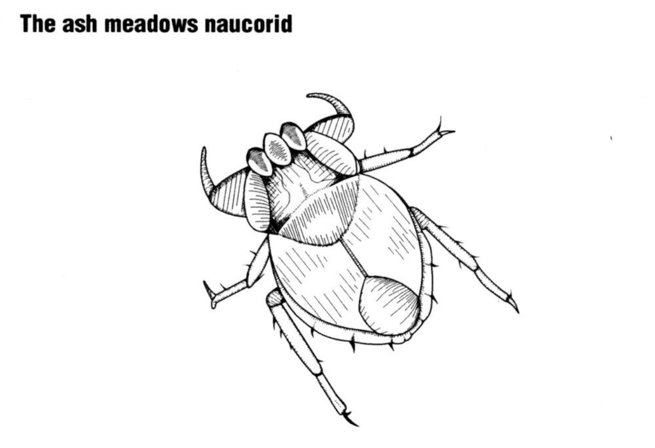

SOURCE: E. Tobin, artist, "Ash Meadows Naucorid *Pelecorus shoshone*," in *Planning Update 1: Ash Meadows National Wildlife Refuge*, U.S. Department of the Interior, U.S. Fish and Wildlife Service, January 1999, http://www.fws.gov/pacific/planning/am_pu1.pdf (accessed November 12, 2011)

TABLE 9.2

Insect species with the highest expenditures under the Endangered Species Act, fiscal year 2010

Ranking	Species	Expenditure
1	Beetle, valley elderberry longhorn (Desmocerus californicus dimorphus)	$2,495,323
2	Butterfly, Karner blue (Lycaeides melissa samuelis)	$1,729,106
3	Butterfly, Oregon silverspot (Speyeria zerene hippolyta)	$875,444
4	Beetle, American burying (Nicrophorus americanus)	$639,848
5	Butterfly, Fender's blue (Icaricia icarioides fenderi)	$612,642
6	Butterfly, Quino checkerspot (Euphydryas editha quino (=E. e. wrighti))	$555,403
7	Butterfly, El Segundo blue (Euphilotes battoides allyni)	$525,715
8	Pomace fly, [unnamed] (Drosophila montgomeryi)	$289,678
9	Pomace fly, [unnamed] (Drosophila substenoptera)	$289,678
10	Dragonfly, Hine's emerald (Somatochlora hineana)	$287,600

SOURCE: Adapted from "Table 2. Species Ranked in Descending Order of Total FY 2010 Reported Expenditures, Not Including Land Acquisition Costs," in *Federal and State Endangered and Threatened Species Expenditures: Fiscal Year 2010*, U.S. Department of the Interior, U.S. Fish and Wildlife Service, 2010, http://www.fws.gov/endangered/esa-library/pdf/2010.EXP.FINAL.pdf (accessed October 25, 2011)

species are extremely sensitive to changing environmental conditions, butterflies and moths are carefully monitored by scientists and conservationists around the world.

The major threats to butterflies include:

- Habitat destruction
- Mowing of pastures, ditches, and highway rights-of-way
- Collisions with moving automobiles
- Insecticides

KARNER BLUE BUTTERFLIES. The Karner blue butterfly was listed under the ESA in 1992 as an endangered species. The USFWS states in "Restoring Habitat for Karner Blue Butterfly on Private Lands in Wisconsin" (April 8, 2009, http://www.fws.gov/Midwest/WisconsinPartners/butterfly.html) that the butterfly is no bigger than a postage stamp. Historically, it occupied habitats in the eastern United States from Minnesota to Maine as well as in Ontario, Canada. The USFWS notes that the species' historic range has declined by 99% due to loss and fragmentation of habitat. It is now found only in portions of Indiana, Michigan, Minnesota, New Hampshire, New York, Ohio, and Wisconsin. Wisconsin has the largest population. Most Karner blue butterfly populations are extremely small and in danger of extinction.

The caterpillars of the Karner blue butterfly feed only on a species of wild lupine. The butterflies are weak flyers and prefer to stay near their favorite lupine patches. In September 2003 the USFWS published *Karner Blue Butterfly (*Lycaeides melissa samuelis*) Recovery Plan* (http://ecos.fws.gov/docs/recovery_plan/030919.pdf), which outlined the status of the species and the steps required to conserve it. In February 2011 the agency (http://www.fws.gov/midwest/endangered/insects/kbb/pdf/kbbRecPlanRevision2011.pdf) issued a memorandum that made a minor update to the recovery plan by adding a new potential recovery unit. Figure 9.2 shows the updated map of recovery units, or populations, of the species, sites for potential recovery units, and other sites where the species has historically been found.

In March 2009 the USFWS (http://www.gpo.gov/fdsys/search/citation.result.FR.action?federalRegister.volume=2009&federalRegister.page=11600&publication=FR) initiated a five-year review for several midwestern species, including the Karner blue butterfly. As of January 2012, the results of the review had not been published.

California Insects

There are 21 insects found only or primarily in California on the list of endangered and threatened species:

- Caseys June beetle
- Delta green ground beetle
- Mount Hermon June beetle
- Ohlone tiger beetle
- Valley elderberry longhorn beetle
- Bay checkerspot butterfly
- Behren's silverspot butterfly
- Callippe silverspot butterfly
- El Segundo blue butterfly
- Lange's metalmark butterfly
- Lotis blue butterfly

FIGURE 9.2

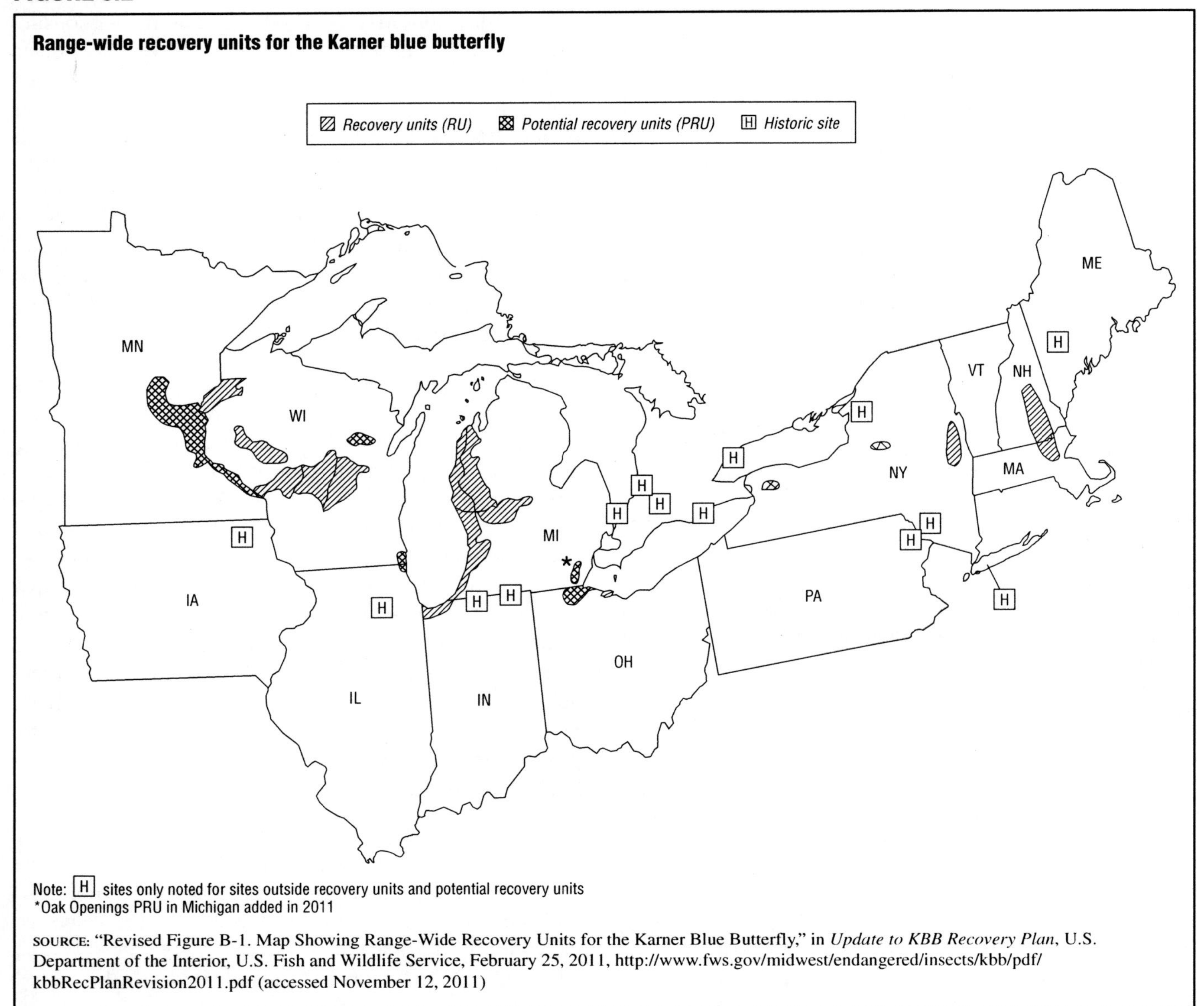

Range-wide recovery units for the Karner blue butterfly

Note: H sites only noted for sites outside recovery units and potential recovery units
*Oak Openings PRU in Michigan added in 2011

SOURCE: "Revised Figure B-1. Map Showing Range-Wide Recovery Units for the Karner Blue Butterfly," in *Update to KBB Recovery Plan*, U.S. Department of the Interior, U.S. Fish and Wildlife Service, February 25, 2011, http://www.fws.gov/midwest/endangered/insects/kbb/pdf/kbbRecPlanRevision2011.pdf (accessed November 12, 2011)

- Mission blue butterfly
- Myrtle's silverspot butterfly
- Palos Verdes blue butterfly
- Quino checkerspot butterfly
- San Bruno elfin butterfly
- Smith's blue butterfly
- Delhi Sands flower-loving fly
- Zayante band-winged grasshopper
- Kern primrose sphinx moth
- Laguna Mountains skipper

VALLEY ELDERBERRY LONGHORN BEETLES. As noted earlier, nearly $2.5 million was spent under the ESA on the threatened valley elderberry longhorn beetle in FY 2010. (See Table 9.2.) It is a stout-bodied beetle nearly 1 inch (2.5 cm) long when fully grown. (See Figure 9.3.) The species overwhelmingly prefers only one type of host plant: elderberry shrubs along creeks and rivers in California's Central Valley. This is an area that experienced extensive agricultural and urban development during the 20th century. Long-term destruction and fragmentation of riparian (river and stream) ecosystems imperiled the beetle population. It was first listed under the ESA in 1980 as a threatened species. Two areas in Sacramento County were designated critical habitat. A recovery plan was finalized in 1984.

In September 2006 the USFWS published *Valley Elderberry Longhorn Beetle (*Desmocerus californicus dimorphus*): 5-Year Review—Summary and Evaluation*

FIGURE 9.3

Valley elderberry longhorn beetle

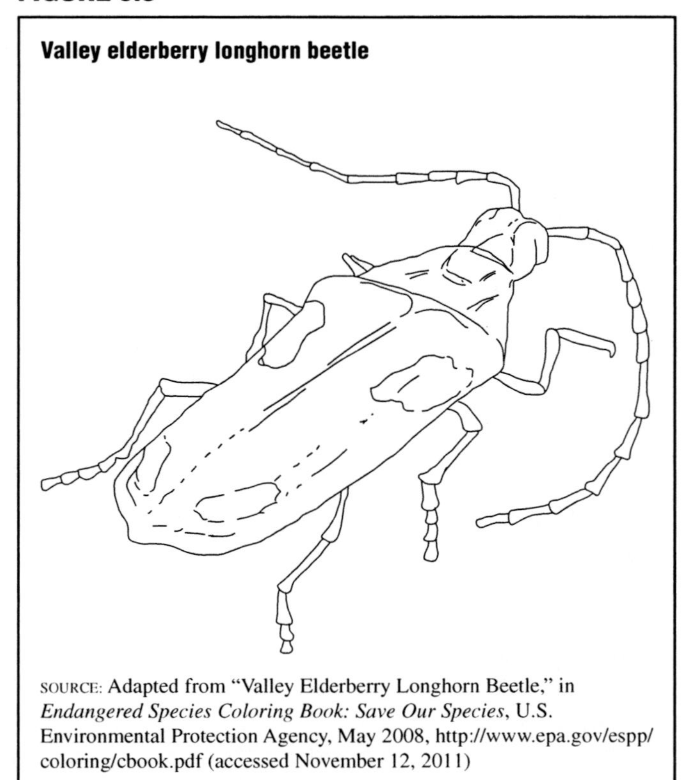

SOURCE: Adapted from "Valley Elderberry Longhorn Beetle," in *Endangered Species Coloring Book: Save Our Species*, U.S. Environmental Protection Agency, May 2008, http://www.epa.gov/espp/coloring/cbook.pdf (accessed November 12, 2011)

(http://ecos.fws.gov/docs/five_year_review/doc779.pdf) and recommended that the species be delisted. The agency noted that 50,000 acres (20,200 ha) of the riparian habitat preferred by the beetle had been given protected status and that over 5,000 acres (2,020 ha) had been restored.

In September 2010 the Pacific Legal Foundation (PLF), a private organization that typically litigates against the ESA, petitioned the USFWS to conduct the delisting based on the information contained in the five-year review. In "The Elderberry Longhorn Beetle Doesn't Belong on Any ESA List" (2012, http://www.pacificlegal.org/page.aspx?pid=1136), the PLF complains about the agency's failure to act on its own delisting recommendation, noting that "landowners, flood control districts, and others must still devote their limited resources to protect a beetle that in reality no longer requires federal protection." As described in Chapter 2, the agency must determine within 90 days whether a delisting petition contains "substantial information" suggesting that a species may require delisting under the ESA. When the agency failed to meet this deadline, the PLF sued the USFWS. In January 2011 the agency (http://www.gpo.gov/fdsys/search/citation.result.FR.action?federalRegister.volume=2011&federalRegister.page=3069&publication=FR) published its petition finding that delisting was warranted. This triggered a mandatory status review, which had to be completed within 12 months. As of January 2012, the status review had not been published.

LANGE'S METALMARK BUTTERFLIES. The Antioch Dunes National Wildlife Refuge is a small strip of sandy land lying between a large industrial complex and the San Joaquin River in the heavily populated San Francisco Bay-Delta area of California. (See Figure 9.4.) According to the USFWS (2012, http://www.fws.gov/refuges/profiles/index.cfm?id=81646), the land on which the refuge sits was once a 5.6-mile (9-km) stretch of unique sand dune habitat formed by glacial activity tens of thousands of years ago. The refuge is home to the critically endangered Lange's metalmark butterfly, a colorful butterfly with metallic-looking markings on its wings. Susan Morse describes the species' fight for survival in "Bringing a Butterfly Back from the Brink" (*Endangered Species Bulletin*, vol. 34, no. 3, Fall 2009). Morse notes that the once billowing Antioch sand dunes were decimated by more than a century of mining by the brickmaking industry. The remaining hardpan sand does not provide an ideal habitat for the butterfly. The species breeds once per year and prefers only one plant for its home: the naked stem buckwheat. The butterfly was first listed under the ESA in 1976. Its population numbered 5,976 in 1997, but dropped to 158 by 2006. Alarmed scientists captured some of the remaining butterflies and began a successful captive-breeding program that has allowed them to release dozens of pupae (cocoons), larvae, and adults back into the wild. Refuge staff and volunteers have also planted thousands of buckwheat seedlings and removed nonnative plants, such as vetch, that threaten to smother the butterfly habitat.

In May 2011 the USFWS (http://www.gpo.gov/fdsys/pkg/FR-2011-05-25/pdf/2011-12861.pdf) initiated a five-year review for dozens of West Coast species, including the Lange's metalmark butterfly. The purpose of the review is to examine the current status of the species and determine if any changes are warranted in their listing statuses. As of January 2012, the results of the review had not been published.

Hawaiian Insects

As of November 2011, there were 16 Hawaiian insects listed under the ESA: 12 subspecies of the pomace fly, two damselflies, a Hawaiian picture-wing fly, and the Blackburn's sphinx moth. (See Table 9.1.)

POMACE FLIES. Pomace flies (also known as Hawaiian picture-wing flies) belong to the genus *Drosophila*, a group commonly called fruit flies. There are more than 100 species of *Drosophila*. They are relatively large flies with elaborate markings on their wings. The Hawaiian subspecies are renowned for their colorful wing patterns. They prefer mostly mesic (adequately moist) forest and wet forest habitats. Each fly is dependent on one or more specific host plants. This dependence is one of the factors that has imperiled the flies. All the host plants face

FIGURE 9.4

Antioch Dunes National Wildlife Refuge

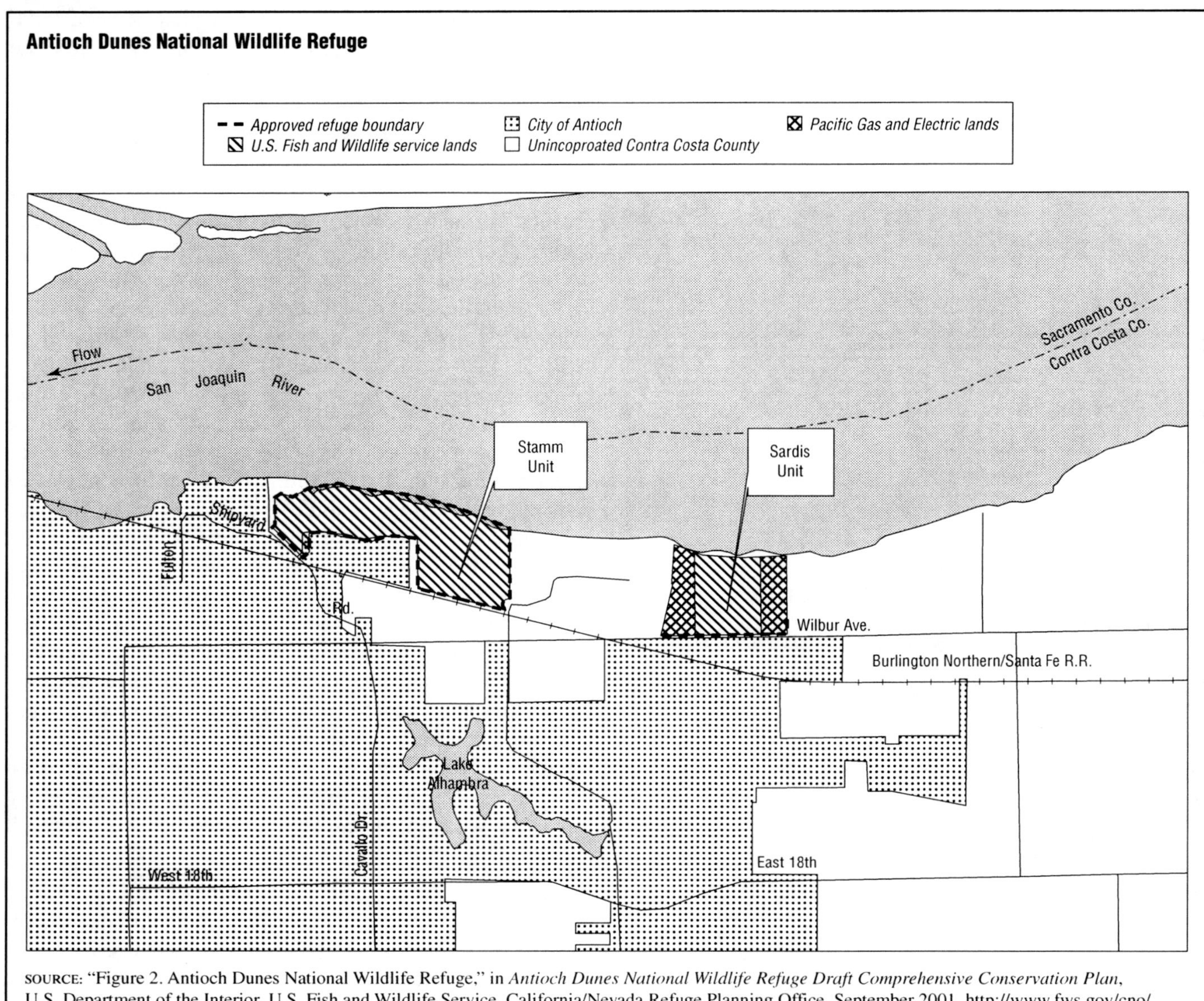

SOURCE: "Figure 2. Antioch Dunes National Wildlife Refuge," in *Antioch Dunes National Wildlife Refuge Draft Comprehensive Conservation Plan*, U.S. Department of the Interior, U.S. Fish and Wildlife Service, California/Nevada Refuge Planning Office, September 2001, http://www.fws.gov/cno/refuges/antioch/CCP.pdf (accessed November 12, 2011)

threats to their survival due to competition with nonnative plants and trampling and ingestion by livestock and wild animals.

In May 2006 all 12 subspecies were listed under the ESA. All but one were designated as endangered. (See Table 9.1.) *Drosophila mulli* was designated as threatened. As of January 2012, a recovery plan had not been issued for the Hawaiian pomace flies.

In April 2010 the USFWS (http://www.gpo.gov/fdsys/pkg/FR-2010-04-08/pdf/2010-7915.pdf#page=1) initiated a five-year review for the pomace flies and dozens of other listed species found in Hawaii, Idaho, Washington, Guam, and the Commonwealth of the Northern Mariana Islands. As of January 2012, the results of the review had not been published.

THREATENED AND ENDANGERED FOREIGN SPECIES OF INSECTS

In *Red List of Threatened Species Version 2011.2*, the IUCN indicates that in 2011, 741 species of insects were threatened. This number accounted for 19% of the 3,844 evaluated species and less than 0.1% of the 1 million described species.

As of November 2011, there were four totally foreign insects listed under the ESA. (See Table 9.3.) All four were butterfly species and were listed as endangered.

THREATENED AND ENDANGERED ARACHNID SPECIES IN THE UNITED STATES

Arachnids are invertebrates with eight legs. They include spiders, mites, ticks, scorpions, and harvestmen.

TABLE 9.3

Foreign endangered and threatened insect species, November 2011

Common name	Scientific name	Listing status	Historic range
Corsican swallowtail butterfly	Papilio hospiton	Endangered	Corsica, Sardinia
Homerus swallowtail butterfly	Papilio homerus	Endangered	Jamaica
Luzon peacock swallowtail butterfly	Papilio chikae	Endangered	Philippines
Queen Alexandra's birdwing butterfly	Troides (=Ornithoptara) alexandrae	Endangered	Papua New Guinea

SOURCE: Adapted from "Generate Species List," in *Species Reports*, U.S. Department of the Interior, U.S. Fish & Wildlife Service, November 2011, http://ecos.fws.gov/tess_public/pub/adHocSpeciesForm.jsp (accessed November 8, 2011)

TABLE 9.4

Endangered arachnid species, November 2011

Common name	Scientific name	Listing status	U.S. or U.S./foreign listed	Recovery plan date	Recovery plan stage
Bee Creek Cave harvestman	Texella reddelli	Endangered	US	8/25/1994	Final
Bone Cave harvestman	Texella reyesi	Endangered	US	8/25/1994	Final
Braken Bat Cave Meshweaver	Cicurina venii	Endangered	US	10/4/2011	Final
Cokendolpher Cave Harvestman	Texella cokendolpheri	Endangered	US	10/4/2011	Final
Government Canyon Bat Cave Meshweaver	Cicurina vespera	Endangered	US	10/4/2011	Final
Government Canyon Bat Cave spider	Neoleptoneta microps	Endangered	US	10/4/2011	Final
Kauai cave wolf or pe`e pe`e maka `ole spider	Adelocosa anops	Endangered	US	7/19/2006	Final
Madla's Cave Meshweaver	Cicurina madla	Endangered	US	10/4/2011	Final
Robber Baron Cave Meshweaver	Cicurina baronia	Endangered	US	10/4/2011	Final
Spruce-fir moss spider	Microhexura montivaga	Endangered	US	9/11/1998	Final
Tooth Cave pseudoscorpion	Tartarocreagris texana	Endangered	US	8/25/1994	Final
Tooth Cave spider	Leptoneta myopica	Endangered	US	8/25/1994	Final

SOURCE: Adapted from "Generate Species List," in *Species Reports*, U.S. Department of the Interior, U.S. Fish & Wildlife Service, November 2011, http://ecos.fws.gov/tess_public/pub/adHocSpeciesForm.jsp (accessed November 8, 2011), and "Listed FWS/Joint FWS and NMFS Species and Populations with Recovery Plans (Sorted by Listed Entity)," in *Recovery Plans Search*, U.S. Department of the Interior, U.S. Fish & Wildlife Service, November 2011, http://ecos.fws.gov/tess_public/pub/speciesRecovery.jsp?sort=1 (accessed November 8, 2011)

As of November 2011, there were 12 U.S. species of arachnids listed under the ESA. (See Table 9.4.) All the arachnids had endangered status, and all had recovery plans in place. The imperiled arachnids fall into four species types:

- Harvestmen—three species
- Meshweaver—four species
- Pseudoscorpion—one species
- Spider—four species

Ten of the arachnids are cave-dwelling species found only in Texas. The only imperiled arachnids outside of Texas are the Kauai Cave wolf spider, which inhabits Hawaii, and the spruce-fir moss spider, which is found in North Carolina and Tennessee.

Table 9.5 shows the 10 arachnid species with the highest expenditures under the ESA during FY 2010. More than $242,000 was spent conserving these species.

TABLE 9.5

Arachnid species with the highest expenditures under the Endangered Species Act, fiscal year 2010

Ranking	Species	Expenditure
1	Meshweaver, Madla's Cave (Cicurina madla)	$63,760
2	Meshweaver, Braken Bat Cave (Cicurina venii)	$43,600
3	Spider, spruce-fir moss (Microhexura montivaga)	$33,600
4	Spider, Government Canyon Bat Cave (Neoleptoneta microps)	$21,260
5	Spider, Tooth Cave (Leptoneta myopica)	$15,600
6	Pseudoscorpion, Tooth Cave (Tartarocreagris texana)	$13,600
7	Meshweaver, Robber Baron Cave (Cicurina baronia)	$13,600
8	Meshweaver, Government Canyon Bat Cave (Cicurina vespera)	$13,260
9	Harvestman, Cokendolpher Cave (Texella cokendolpheri)	$13,000
10	Spider, Kauai cave wolf or pe`e pe`e maka `ole (Adelocosa anops)	$11,000

SOURCE: Adapted from "Table 2. Species Ranked in Descending Order of Total FY 2010 Reported Expenditures, Not Including Land Acquisition Costs," in *Federal and State Endangered and Threatened Species Expenditures: Fiscal Year 2010*, U.S. Department of the Interior, U.S. Fish and Wildlife Service, 2010, http://www.fws.gov/endangered/esa-library/pdf/2010.EXP.FINAL.pdf (accessed October 25, 2011)

Texas Cave Arachnids

Ten of the listed arachnids are found only in underground karst caves in three counties in Texas. (See Figure 9.5.) Karst is a geological term referring to a type of underground terrain resulting when limestone bedrock is exposed to mildly acidic groundwater over a long period. Eventually, the bedrock becomes a honeycomb of cracks, fissures,

FIGURE 9.5

A karst cave provides habitat for endangered invertebrates

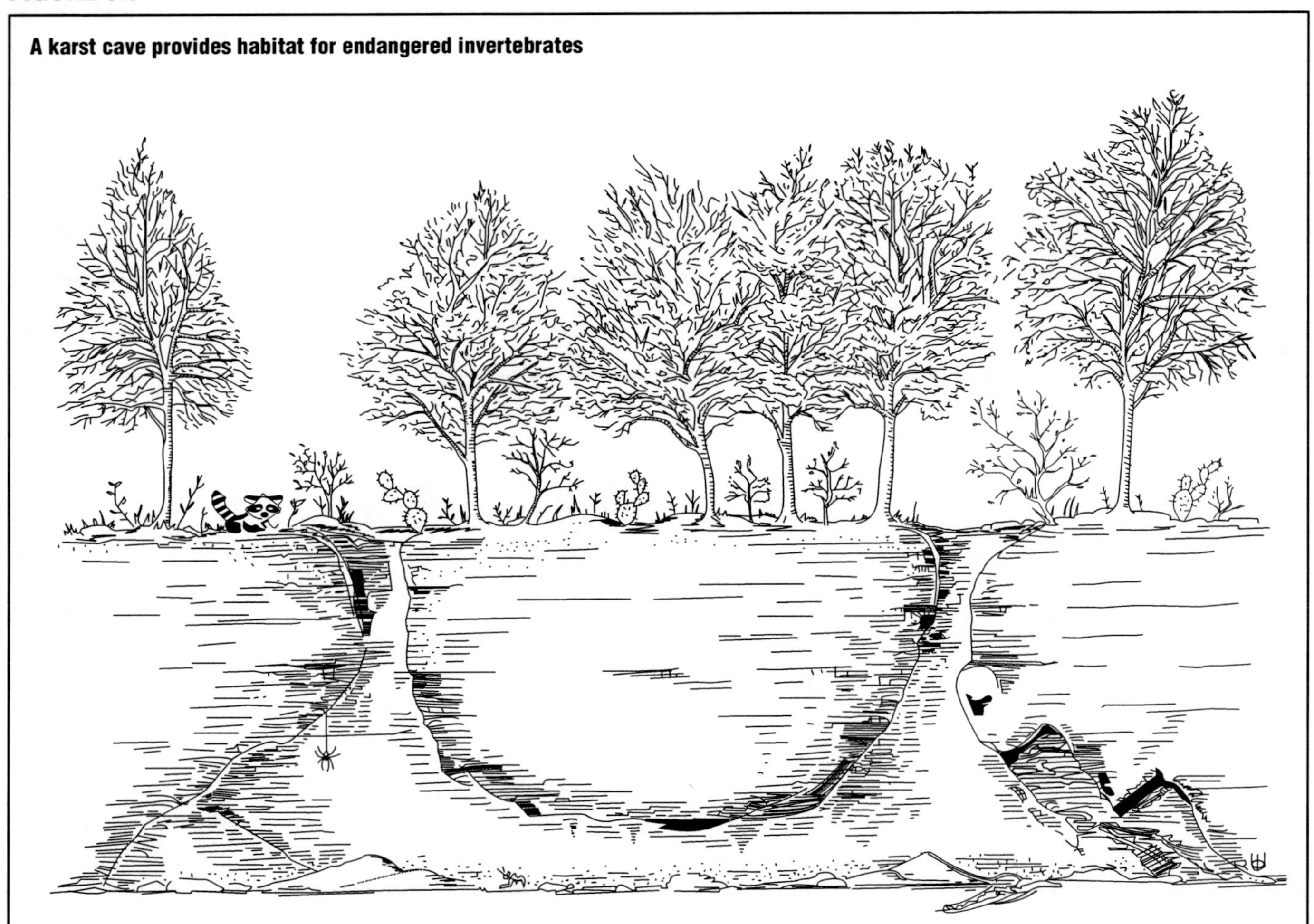

SOURCE: Lisa O'Donnell, William R. Elliott, and Ruth A. Stanford, "Front Cover," in *Recovery Plan for Endangered Karst Invertebrates in Travis and Williamson Counties, Texas*, U.S. Department of the Interior, U.S. Fish and Wildlife Service, 1994, http://ecos.fws.gov/docs/recovery_plans/1994/940825.pdf (accessed November 12, 2011)

holes, and other openings. There are dozens of these karst caves located in Bexar, Travis, and Williamson Counties in Texas. In recent decades scientists have discovered unusual invertebrate species living in these caves. The tiny cave dwellers are eyeless and have no pigment (color) to their bodies. Ten of the creatures have been added to the endangered species list. They include four meshweavers (tiny web-making arachnids), three harvestmen (commonly known as daddy longlegs or granddaddy longlegs), two true spiders, and one pseudoscorpion.

The species were listed under the ESA after a collection of conservation groups petitioned the USFWS in 1992. The creatures were listed as endangered in 2000. In 2003 approximately 1,000 acres (405 ha) were designated as critical habitat for six of the arachnids. In addition, four of the species are included in the USFWS's *Recovery Plan for Endangered Karst Invertebrates in Travis and Williamson Counties, Texas* (August 1994, http://ecos.fws.gov/docs/recovery_plan/940825.pdf), which also covers other imperiled invertebrate species living in the caves.

In December 2009 the USFWS published five-year reviews for four of the arachnids: Bee Creek Cave harvestman, Bone Cave harvestman, Tooth Cave pseudoscorpion, and Tooth Cave spider. The agency recommended maintaining the endangered listings for all of the species, noting that no significant steps had been taken toward protecting the caves in which the spiders occur.

In *Bexar County Karst Invertebrates Draft Recovery Plan* (March 2008, http://ecos.fws.gov/docs/recovery_plan/080516.pdf), the USFWS covers nine invertebrate species that are found in the karst caves of Bexar County, Texas. Included in the plan are six endangered arachnids: Braken Bat Cave meshweaver, Cokendolpher Cave harvestman, Government Canyon Bat Cave meshweaver, Government Canyon Bat Cave spider, Madla's Cave meshweaver, and Robber Baron Cave meshweaver. The USFWS notes that the health of the species' cave habitat is very dependent on surface conditions above the cave. This is because the invertebrates depend on surface plants and animals to provide nutrients, for example, through leafs and animal droppings. The recovery plan calls for scientists to select a limited number of the caves (or clusters of caves) and preserving them and their surface environments for the benefit of the imperiled species. As of January 2012, a final recovery plan had not been published for the Bexar County Cave species.

In February 2011 the USFWS (http://www.gpo.gov/fdsys/pkg/FR-2011-02-22/pdf/2011-3038.pdf) proposed revising critical habitat for the Braken Bat Cave meshweaver, the Cokendolpher Cave harvestman, the Madla's Cave meshweaver, and the Robber Baron Cave meshweaver and suggested establishing critical habitat for the Government Canyon Bat Cave meshweaver and the Government Canyon Bat Cave spider. The agency called for approximately 6,906 acres (2,795 ha) to be established as critical habitat for these arachnids in Bexar County. As of January 2012, a final decision on the proposal had not been published.

FIGURE 9.6

Spruce-fir moss spider

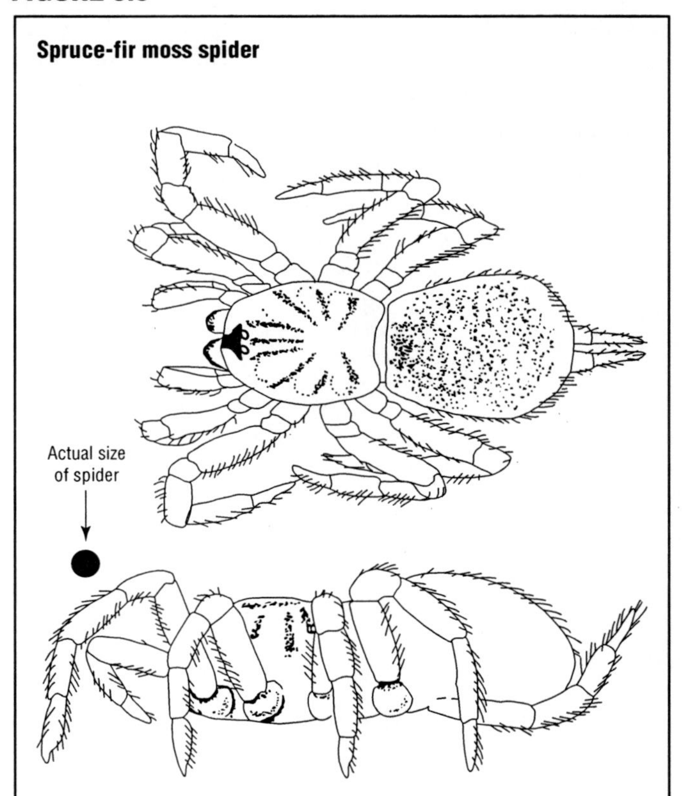

SOURCE: "Untitled," in *Spruce-Fir Moss Spider*, U.S. Department of the Interior, U.S. Fish and Wildlife Service, August 2000, http://www.fws.gov/asheville/pdfs/Spruce%20Fir%20Moss%20Spider.pdf (accessed November 12, 2011)

Kauai Cave Wolf Spiders

The Kauai Cave wolf spider, which ranges from about 0.5 to 0.75 of an inch (1.3 to 1.9 cm) in length, is a blind species found only in special caves on the southern part of the island of Kauai in Hawaii. These caves were formed by young lava flows. Unlike most other spiders, which trap their prey in webs, the Kauai Cave wolf spider hunts its prey directly. Its prey includes the Kauai Cave amphipod, a species that is also highly endangered. The USFWS originally listed both species as endangered in January 2000. Female Kauai Cave wolf spiders lay 15 to 30 eggs per clutch and carry the young on their back after hatching. Cave species are extremely sensitive to changes in temperature and light. In April 2006 the USFWS published *Recovery Plan for the Kauai Cave Arthropods: The Kauai Cave Wolf Spider (*Adelocosa anops*) and the Kauai Cave Amphipod (*Spelaeorchestia koloana*)* (http://ecos.fws.gov/docs/recovery_plan/060719.pdf). At that time fewer than 28 individual wolf spiders were known to exist. The critical habitat established for the species in 2003 includes 14 units totaling 272 acres (110 ha) in the southern part of the island.

Spruce-Fir Moss Spiders

The spruce-fir moss spider is a member of the tarantula family, but is very tiny, measuring only 0.1 to 0.15 of an inch (0.3 to 0.4 cm) in length when fully grown. (See Figure 9.6.) According to the USFWS, in "Spruce-Fir Moss Spider" (August 2000, http://www.fws.gov/asheville/pdfs/Spruce%20Fir%20Moss%20Spider.pdf), the spider is found only in the highest mountain forests of western North Carolina and eastern Tennessee. These forests contain Fraser fir and red spruce trees. The spider lives in damp moss mats that grow on rocks sheltered beneath the trees.

The spider was placed on the list of endangered species in 1995. Its populations have declined largely due to the introduction of an invasive European insect species: the balsam-woolly adelgid. The balsam-woolly adelgid infests Fraser fir trees, causing them to die within two to seven years. The death and subsequent toppling of many fir trees thins the forest, allowing winds to blow over other types of trees. The resulting increase in light level and temperature causes the moss mats on the forest floor to dry up.

In July 2001 the USFWS designated critical habitat for the species, including areas in the Great Smoky Mountains National Park and the Pisgah and Cherokee National Forests, as well as a preserve managed by the

Nature Conservancy. This designation of critical habitat was prompted by a lawsuit against the USFWS, which had previously deemed designating critical habitat "not prudent" because it believed the spider would be more vulnerable to collectors.

In July 2008 the USFWS initiated a five-year status review of 20 southeastern species, including the spruce-fir moss spider. As of January 2012, the results of the review had not been published.

THREATENED AND ENDANGERED FOREIGN SPECIES OF SPIDERS

The IUCN notes in *Red List of Threatened Species Version 2011.2* that 19 species of arachnids were threatened in 2011. This number accounted for 58% of the 33 evaluated species, but less than 0.1% of the more than 102,000 described arachnid species.

As of November 2011, there were no foreign spiders listed under the ESA.

CHAPTER 10
PLANTS

Plants belong to the Plantae kingdom. In general, there are two types of land-growing plants: vascular and nonvascular. Vascular plants have specially developed organs similar to veins that move liquids through their system. This category includes trees, shrubs, flowers, and grasses. Nonvascular plants are mosses, liverworts, and hornworts. The vast majority of plant species on the earth are vascular plants that reproduce through their flowers.

In science, plants are more often identified by their scientific name than are animals. Plant species are so abundant and diverse that many plants have multiple common names. However, there are plants that have no common names because they are rare or are geographically remote. To avoid confusion, this chapter will include the scientific name for any specific common name given.

Several factors contribute to the endangerment of plant species. Many species are the victims of habitat loss due to land and agricultural development. Others have declined due to pollution or habitat damage, or as a result of competition with invasive species. Still other imperiled plants have succumbed to introduced or unknown plant diseases. Finally, collectors or dealers often illegally seek rare, showy, or unusual plants and have depleted populations through overcollection.

The preservation of plant species is important for many reasons. Not only are plants of aesthetic value but also they are crucial components of every ecosystem on the earth. Furthermore, plants serve several functions that are directly beneficial to humans. First, they provide genetic variation that is used in the breeding of new crop varieties—native plants provide genes that allow for adaptation to local environments and resistance to pests, disease, or drought. In addition, plants are the source of many human medicines.

AMERICAN CHESTNUT TREE: MAKING A COMEBACK?

During the 1800s the American chestnut (*Castanea dentata*) was the predominant tree of many forests in the eastern United States. Its range extended from Maine to Mississippi. (See Figure 10.1.) In *American Chestnut... an American Wood* (February 1973, http://www.fpl.fs.fed.us/documnts/usda/amwood/230chest.pdf), the U.S. Forest Service indicates that the heaviest concentrations were in the southern Appalachian Mountains, where the tree made up more than one-third of the overstory trees (the topmost layer of foliage in a forest). Mature trees reached 3 to 5 feet (0.9 to 1.5 m) in diameter and rose to 90 feet (27 m) in height with a huge canopy. The species was fast growing and produced a light, durable wood that was extremely popular for firewood and for making furniture, shingles, caskets, telephone poles, railroad ties, and other products. The trees were also valued for their chestnuts and tannin content. Tannin is an extract used in the leather industry.

In 1904 observers in New York City reported that an unknown blight (disease) was killing American chestnut trees at the Bronx Zoo. By 1940 the blight had spread through the entire range of the species, leaving all the trees dead or dying. The tree structure was not damaged by the disease, so the harvesting of dead trees continued for several more decades. Even though sprouts would grow from the stumps that were left behind, they eventually succumbed to the blight. By the 1970s the American chestnut had been virtually eliminated. Approximately 3 billion to 4 billion trees had been killed. The culprit was a fungus originally called *Endothia parasitica*, but later renamed *Cryphonectria parasitica*. Scientists believe the disease came into the United States with ornamental chestnut trees that were imported from Japan or China. Oriental trees could carry the disease, but not succumb to it, because they had a natural immunity to it.

FIGURE 10.1

Historical distribution of the American chestnut

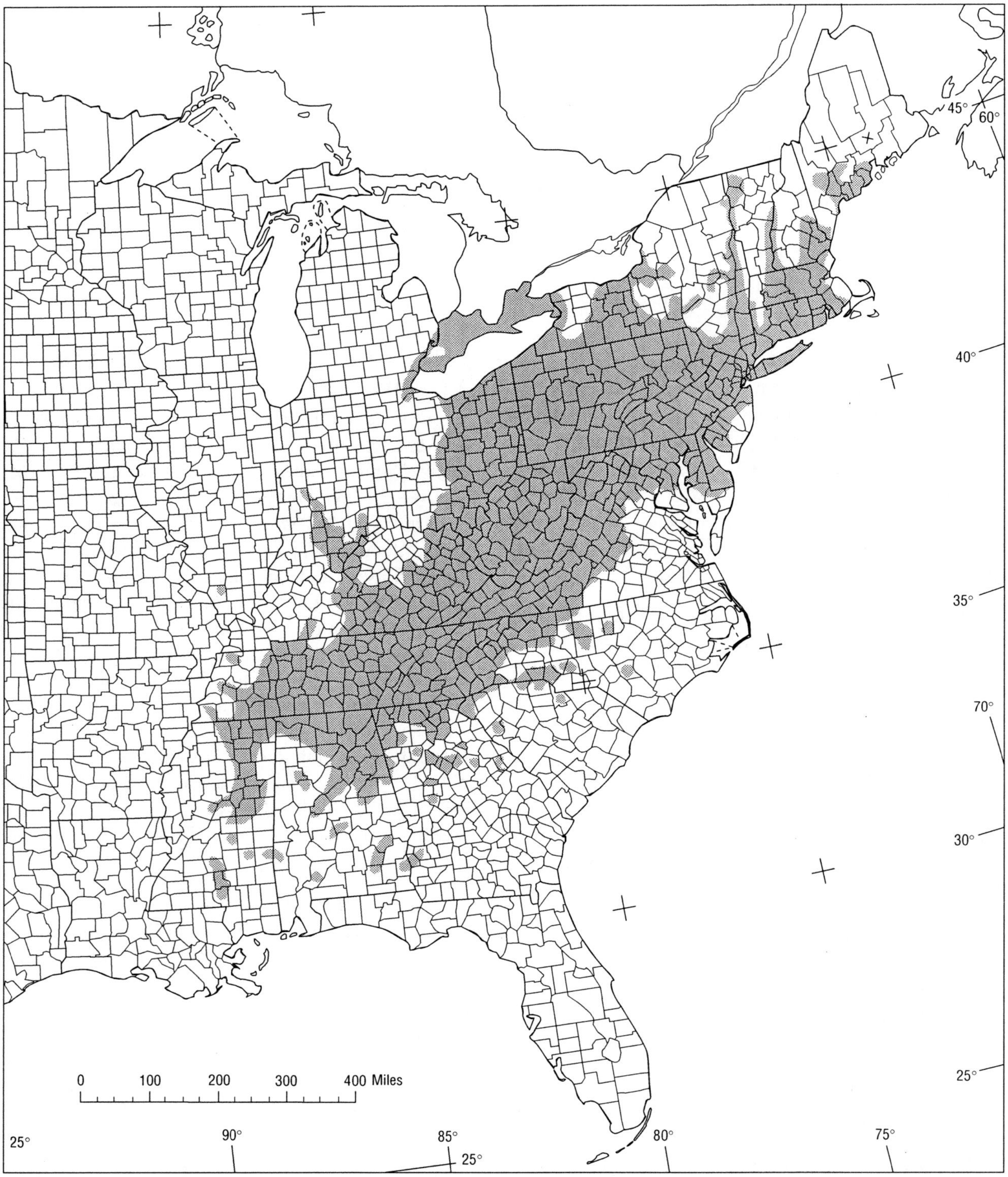

SOURCE: Joseph R. Saucier, "Figure 1. Natural Range of American Chestnut," in *American Chestnut (Castanea Dentata). An American Wood*, U.S. Department of Agriculture, U.S. Forest Service, February 1973, http://www.fpl.fs.fed.us/documnts/usda/amwood/230chest.pdf (accessed November 14, 2011)

During the 1920s scientists began crossing the remaining American chestnut trees with the oriental species. Even though the hybrid trees resulted with some resistance to the blight, they were inferior in quality to the original American species. Advances in genetic research and forestry techniques have gradually led to better hybrids. Three organizations have played a key role in the research: the American Chestnut Foundation (TACF), the American Chestnut Research and Restoration Center (ACRRC) at the State University of New York, and the American Chestnut Cooperators' Foundation (ACCF) at Virginia Tech University. The TACF and the ACRRC focus on crossing naturally blight-resistant Asiatic species with American species, whereas the ACCF produces crosses between American chestnut trees that are found to have some resistance to the blight in hopes of eventually producing offspring with higher resistance. All three organizations are confident that vigorous blight-resistant American chestnut trees can be developed during the 21st century.

In "American Chestnut (*Castanea dentata*) Restoration Research" (2012, http://www.srs.fs.usda.gov/uplandhardwood/americanchestnut.html), the Forest Service indicates that it does not conduct breeding itself, but provides support for TACF activities through research and by allowing plantings of the hybrid trees in the national forests. According to the Forest Service, these hybrids include crosses that are 15/16 (nearly 94%) American chestnut. Since 2005 hundreds of the hybrids have been planted in the national forests in the South and "are growing successfully." The agency notes that it also provides support for studies that are conducted by other organizations that use "pure American seedlings and less advanced genetic material."

Jordan Schrader reports in "Cutting-Edge Science Restoring American Chestnut Tree" (*USA Today*, October 20, 2009) that the genetic manipulation of chestnut trees is criticized by some conservation groups, including the Sierra Club. Neil Carman, a plant biologist for the Sierra Club, observes that genetically modified American chestnut trees "could have unintended consequences, for example, for the insects that pollinate them." However, the TACF believes genetic techniques hold great promise for other plant species that have been decimated by invasive pests.

PROTECTION OF PLANTS UNDER THE ENDANGERED SPECIES ACT

The Endangered Species Act (ESA) of 1973 protects listed plants from deliberate destruction or vandalism. Plants also receive protection under the consultation requirements of the act—that is, all federal agencies must consult with the U.S. Fish and Wildlife Service (USFWS) to determine how best to conserve species and to ensure that no issued permits will jeopardize listed species or harm their habitat.

Regardless, many conservationists believe plants receive less protection than animals under the ESA. First, the ESA only protects plants that are found on federal lands. It imposes no restrictions on private landowners whose property is home to endangered plants. Critics also complain that the USFWS has been slow to list plant species and that damage to plant habitats is not addressed with the same seriousness as for animal species. However, the USFWS points out that the number of plants listed under the ESA has risen dramatically over the past three decades. As of November 2011, this number was nearly 800. In fact, plants (and lichens) accounted for 57% of all species listed under the ESA at that time. (See Figure 10.2.)

In 2000, in an effort to bolster conservation efforts for plants, the USFWS formed an agreement with the Center for Plant Conservation (CPC), a national association of botanical gardens and arboreta. The two organizations are cooperating in developing conservation measures to help save North American plant species, particularly those that are listed as threatened or endangered. Central to the effort is the creation of educational programs that are aimed at informing the public about the importance of plant species for aesthetic, economic, biological, and medical reasons. The CPC also aids in developing recovery plans for listed plant species and collects

FIGURE 10.2

Categories of endangered and threatened species, November 1, 2011

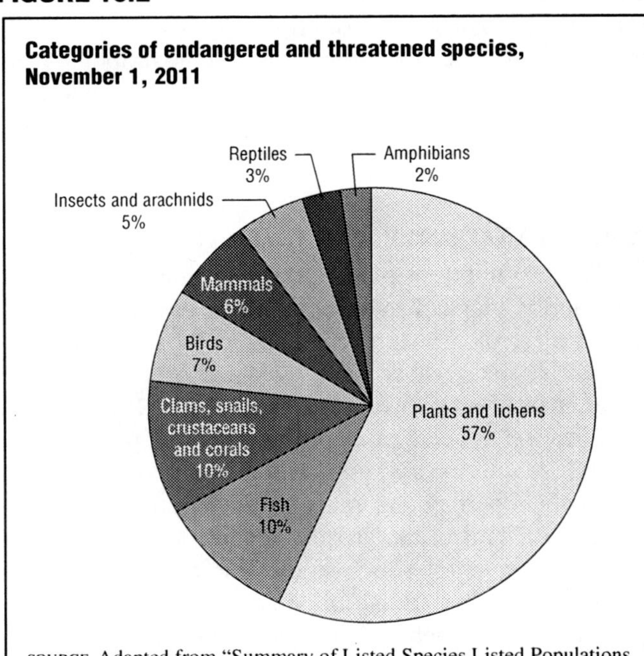

SOURCE: Adapted from "Summary of Listed Species Listed Populations and Recovery Plans as of Tue, 1 Nov 2011 14:29:49 UTC," in *Species Reports*, U.S. Department of the Interior, U.S. Fish and Wildlife Service, November 1, 2011, http://ecos.fws.gov/tess_public/pub/Boxscore.do (accessed November 1, 2011)

listed species for preservation. As of January 2012, the CPC's (http://www.centerforplantconservation.org/) collection included over 750 "of America's most imperiled native plants."

THREATENED AND ENDANGERED U.S. PLANT SPECIES

As of November 2011, there were 794 U.S. plant species listed under the ESA. (See Table 10.1.) Nearly all the plants had recovery plans in place. Because several species of imperiled plants are often found in the same ecosystem, many recovery plans cover multiple plant species.

The status of most plant species has not been studied in detail. Thus, many more plants are probably in danger of extinction than are listed under the ESA.

The 10 plants with the highest expenditures under the ESA during fiscal year (FY) 2010 are listed in Table 10.2. Ute ladies'-tresses (*Spiranthes diluvialis*) dominated the list with nearly $2.8 million in spending. (See Figure 10.3.) It is an orchid found in the interior western states. Almost $1.9 million was spent on Contra Costa goldfields (*Lasthenia conjugens*), a wildflower that grows only in vernal (temporary rainy-season) pools in California and southern Oregon. (See Figure 10.4.) Johnson's seagrass (*Halophila johnsonii*) is a plant that lives underwater in salty lagoons in southeastern Florida. (See Figure 10.5.) It cost more than $1.4 million in FY 2010.

As of November 2011, there were recovery plans for several hundreds of listed plant species. (See Table 10.1.) Many of the plans cover multiple species. The USFWS provides links to the plans and details about them in "Listed FWS/Joint FWS and NMFS Species and Populations with Recovery Plans" (January 7, 2012, http://ecos.fws.gov/tess_public/TESSWebpageRecovery?sort=1).

PLANT TAXONOMY AND CATEGORIZATION

The taxonomy of plant species can be quite complicated and is plagued by disagreements among scientists. Historically, plants were categorized by morphology—physical characteristics, such as shape or color of their leaves, fruit, bark, and so on. During the 1960s a new classification scheme emerged that groups plants based on their evolutionary similarities, for example, their chemical properties and reproductive mechanisms. This taxonomy is part of the broader science known as phylogenetic systematics, which studies the evolutionary relationships between living organisms. In the future the systematics approach is expected to be used to classify all life forms.

In general, plants are assigned to the same taxonomic levels that are used to classify animals. This hierarchical structure includes kingdom, phylum, class, order, family, genus, and species. Beneath the species level, plants can be classified to subspecies, just as in animal taxonomy. There is an additional classification for plants at this level called variety (abbreviated as "var."). Varieties are subgroups with unique differences between them. For example, the invasive species known as kudzu has the scientific name *Pueraria montana*. There are two varieties: *Pueraria montana* var. *lobata* and *Pueraria montana* var. *montana*. The *lobata* variety is commonly found in the United States, whereas the *montana* variety is not.

Plant taxonomy also includes additional taxa (groups) between kingdom and phylum called subkingdom, superdivision, and division that distinguish between broad categories of plants. The subkingdom level distinguishes between vascular and nonvascular plants. Within vascular plants, there are two superdivisions: seed plants and seedless plants. Seed plants are divided into various divisions, the largest of which is flowering plants.

The USFWS uses four broad categories for plant types: conifers and cycads, ferns and allies, lichens, and flowering plants. These categories are division levels or similar groupings.

Conifers and Cycads

Conifers are cone-bearing, woody plants. Most are trees; only a few species are shrubs. Common tree types include cedar, cypress, fir, pine, redwood, and spruce. Cycads are unusual plants often mistaken for palms or ferns. They have thick, soft trunks and large, leaflike crowns. Cycads are found in tropical or semitropical regions. Their rarity makes them popular with collectors. As of November 2011, there were only three conifers or cycads listed under the ESA. (See Table 10.1.)

The Santa Cruz cypress (*Cupressus abramsiana*) and the gowen cypress (*Cupressus goveniana*) are conifers found only in Southern California. The Santa Cruz cypress was listed as endangered, and the gowen cypress was listed as threatened. Both species are imperiled because they have a limited range of distribution and are threatened by alteration and loss of habitat.

The Florida torreya (*Torreya taxifolia*) is a cycad. This scrubby tree is extremely rare and is found only on the bluffs along the Apalachicola River in the Florida Panhandle. An unknown disease virtually wiped out the species in the wild during the 1950s. Thus, it has been given an endangered listing.

Ferns and Allies

According to the International Union for Conservation of Nature (IUCN), in *Red List of Threatened Species Version 2011.2* (2011, http://www.iucnredlist.org/documents/summarystatistics/2011_2_RL_Stats_Table1.pdf), there are approximately 12,000 described species considered to be ferns and allies. Ferns are an abundant and diverse

TABLE 10.1

Endangered and threatened plant species, November 2011

Common name	Scientific name	Species group	Listing status[a]	U.S. or U.S./foreign listed	Recovery plan date	Recovery plan stage[b]
A`e	Zanthoxylum dipetalum var. tomentosum	Flowering plants	E	US	5/11/1998	F
A`e	Zanthoxylum hawaiiense	Flowering plants	E	US	5/11/1998	F
`Ahinahina	Argyroxiphium sandwicense ssp. macrocephalum	Flowering plants	T	US	7/29/1997	F
`Ahinahina	Argyroxiphium sandwicense ssp. sandwicense	Flowering plants	E	US	9/30/1993	F
`Aiakeakua, popolo	Solanum sandwicense	Flowering plants	E	US	9/20/1995	F
`Aiea	Nothocestrum breviflorum	Flowering plants	E	US	5/11/1998	F
`Aiea	Nothocestrum peltatum	Flowering plants	E	US	9/20/1995	F
`Akoko	Chamaesyce celastroides var. kaenana	Flowering plants	E	US	8/10/1998	F
`Akoko	Chamaesyce deppeana	Flowering plants	E	US	8/10/1998	F
`Akoko	Chamaesyce eleanoriae	Flowering plants	E	US	6/17/2010	O
`Akoko	Chamaesyce herbstii	Flowering plants	E	US	8/10/1998	F
`Akoko	Chamaesyce kuwaleana	Flowering plants	E	US	8/10/1998	F
`Akoko	Chamaesyce remyi var. kauaiensis	Flowering plants	E	US	6/17/2010	O
`Akoko	Chamaesyce remyi var. remyi	Flowering plants	E	US	6/17/2010	O
`Akoko	Chamaesyce rockii	Flowering plants	E	US	8/10/1998	F
`Akoko	Euphorbia haeleeleana	Flowering plants	E	US	12/10/2002	F
Alabama canebrake pitcher-plant	Sarracenia rubra alabamensis	Flowering plants	E	US	10/8/1992	F
Alabama leather flower	Clematis socialis	Flowering plants	E	US	12/27/1989	F
Alabama streak-sorus fern	Thelypteris pilosa var. alabamensis	Ferns and Allies	T	US	10/25/1996	F
Alani	Melicope adscendens	Flowering plants	E	US	7/29/1997	F
Alani	Melicope balloui	Flowering plants	E	US	7/29/1997	F
Alani	Melicope degeneri	Flowering plants	E	US	6/17/2010	O
Alani	Melicope haupuensis	Flowering plants	E	US	9/20/1995	F
Alani	Melicope knudsenii	Flowering plants	E	US	9/20/1995	F
Alani	Melicope lydgatei	Flowering plants	E	US	8/10/1998	F
Alani	Melicope mucronulata	Flowering plants	E	US	7/29/1997	F
Alani	Melicope munroi	Flowering plants	E	US	12/10/2002	F
Alani	Melicope ovalis	Flowering plants	E	US	7/29/1997	F
Alani	Melicope pallida	Flowering plants	E	US	8/23/1998	F
Alani	Melicope paniculata	Flowering plants	E	US	6/17/2010	O
Alani	Melicope puberula	Flowering plants	E	US	6/17/2010	O
Alani	Melicope quadrangularis	Flowering plants	E	US	8/23/1998	F
Alani	Melicope reflexa	Flowering plants	E	US	9/26/1996	F
Alani	Melicope saint-johnii	Flowering plants	E	US	8/10/1998	F
Alani	Melicope zahlbruckneri	Flowering plants	E	US	5/11/1998	F
Aleutian shield fern	Polystichum aleuticum	Ferns and Allies	E	US	9/30/1992	F
Amargosa niterwort	Nitrophila mohavensis	Flowering plants	E	US	9/28/1990	F
American chaffseed	Schwalbea americana	Flowering plants	E	US	9/29/1995	F
American hart's-tongue fern	Asplenium scolopendrium var. americanum	Ferns and Allies	T	US/foreign	9/15/1993	F
`Anaunau	Lepidium arbuscula	Flowering plants	E	US	8/10/1998	F
Antioch Dunes evening-primrose	Oenothera deltoides ssp. howellii	Flowering plants	E	US	4/25/1984	RF(1)
`Anunu	Sicyos alba	Flowering plants	E	US	5/11/1998	F
Apalachicola rosemary	Conradina glabra	Flowering plants	E	US	9/27/1994	F
Applegate's milk-vetch	Astragalus applegatei	Flowering plants	E	US	4/10/1998	F
Arizona Cliff-rose	Purshia (=Cowania) subintegra	Flowering plants	E	US	6/16/1995	F
Arizona hedgehog cactus	Echinocereus triglochidiatus var. arizonicus	Flowering plants	E	US	9/30/1984	D
Ash-grey paintbrush	Castilleja cinerea	Flowering plants	T	US	None	—

TABLE 10.1

Endangered and threatened plant species, November 2011 [CONTINUED]

Common name	Scientific name	Species group	Listing status[a]	U.S. or U.S./foreign listed	Recovery plan date	Recovery plan stage[b]
Ash Meadows blazingstar	Mentzelia leucophylla	Flowering plants	T	US	9/28/1990	F
Ash Meadows gumplant	Grindelia fraxino-pratensis	Flowering plants	T	US	9/28/1990	F
Ash Meadows ivesia	Ivesia kingii var. eremica	Flowering plants	T	US	9/28/1990	F
Ash meadows milk-vetch	Astragalus phoenix	Flowering plants	T	US	9/28/1990	F
Ash Meadows sunray	Enceliopsis nudicaulis var. corrugata	Flowering plants	T	US	9/28/1990	F
Ashy dogweed	Thymophylla tephroleuca	Flowering plants	E	US	7/29/1988	F
Asplenium-leaved diellia	Diellia erecta	Ferns and Allies	E	US	7/10/1999 and 12/10/2002	F and F
Aupaka	Isodendrion hosakae	Flowering plants	E	US	5/23/1994	F
Aupaka	Isodendrion laurifolium	Flowering plants	E	US	7/10/1999	F
Aupaka	Isodendrion longifolium	Flowering plants	T	US	12/10/2002	F
Autumn Buttercup	Ranunculus aestivalis (=acriformis)	Flowering plants	E	US	9/16/1991	F
Avon Park harebells	Crotalaria avonensis	Flowering plants	E	US	5/18/1999	F
`Awikiwiki	Canavalia molokaiensis	Flowering plants	E	US	6/2/2011	F
`Awikiwiki	Canavalia napaliensis	Flowering plants	E	US	6/17/2010	O
Awiwi	Centaurium sebaeoides	Flowering plants	E	US	7/10/1999	F
Awiwi	Hedyotis cookiana	Flowering plants	E	US	8/23/1998	F
Bakersfield cactus	Opuntia treleasei	Flowering plants	E	US	9/30/1998	F
Baker's larkspur	Delphinium bakeri	Flowering plants	E	US	None	—
Bariaco	Trichilia triacantha	Flowering plants	E	US	8/20/1991	F
Barneby reed-mustard	Schoenocrambe barnebyi	Flowering plants	E	US	9/14/1994	F
Barneby ridge-cress	Lepidium barnebyanum	Flowering plants	E	US	7/23/1993	F
Beach jacquemontia	Jacquemontia reclinata	Flowering plants	E	US	5/18/1999	F
Beach layia	Layia carnosa	Flowering plants	E	US	9/29/1998	F
Bear Valley sandwort	Arenaria ursina	Flowering plants	T	US	None	—
Beautiful goetzea	Goetzea elegans	Flowering plants	E	US	4/28/1987	F
Beautiful pawpaw	Deeringothamnus pulchellus	Flowering plants	E	US	5/18/1999	F
Ben Lomond spineflower	Chorizanthe pungens var. hartwegiana	Flowering plants	E	US	9/28/1998	F
Ben Lomond wallflower	Erysimum teretifolium	Flowering plants	E	US	9/28/1998	F
Big-leaved crownbeard	Verbesina dissita	Flowering plants	T	US/foreign	None	—
Black lace cactus	Echinocereus reichenbachii var. albertii	Flowering plants	E	US	3/18/1987	F
Black spored quillwort	Isoetes melanospora	Ferns and Allies	E	US	None	—
Blowout penstemon	Penstemon haydenii	Flowering plants	E	US	7/17/1992	F
Blue Ridge goldenrod	Solidago spithamaea	Flowering plants	T	US	10/28/1987	F
Bradshaw's desert-parsley	Lomatium bradshawii	Flowering plants	E	US	6/29/2010	F
Brady pincushion cactus	Pediocactus bradyi	Flowering plants	E	US	3/28/1985	F
Braun's rock-cress	Arabis perstellata	Flowering plants	E	US	7/22/1997	F
Braunton's milk-vetch	Astragalus brauntonii	Flowering plants	E	US	9/30/1999	F
Britton's beargrass	Nolina brittoniana	Flowering plants	E	US	6/20/1996	RF(1)
Brooksville bellflower	Campanula robinsiae	Flowering plants	E	US	6/20/1994	F
Bunched arrowhead	Sagittaria fasciculata	Flowering plants	E	US	9/8/1983	F
Bunched cory cactus	Coryphantha ramillosa	Flowering plants	T	US/foreign	4/13/1990	F
Burke's goldfields	Lasthenia burkei	Flowering plants	E	US	None	—
Butte County meadowfoam	Limnanthes floccosa ssp. californica	Flowering plants	E	US	12/15/2005	F
California jewelflower	Caulanthus californicus	Flowering plants	E	US	9/30/1998	F
California Orcutt grass	Orcuttia californica	Flowering plants	E	US	9/3/1998	F
California seablite	Suaeda californica	Flowering plants	E	US	2/10/2010	D
California taraxacum	Taraxacum californicum	Flowering plants	E	US	None	—
Calistoga allocarya	Plagiobothrys strictus	Flowering plants	E	US	None	—
Canby's dropwort	Oxypolis canbyi	Flowering plants	E	US	4/10/1990	F
Canelo Hills ladies'-tresses	Spiranthes delitescens	Flowering plants	E	US	None	—
Capa rosa	Callicarpa ampla	Flowering plants	E	US	7/31/1995	F
Carter's mustard	Warea carteri	Flowering plants	E	US	5/18/1999	F

TABLE 10.1
Endangered and threatened plant species, November 2011 [CONTINUED]

Common name	Scientific name	Species group	Listing status[a]	U.S. or U.S./foreign listed	Recovery plan date	Recovery plan stage[b]
Carter's panicgrass	Panicum fauriei var. carteri	Flowering plants	E	US	12/9/1993	D
Catalina Island mountain-mahogany	Cercocarpus traskiae	Flowering plants	E	US	None	—
Chapman rhododendron	Rhododendron chapmanii	Flowering plants	E	US	9/8/1983	F
Chinese Camp brodiaea	Brodiaea pallida	Flowering plants	T	US	None	—
Chisos Mountain hedgehog Cactus	Echinocereus chisoensis var. chisoensis	Flowering plants	T	US	12/8/1993	F
Chorro Creek bog thistle	Cirsium fontinale var. obispoense	Flowering plants	E	US	9/28/1998	F
Chupacallos	Pleodendron macranthum	Flowering plants	E	US	9/11/1998	F
Clara Hunt's milk-vetch	Astragalus clarianus	Flowering plants	E	US	None	—
clay-loving wild buckwheat	Eriogonum pelinophilum	Flowering plants	E	US	11/10/1988	F
Clay phacelia	Phacelia argillacea	Flowering plants	E	US	4/12/1982	F
Clay reed-mustard	Schoenocrambe argillacea	Flowering plants	T	US	9/14/1994	F
Clay's hibiscus	Hibiscus clayi	Flowering plants	E	US	9/20/1995	F
Clover lupine	Lupinus tidestromii	Flowering plants	E	US	9/29/1998	F
Coachella Valley milk-vetch	Astragalus lentiginosus var. coachellae	Flowering plants	E	US	None	—
Coastal dunes milk-vetch	Astragalus tener var. titi	Flowering plants	E	US	6/17/2005	F
Cobana negra	Stahlia monosperma	Flowering plants	T	US/foreign	11/1/1996	F
Cochise pincushion cactus	Coryphantha robbinsorum	Flowering plants	T	US/foreign	9/27/1993	F
Colorado Butterfly plant	Gaura neomexicana var. coloradensis	Flowering plants	T	US	5/25/2010	O
Colorado hookless Cactus	Sclerocactus glaucus	Flowering plants	T	US	4/14/2010	O
Colusa grass	Neostapfia colusana	Flowering plants	T	US	12/15/2005	F
Conejo dudleya	Dudleya abramsii ssp. parva	Flowering plants	T	US	9/30/1999	F
Contra Costa goldfields	Lasthenia conjugens	Flowering plants	E	US	12/15/2005	F
Contra Costa wallflower	Erysimum capitatum var. angustatum	Flowering plants	E	US	4/25/1984	RF(1)
Cooke's koki`o	Kokia cookei	Flowering plants	E	US	5/27/1998	F
Cook's holly	Ilex cookii	Flowering plants	E	US	1/31/1991	F
Cook's lomatium	Lomatium cookii	Flowering plants	E	US	9/22/2006	D
Cooley's meadowrue	Thalictrum cooleyi	Flowering plants	E	US	4/21/1994	F
Cooley's water-willow	Justicia cooleyi	Flowering plants	E	US	6/20/1994	F
Coyote ceanothus	Ceanothus ferrisae	Flowering plants	E	US	9/30/1998	F
Crenulate lead-plant	Amorpha crenulata	Flowering plants	E	US	5/18/1999	F
Cumberland rosemary	Conradina verticillata	Flowering plants	T	US	7/12/1996	F
Cumberland sandwort	Arenaria cumberlandensis	Flowering plants	E	US	6/20/1996	F
Cushenbury buckwheat	Eriogonum ovalifolium var. vineum	Flowering plants	E	US	9/30/1997	D
Cushenbury milk-vetch	Astragalus albens	Flowering plants	E	US	9/30/1997	D
Cushenbury oxytheca	Oxytheca parishii var. goodmaniana	Flowering plants	E	US	9/30/1997	D
Davis' green pitaya	Echinocereus viridiflorus var. davisii	Flowering plants	E	US	9/20/1984	F
DeBeque phacelia	Phacelia submutica	Flowering plants	T	US	None (Listed in 2011)	—
Decurrent false aster	Boltonia decurrens	Flowering plants	T	US	9/28/1990	F
Del Mar manzanita	Arctostaphylos glandulosa ssp. crassifolia	Flowering plants	E	US/foreign	None	—
Deltoid spurge	Chamaesyce deltoidea ssp. deltoidea	Flowering plants	E	US	5/18/1999	F
Deseret milk-vetch	Astragalus desereticus	Flowering plants	T	US	None	—
Desert yellowhead	Yermo xanthocephalus	Flowering plants	T	US	2/22/2010	O
Diamond Head schiedea	Schiedea adamantis	Flowering plants	E	US	2/2/1994	F
Dudley Bluffs bladderpod	Lesquerella congesta	Flowering plants	T	US	8/13/1993	F
Dudley Bluffs twinpod	Physaria obcordata	Flowering plants	T	US	8/13/1993	F
Dwarf Bear-poppy	Arctomecon humilis	Flowering plants	E	US	12/31/1985	F
Dwarf-flowered heartleaf	Hexastylis naniflora	Flowering plants	T	US	None	—
Dwarf iliau	Wilkesia hobdyi	Flowering plants	E	US	8/23/1998	F
Dwarf lake iris	Iris lacustris	Flowering plants	T	US/foreign	None	—
Dwarf naupaka	Scaevola coriacea	Flowering plants	E	US	7/29/1997	F

TABLE 10.1

Endangered and threatened plant species, November 2011 [CONTINUED]

Common name	Scientific name	Species group	Listing status[a]	U.S. or U.S./foreign listed	Recovery plan date	Recovery plan stage[b]
Eastern prairie fringed orchid	Platanthera leucophaea	Flowering plants	T	US/foreign	9/29/1999	F
El Dorado bedstraw	Galium californicum ssp. sierrae	Flowering plants	E	US	8/30/2002	F
Elfin tree fern	Cyathea dryopteroides	Ferns and Allies	E	US	1/31/1991	F
Encinitas baccharis	Baccharis vanessae	Flowering plants	T	US	None	—
Erubia	Solanum drymophilum	Flowering plants	E	US	7/9/1992	F
Etonia rosemary	Conradina etonia	Flowering plants	E	US	9/27/1994	F
Eureka Dune grass	Swallenia alexandrae	Flowering plants	E	US	12/13/1982	F
Eureka Valley evening-primrose	Oenothera avita ssp. eurekensis	Flowering plants	E	US	12/13/1982	F
Ewa Plains `akoko	Chamaesyce skottsbergii var. kalaeloana	Flowering plants	E	US	10/5/1993	D
Fassett's locoweed	Oxytropis campestris var. chartacea	Flowering plants	T	US	3/29/1991	F
Few-flowered navarretia	Navarretia leucocephala ssp. pauciflora (=N. pauciflora)	Flowering plants	E	US	12/15/2005	F
Fish Slough milk-vetch	Astragalus lentiginosus var. piscinensis	Flowering plants	T	US	9/30/1998	F
Fleshy owl's-clover	Castilleja campestris ssp. succulenta	Flowering plants	T	US	12/15/2005	F
Florida bonamia	Bonamia grandiflora	Flowering plants	T	US	6/20/1996	RF(1)
Florida golden aster	Chrysopsis floridana	Flowering plants	E	US	8/29/1988	F
Florida perforate cladonia	Cladonia perforata	Lichens	E	US	5/18/1999	F
Florida skullcap	Scutellaria floridana	Flowering plants	T	US	6/22/1994	F
Florida torreya	Torreya taxifolia	Conifers and cycads	E	US	9/9/1986	F
Florida ziziphus	Ziziphus celata	Flowering plants	E	US	5/18/1999	F
Fosberg's love grass	Eragrostis fosbergii	Flowering plants	E	US	8/10/1998	F
Fountain thistle	Cirsium fontinale var. fontinale	Flowering plants	E	US	9/30/1998	F
Four-petal pawpaw	Asimina tetramera	Flowering plants	E	US	5/18/1999	F
Fragrant prickly-apple	Cereus eriophorus var. fragrans	Flowering plants	E	US	5/18/1999	F
Fringed campion	Silene polypetala	Flowering plants	E	US	10/1/1996	D
Furbish lousewort	Pedicularis furbishiae	Flowering plants	E	US/foreign	7/2/1991	RF(1)
Gambel's watercress	Rorippa gambellii	Flowering plants	E	US	9/28/1998	F
Garber's spurge	Chamaesyce garberi	Flowering plants	T	US	5/18/1999	F
Garrett's mint	Dicerandra christmanii	Flowering plants	E	US	5/18/1999	F
Gaviota Tarplant	Deinandra increscens ssp. villosa	Flowering plants	E	US	None	—
Gentian pinkroot	Spigelia gentianoides	Flowering plants	E	US	3/23/2011	D
Gentner's Fritillary	Fritillaria gentneri	Flowering plants	E	US	8/28/2003	F
Godfrey's butterwort	Pinguicula ionantha	Flowering plants	T	US	6/22/1994	F
Golden Paintbrush	Castilleja levisecta	Flowering plants	T	US/foreign	6/29/2010	F
Golden sedge	Carex lutea	Flowering plants	E	US	None	—
Gowen cypress	Cupressus goveniana ssp. goveniana	Conifers and cycads	T	US	6/17/2005	F
Greene's tuctoria	Tuctoria greenei	Flowering plants	E	US	12/15/2005	F
Green pitcher-plant	Sarracenia oreophila	Flowering plants	E	US	12/12/1994	RF(2)
Guthrie's (=Pyne's) ground-plum	Astragalus bibullatus	Flowering plants	E	US	8/9/2011	F
Gypsum wild-buckwheat	Eriogonum gypsophilum	Flowering plants	T	US	3/30/1984	F
Haha	Cyanea acuminata	Flowering plants	E	US	8/10/1998	F
Haha	Cyanea asarifolia	Flowering plants	E	US	8/23/1998	F
Haha	Cyanea copelandii ssp. copelandii	Flowering plants	E	US	9/26/1996	F
Haha	Cyanea copelandii ssp. haleakalaensis	Flowering plants	E	US	7/10/1999	F
Haha	Cyanea dolichopoda	Flowering plants	E	US	6/17/2010	O
Haha	Cyanea dunbarii	Flowering plants	E	US	None	—
Haha	Cyanea eleeleensis	Flowering plants	E	US	6/17/2010	O
Haha	Cyanea glabra	Flowering plants	E	US	12/10/2002	F
Haha	Cyanea grimesiana ssp. grimesiana	Flowering plants	E	US	7/10/1999	F
Haha	Cyanea grimesiana ssp. obatae	Flowering plants	E	US	8/10/1998	F
Haha	Cyanea hamatiflora ssp. carlsonii	Flowering plants	E	US	5/11/1998	F
Haha	Cyanea hamatiflora ssp. hamatiflora	Flowering plants	E	US	7/10/1999	F

TABLE 10.1

Endangered and threatened plant species, November 2011 [CONTINUED]

Common name	Scientific name	Species group	Listing status[a]	U.S. or U.S./foreign listed	Recovery plan date	Recovery plan stage[b]
Haha	Cyanea humboldtiana	Flowering plants	E	US	8/10/1998	F
Haha	Cyanea kolekoleensis	Flowering plants	E	US	6/17/2010	O
Haha	Cyanea koolauensis	Flowering plants	E	US	8/10/1998	F
Haha	Cyanea kuhihewa	Flowering plants	E	US	6/17/2010	O
Haha	Cyanea lobata	Flowering plants	E	US	7/29/1997	F
Haha	Cyanea longiflora	Flowering plants	E	US	8/10/1998	F
Haha	Cyanea macrostegia ssp. gibsonii	Flowering plants	E	US	9/29/1995	F
Haha	Cyanea mannii	Flowering plants	E	US	9/26/1996	F
Haha	Cyanea mceldowneyi	Flowering plants	E	US	7/29/1997	F
Haha	Cyanea pinnatifida	Flowering plants	E	US	8/10/1998	F
Haha	Cyanea platyphylla	Flowering plants	E	US	5/11/1998	F
Haha	Cyanea procera	Flowering plants	E	US	9/26/1996	F
Haha	Cyanea recta	Flowering plants	T	US	8/23/1998	F
Haha	Cyanea remyi	Flowering plants	E	US	9/20/1995	F
Haha	Cyanea shipmanii	Flowering plants	E	US	9/26/1996	F
Haha	Cyanea st.-johnii	Flowering plants	E	US	8/10/1998	F
Haha	Cyanea stictophylla	Flowering plants	E	US	9/26/1996	F
Haha	Cyanea superba	Flowering plants	E	US	8/10/1998	F
Haha	Cyanea truncata	Flowering plants	E	US	8/10/1998	F
Haha	Cyanea undulata	Flowering plants	E	US	5/31/1994	F
Hairy Orcutt grass	Orcuttia pilosa	Flowering plants	E	US	12/15/2005	F
Hairy rattleweed	Baptisia arachnifera	Flowering plants	E	US	3/19/1984	F
Haiwale	Cyrtandra paliku	Flowering plants	E	US	6/17/2010	O
Ha`iwale	Cyrtandra crenata	Flowering plants	E	US	8/10/1998	F
Ha`iwale	Cyrtandra dentata	Flowering plants	E	US	8/10/1998	F
Ha`iwale	Cyrtandra giffardii	Flowering plants	E	US	9/26/1996	F
Ha`iwale	Cyrtandra limahuliensis	Flowering plants	T	US	8/23/1998	F
Ha`iwale	Cyrtandra munroi	Flowering plants	E	US	9/29/1995	F
Ha`iwale	Cyrtandra oenobarba	Flowering plants	E	US	6/17/2010	O
Ha`iwale	Cyrtandra polyantha	Flowering plants	E	US	8/10/1998	F
Ha`iwale	Cyrtandra subumbellata	Flowering plants	E	US	8/10/1998	F
Ha`iwale	Cyrtandra tintinnabula	Flowering plants	E	US	5/11/1998	F
Ha`iwale	Cyrtandra viridiflora	Flowering plants	E	US	8/10/1998	F
Hala pepe	Pleomele hawaiiensis	Flowering plants	E	US	9/26/1996	F
Harperella	Ptilimnium nodosum	Flowering plants	E	US	3/5/1991	F
Harper's beauty	Harperocallis flava	Flowering plants	E	US	9/14/1983	F
Hartweg's golden sunburst	Pseudobahia bahiifolia	Flowering plants	E	US	None	—
Hau kuahiwi	Hibiscadelphus giffardianus	Flowering plants	E	US	9/26/1996	F
Hau kuahiwi	Hibiscadelphus hualalaiensis	Flowering plants	E	US	5/11/1998	F
Hau kuahiwi	Hibiscadelphus woodii	Flowering plants	E	US	8/23/1998	F
Hawaiian bluegrass	Poa sandvicensis	Flowering plants	E	US	8/23/1998	F
Hawaiian gardenia (=Na`u)	Gardenia brighamii	Flowering plants	E	US	9/30/1993	F
Hawaiian red-flowered geranium	Geranium arboreum	Flowering plants	E	US	7/29/1997	F
Hawaiian vetch	Vicia menziesii	Flowering plants	E	US	5/18/1984	F
Hayun lagu (=(Guam), Tronkon guafi (Rota))	Serianthes nelsonii	Flowering plants	E	US	2/2/1994	F
Heau	Exocarpos luteolus	Flowering plants	E	US	9/20/1995	F
Heliotrope milk-vetch	Astragalus montii	Flowering plants	T	US	9/27/1995	D
Heller's blazingstar	Liatris helleri	Flowering plants	T	US	1/28/2000	RF(1)
Hickman's potentilla	Potentilla hickmanii	Flowering plants	E	US	6/17/2005	F
Hidden Lake bluecurls	Trichostema austromontanum ssp. compactum	Flowering plants	T	US	None	—

TABLE 10.1

Endangered and threatened plant species, November 2011 [CONTINUED]

Common name	Scientific name	Species group	Listing status[a]	U.S. or U.S./foreign listed	Recovery plan date	Recovery plan stage[b]
Highlands scrub hypericum	Hypericum cumulicola	Flowering plants	E	US	5/18/1999	F
Higo Chumbo	Harrisia portoricensis	Flowering plants	T	US	11/12/1996	F
Higuero de sierra	Crescentia portoricensis	Flowering plants	E	US	9/23/1991	F
Hilo ischaemum	Ischaemum byrone	Flowering plants	E	US	9/26/1996	F
Hinckley oak	Quercus hinckleyi	Flowering plants	T	US	9/30/1992	F
Ho`awa	Pittosporum napaliense	Flowering plants	E	US	6/17/2010	O
Hoffmann's rock-cress	Arabis hoffmannii	Flowering plants	E	US	9/26/2000	F
Hoffmann's slender-flowered gilia	Gilia tenuiflora ssp. hoffmannii	Flowering plants	E	US	9/26/2000	F
Holei	Ochrosia kilaueaensis	Flowering plants	E	US	5/11/1998	F
Holmgren milk-vetch	Astragalus holmgreniorum	Flowering plants	E	US	9/29/2006	F
Holy Ghost ipomopsis	Ipomopsis sancti-spiritus	Flowering plants	E	US	9/26/2002	F
Honohono	Haplostachys haplostachya	Flowering plants	E	US	9/20/1993	D
Hoover's spurge	Chamaesyce hooveri	Flowering plants	T	US	12/15/2005	F
Houghton's goldenrod	Solidago houghtonii	Flowering plants	T	US/foreign	9/17/1997	F
Howell's spectacular thelypody	Thelypodium howellii spectabilis	Flowering plants	T	US	6/3/2002	F
Howell's spineflower	Chorizanthe howellii	Flowering plants	E	US	9/29/1998	F
Huachuca water-umbel	Lilaeopsis schaffneriana var. recurva	Flowering plants	E	US/foreign	None	—
Ihi`ihi	Marsilea villosa	Ferns and Allies	E	US	4/18/1996	F
Indian Knob mountain balm	Eriodictyon altissimum	Flowering plants	E	US	9/28/1998	F
Ione (incl. Irish Hill) buckwheat	Eriogonum apricum (incl. var. prostratum)	Flowering plants	E	US	None	—
Ione manzanita	Arctostaphylos myrtifolia	Flowering plants	T	US	None	—
Island Barberry	Berberis pinnata ssp. insularis	Flowering plants	E	US	9/26/2000	F
Island bedstraw	Galium buxifolium	Flowering plants	E	US	9/26/2000	F
Island malacothrix	Malacothrix squalida	Flowering plants	E	US	9/26/2000	F
Island phacelia	Phacelia insularis ssp. insularis	Flowering plants	E	US	9/26/2000	F
Island rush-rose	Helianthemum greenei	Flowering plants	T	US	9/26/2000	F
Jesup's milk-vetch	Astragalus robbinsii var. jesupi	Flowering plants	E	US	11/21/1989	F
Johnson's seagrass	Halophila johnsonii	Flowering plants	T	US	10/4/2002	F
Johnston's frankenia	Frankenia johnstonii	Flowering plants	E	US/foreign	5/24/1988	F
Jones Cycladenia	Cycladenia jonesii (=humilis)	Flowering plants	T	US	12/30/2008	O
Kamakahala	Labordia cyrtandrae	Flowering plants	E	US	8/10/1998	F
Kamakahala	Labordia helleri	Flowering plants	E	US	6/17/2010	O
Kamakahala	Labordia lydgatei	Flowering plants	E	US	5/31/1994	F
Kamakahala	Labordia pumila	Flowering plants	E	US	6/17/2010	O
Kamakahala	Labordia tinifolia var. lanaiensis	Flowering plants	E	US	12/10/2002	F
Kamakahala	Labordia tinifolia var. wahiawaensis	Flowering plants	E	US	8/23/1998	F
Kamakahala	Labordia triflora	Flowering plants	E	US	7/10/1999	F
Kamanomano	Cenchrus agrimonioides	Flowering plants	E	US	12/10/2002	F
Kauai hau kuahiwi	Hibiscadelphus distans	Flowering plants	E	US	6/5/1996	F
Kauila	Colubrina oppositifolia	Flowering plants	E	US	9/26/1996	F
Kaulu	Pteralyxia kauaiensis	Flowering plants	E	US	8/23/1998	F
Kearney's blue-star	Amsonia kearneyana	Flowering plants	E	US	5/24/1993	F
Keck's Checker-mallow	Sidalcea keckii	Flowering plants	E	US	None	—
Kenwood Marsh checker-mallow	Sidalcea oregana ssp. valida	Flowering plants	E	US	None	—
Kern mallow	Eremalche kernensis	Flowering plants	E	US	9/30/1998	F
Key tree cactus	Pilosocereus robinii	Flowering plants	E	US/foreign	5/18/1999	F
Kincaid's Lupine	Lupinus sulphureus (=oreganus) ssp. kincaidii (=var. kincaidii)	Flowering plants	T	US	6/29/2010	F
Kio`ele	Hedyotis coriacea	Flowering plants	E	US	7/29/1997	F
Kiponapona	Phyllostegia racemosa	Flowering plants	E	US	5/11/1998	F

TABLE 10.1

Endangered and threatened plant species, November 2011 [CONTINUED]

Common name	Scientific name	Species group	Listing status[a]	U.S. or U.S./foreign listed	Recovery plan date	Recovery plan stage[b]
Kneeland Prairie penny-cress	Thlaspi californicum	Flowering plants	E	US	8/14/2003	F
Knieskern's Beaked-rush	Rhynchospora knieskernii	Flowering plants	T	US	9/29/1993	F
Knowlton's cactus	Pediocactus knowltonii	Flowering plants	E	US	3/29/1985	F
Kodachrome bladderpod	Lesquerella tumulosa	Flowering plants	E	US	10/27/2009	O
Kohe malama malama o kanaloa	Kanaloa kahoolawensis	Flowering plants	E	US	12/10/2002	F
Koki`o	Kokia drynarioides	Flowering plants	E	US	5/6/1994	F
Koki`o	Kokia kauaiensis	Flowering plants	E	US	8/23/1998	F
Koki`o ke`oke`o	Hibiscus arnottianus ssp. immaculatus	Flowering plants	E	US	9/26/1996	F
Koki`o ke`oke`o	Hibiscus waimeae ssp. hannerae	Flowering plants	E	US	9/20/1995	F
Kolea	Myrsine juddii	Flowering plants	E	US	8/10/1998	F
Kolea	Myrsine knudsenii	Flowering plants	E	US	6/17/2010	O
Kolea	Myrsine linearifolia	Flowering plants	T	US	8/23/1998	F
Kolea	Myrsine mezii	Flowering plants	E	US	6/17/2010	O
Ko`oko`olau	Bidens micrantha ssp. kalealaha	Flowering plants	E	US	7/29/1997	F
Ko`oko`olau	Bidens wiebkei	Flowering plants	E	US	5/20/1998	F
Ko`oloa`ula	Abutilon menziesii	Flowering plants	E	US	9/29/1995	F
Kopa	Hedyotis schlechtendahliana var. remyi	Flowering plants	E	US	12/10/2002	F
Kopiko	Psychotria grandiflora	Flowering plants	E	US	6/17/2010	O
Kopiko	Psychotria hobdyi	Flowering plants	E	US	6/17/2010	O
Kral's water-plantain	Sagittaria secundifolia	Flowering plants	T	US	8/12/1991	F
Kuahiwi laukahi	Plantago hawaiensis	Flowering plants	E	US	9/26/1996	F
Kuahiwi laukahi	Plantago princeps	Flowering plants	E	US	7/10/1999	F
Kuawawaenohu	Alsinidendron lychnoides	Flowering plants	E	US	8/23/1998	F
Kuenzler hedgehog cactus	Echinocereus fendleri var. kuenzleri	Flowering plants	E	US	3/28/1985	F
Kula wahine noho	Isodendrion pyrifolium	Flowering plants	E	US	9/26/1996	F
Kulu`i	Nototrichium humile	Flowering plants	E	US	8/10/1998	F
La Graciosa thistle	Cirsium loncholepis	Flowering plants	E	US	None	—
Laguna Beach liveforever	Dudleya stolonifera	Flowering plants	T	US	None	—
Lake County stonecrop	Parvisedum leiocarpum	Flowering plants	E	US	12/15/2005	F
Lakela's mint	Dicerandra immaculata	Flowering plants	E	US	5/18/1999	F
Lakeside daisy	Hymenoxys herbacea	Flowering plants	T	US/foreign	9/19/1990	F
Lanai sandalwood (= `iliahi)	Santalum freycinetianum var. lanaiense	Flowering plants	E	US	9/29/1995	F
Lane Mountain milk-vetch	Astragalus jaegerianus	Flowering plants	E	US	None	—
Large-flowered fiddleneck	Amsinckia grandiflora	Flowering plants	E	US	9/29/1997	F
Large-flowered skullcap	Scutellaria montana	Flowering plants	T	US	5/15/1996	F
Large-flowered woolly Meadowfoam	Limnanthes floccosa ssp. grandiflora	Flowering plants	E	US	9/22/2006	D
Large-fruited sand-verbena	Abronia macrocarpa	Flowering plants	E	US	9/30/1992	F
Last Chance townsendia	Townsendia aprica	Flowering plants	T	US	8/20/1993	F
Lau `ehu	Panicum niihauense	Flowering plants	E	US	12/10/2002	F
Laulihilihi	Schiedea stellarioides	Flowering plants	E	US	9/20/1995	F
Layne's butterweed	Senecio layneae	Flowering plants	T	US	8/30/2002	F
Leafy prairie-clover	Dalea foliosa	Flowering plants	E	US	9/30/1996	F
Leedy's roseroot	Rhodiola integrifolia ssp. leedyi	Flowering plants	T	US	9/25/1998	F
Lee pincushion cactus	Coryphantha sneedii var. leei	Flowering plants	T	US	3/21/1986	F
lehua makanoe	Lysimachia daphnoides	Flowering plants	E	US	6/17/2010	O
Lewton's polygala	Polygala lewtonii	Flowering plants	E	US	5/18/1999	F
Liliwai	Acaena exigua	Flowering plants	E	US	7/29/1997	F
Little Aguja (=Creek) Pondweed	Potamogeton clystocarpus	Flowering plants	E	US	6/20/1994	F
Little amphianthus	Amphianthus pusillus	Flowering plants	T	US	7/7/1993	F
Lloyd's Mariposa cactus	Echinomastus mariposensis	Flowering plants	T	US/foreign	4/13/1990	F

TABLE 10.1

Endangered and threatened plant species, November 2011 [CONTINUED]

Common name	Scientific name	Species group	Listing status[a]	U.S. or U.S./foreign listed	Recovery plan date	Recovery plan stage[b]
Loch Lomond coyote thistle	Eryngium constancei	Flowering plants	E	US	12/15/2005	F
Lompoc yerba santa	Eriodictyon capitatum	Flowering plants	E	US	None	—
Longspurred mint	Dicerandra cornutissima	Flowering plants	E	US	7/1/1987	F
Louisiana quillwort	Isoetes louisianensis	Ferns and Allies	E	US	9/30/1996	F
Lo`ulu	Pritchardia affinis	Flowering plants	E	US	9/26/1996	F
Lo`ulu	Pritchardia kaalae	Flowering plants	E	US	8/10/1998	F
Lo`ulu	Pritchardia munroi	Flowering plants	E	US	9/26/1996	F
Lo`ulu	Pritchardia napaliensis	Flowering plants	E	US	8/23/1998	F
Lo`ulu	Pritchardia remota	Flowering plants	E	US	3/31/1998	F
Lo`ulu	Pritchardia schattaueri	Flowering plants	E	US	5/11/1998	F
Lo`ulu	Pritchardia viscosa	Flowering plants	E	US	8/23/1998	F
Lyon's pentachaeta	Pentachaeta lyonii	Flowering plants	E	US	9/30/1999	F
Lyrate bladderpod	Lesquerella lyrata	Flowering plants	T	US	10/17/1996	F
MacFarlane's four-o'clock	Mirabilis macfarlanei	Flowering plants	T	US	6/30/2000	RF(1)
Maguire primrose	Primula maguirei	Flowering plants	T	US	9/27/1990	F
Mahoe	Alectryon macrococcus	Flowering plants	E	US	7/29/1997	F
Makou	Peucedanum sandwicense	Flowering plants	T	US	8/23/1998	F
Malheur wire-lettuce	Stephanomeria malheurensis	Flowering plants	E	US	3/21/1991	F
Mancos milk-vetch	Astragalus humillimus	Flowering plants	E	US	12/20/1989	F
Mann's bluegrass	Poa mannii	Flowering plants	E	US	8/23/1998	F
Many-flowered navarretia	Navarretia leucocephala ssp. plieantha	Flowering plants	E	US	12/15/2005	F
Ma`oli`oli	Schiedea apokremnos	Flowering plants	E	US	9/20/1995	F
Ma`oli`oli	Schiedea kealiae	Flowering plants	E	US	8/10/1998	F
Mapele	Cyrtandra cyaneoides	Flowering plants	E	US	8/23/1998	F
Marcescent dudleya	Dudleya cymosa ssp. marcescens	Flowering plants	T	US	9/30/1999	F
Marin dwarf-flax	Hesperolinon congestum	Flowering plants	T	US	9/30/1998	F
Mariposa pussypaws	Calyptridium pulchellum	Flowering plants	T	US	None	—
Marsh Sandwort	Arenaria paludicola	Flowering plants	E	US	9/28/1998	F
Mat-forming quillwort	Isoetes tegetiformans	Ferns and Allies	E	US	7/7/1993	F
Maui remya	Remya mauiensis	Flowering plants	E	US	7/29/1997	F
Mauna Loa (=Ka'u) silversword	Argyroxiphium kauense	Flowering plants	E	US	11/21/1995	F
McDonald's rock-cress	Arabis macdonaldiana	Flowering plants	E	US	2/28/1984	F
Mead's milkweed	Asclepias meadii	Flowering plants	T	US	9/22/2003	F
Mehamehame	Flueggea neowawraea	Flowering plants	E	US	7/10/1999	F
Menzies' wallflower	Erysimum menziesii	Flowering plants	E	US	9/29/1998	F
Mesa Verde cactus	Sclerocactus mesae-verdae	Flowering plants	T	US	3/30/1984	F
Metcalf Canyon jewelflower	Streptanthus albidus ssp. albidus	Flowering plants	E	US	9/30/1998	F
Mexican flannelbush	Fremontodendron mexicanum	Flowering plants	E	US/foreign	None	—
Miccosukee gooseberry	Ribes echinellum	Flowering plants	T	US	1/0/1900	0
Michaux's sumac	Rhus michauxii	Flowering plants	E	US	4/30/1993	F
Michigan monkey-flower	Mimulus michiganensis	Flowering plants	E	US	9/17/1997	F
Minnesota dwarf trout lily	Erythronium propullans	Flowering plants	E	US	12/16/1987	F
Missouri bladderpod	Physaria filiformis	Flowering plants	T	US	4/7/1988	F
Mohr's Barbara button	Marshallia mohrii	Flowering plants	T	US	11/26/1991	F
Monterey clover	Trifolium trichocalyx	Flowering plants	E	US	6/17/2005	F
Monterey gilia	Gilia tenuiflora ssp. arenaria	Flowering plants	E	US	9/29/1998	F
Monterey spineflower	Chorizanthe pungens var. pungens	Flowering plants	T	US	9/29/1998	F
Morefield's leather flower	Clematis morefieldii	Flowering plants	E	US	5/3/1994	F
Morro manzanita	Arctostaphylos morroensis	Flowering plants	T	US	9/28/1998	F
Mountain golden heather	Hudsonia montana	Flowering plants	T	US	9/14/1983	F
Mountain sweet pitcher-plant	Sarracenia rubra ssp. jonesii	Flowering plants	E	US	8/13/1990	F
Munz's onion	Allium munzii	Flowering plants	E	US	None	—

TABLE 10.1

Endangered and threatened plant species, November 2011 [CONTINUED]

Common name	Scientific name	Species group	Listing status[a]	U.S. or U.S./foreign listed	Recovery plan date	Recovery plan stage[b]
Naenae	Dubautia kalalauensis	Flowering plants	E	US	6/17/2010	O
Naenae	Dubautia kenwoodii	Flowering plants	E	US	6/17/2010	O
Na`ena`e	Dubautia herbstobatae	Flowering plants	E	US	8/10/1998	F
Na`ena`e	Dubautia imbricata imbricata	Flowering plants	E	US	6/17/2010	O
Na`ena`e	Dubautia latifolia	Flowering plants	E	US	8/23/1998	F
Na`ena`e	Dubautia pauciflorula	Flowering plants	E	US	5/31/1994	F
Na`ena`e	Dubautia plantaginea magnifolia	Flowering plants	E	US	6/17/2010	O
Na`ena`e	Dubautia plantaginea ssp. humilis	Flowering plants	E	US	12/10/2002	F
Na`ena`e	Dubautia waialealae	Flowering plants	E	US	6/17/2010	O
(=Na`ena`e) lo`ulu	Pritchardia hardyi	Flowering plants	E	US	6/17/2010	O
Nani wai`ale`ale	Viola kauaiensis var. wahiawaensis	Flowering plants	E	US	9/20/1995	F
Nanu	Gardenia mannii	Flowering plants	E	US	8/10/1998	F
Napa bluegrass	Poa napensis	Flowering plants	E	US	None	—
Na Pali beach hedyotis	Hedyotis st.-johnii	Flowering plants	E	US	9/20/1995	F
(=Native yellow hibiscus) ma`o hau hele	Hibiscus brackenridgei	Flowering plants	E	US	7/10/1999	F
Navajo sedge	Carex specuicola	Flowering plants	T	US	9/24/1987	F
Navasota ladies'-tresses	Spiranthes parksii	Flowering plants	E	US	9/21/1984	F
Nehe	Lipochaeta fauriei	Flowering plants	E	US	8/23/1998	F
Nehe	Lipochaeta kamolensis	Flowering plants	E	US	7/29/1997	F
Nehe	Lipochaeta lobata var. leptophylla	Flowering plants	E	US	8/10/1998	F
Nehe	Lipochaeta micrantha	Flowering plants	E	US	9/20/1995	F
Nehe	Lipochaeta tenuifolia	Flowering plants	E	US	8/10/1998	F
Nehe	Lipochaeta waimeaensis	Flowering plants	E	US	9/20/1995	F
Nellie cory cactus	Coryphantha minima	Flowering plants	E	US	9/20/1984	F
Nelson's checker-mallow	Sidalcea nelsoniana	Flowering plants	T	US	6/29/2010	F
Nevin's barberry	Berberis nevinii	Flowering plants	E	US	None	—
Nichol's Turk's head cactus	Echinocactus horizonthalonius var. nicholii	Flowering plants	E	US	4/14/1986	F
Nioi	Eugenia koolauensis	Flowering plants	E	US	8/10/1998	F
Nipomo Mesa lupine	Lupinus nipomensis	Flowering plants	E	US	None	—
No common name	Abutilon eremitopetalum	Flowering plants	E	US	9/29/1995	F
No common name	Abutilon sandwicense	Flowering plants	E	US	8/10/1998	F
No common name	Achyranthes mutica	Flowering plants	E	US	12/10/2002	F
No common name	Adiantum vivesii	Ferns and Allies	E	US	1/17/1995	F
No common name	Alsinidendron obovatum	Flowering plants	E	US	8/10/1998	F
No common name	Alsinidendron trinerve	Flowering plants	E	US	8/10/1998	F
No common name	Alsinidendron viscosum	Flowering plants	E	US	9/20/1995	F
No common name	Amaranthus brownii	Flowering plants	E	US	3/31/1998	F
No common name	Aristida chaseae	Flowering plants	E	US	7/31/1995	F
No common name	Asplenium fragile var. insulare	Ferns and Allies	E	US	4/10/1998	F
No common name	Auerodendron pauciflorum	Flowering plants	E	US	9/29/1997	F
No common name	Bonamia menziesii	Flowering plants	E	US	7/10/1999	F
No common name	Calyptranthes thomasiana	Flowering plants	E	US/foreign	9/30/1997	F
No common name	Catesbaea melanocarpa	Flowering plants	E	US	8/18/2005	F
No common name	Chamaecrista glandulosa var. mirabilis	Flowering plants	E	US	5/12/1994	F
No common name	Chamaesyce halemanui	Flowering plants	E	US	8/23/1998	F
No common name	Cordia bellonis	Flowering plants	E	US	10/1/1999	F
No common name	Cranichis ricartii	Flowering plants	E	US	7/15/1996	F
No common name	Cyanea (=Rollandia) crispa	Flowering plants	E	US	8/10/1998	F
No common name	Daphnopsis hellerana	Flowering plants	E	US	8/7/1992	F
No common name	Delissea rhytidosperma	Flowering plants	E	US	8/23/1998	F
No common name	Delissea undulata	Flowering plants	E	US	5/11/1998	F
No common name	Diellia falcata	Ferns and Allies	E	US	8/10/1998	F

TABLE 10.1

Endangered and threatened plant species, November 2011 [CONTINUED]

Common name	Scientific name	Species group	Listing status[a]	U.S. or U.S./foreign listed	Recovery plan date	Recovery plan stage[b]
No common name	Diellia mannii	Ferns and Allies	E	US	6/17/2010	O
No common name	Diellia pallida	Ferns and Allies	E	US	8/23/1998	F
No common name	Diellia unisora	Ferns and Allies	E	US	8/10/1998	F
No common name	Diplazium molokaiense	Ferns and Allies	E	US	4/10/1998	F
No common name	Doryopteris angelica	Ferns and Allies	E	US	6/17/2010	O
No common name	Elaphoglossum serpens	Ferns and Allies	E	US	1/17/1995	F
No common name	Eugenia woodburyana	Flowering plants	E	US	10/6/1998	F
No common name	Gahnia lanaiensis	Flowering plants	E	US	9/29/1995	F
No common name	Geocarpon minimum	Flowering plants	T	US	7/26/1993	F
No common name	Gesneria pauciflora	Flowering plants	T	US	10/6/1998	F
No common name	Gouania hillebrandii	Flowering plants	E	US	7/16/1990	F
No common name	Gouania meyenii	Flowering plants	E	US	8/10/1998	F
No common name	Gouania vitifolia	Flowering plants	E	US	8/10/1998	F
No common name	Hedyotis degeneri	Flowering plants	E	US	8/10/1998	F
No common name	Hedyotis parvula	Flowering plants	E	US	8/10/1998	F
No common name	Hesperomannia arborescens	Flowering plants	E	US	8/10/1998	F
No common name	Hesperomannia arbuscula	Flowering plants	E	US	8/10/1998	F
No common name	Hesperomannia lydgatei	Flowering plants	E	US	5/31/1994	F
No common name	Ilex sintenisii	Flowering plants	E	US	7/31/1995	F
No common name	Keysseria (=Lagenifera) erici	Flowering plants	E	US	6/17/2010	O
No common name	Keysseria (=Lagenifera) helenae	Flowering plants	E	US	6/17/2010	O
No common name	Lepanthes eltoroensis	Flowering plants	E	US	7/15/1996	F
No common name	Leptocereus grantianus	Flowering plants	E	US	7/26/1995	F
No common name	Lipochaeta venosa	Flowering plants	E	US	5/23/1994	F
No common name	Lobelia gaudichaudii ssp. koolauensis	Flowering plants	E	US	8/10/1998	F
No common name	Lobelia monostachya	Flowering plants	E	US	8/10/1998	F
No common name	Lobelia niihauensis	Flowering plants	E	US	8/10/1998	F
No common name	Lobelia oahuensis	Flowering plants	E	US	8/10/1998	F
No common name	Lyonia truncata var. proctorii	Flowering plants	E	US	7/31/1995	F
No common name	Lysimachia filifolia	Flowering plants	E	US	9/20/1995	F
No common name	Lysimachia iniki	Flowering plants	E	US	6/17/2010	O
No common name	Lysimachia lydgatei	Flowering plants	E	US	7/29/1997	F
No common name	Lysimachia maxima	Flowering plants	E	US	None	—
No common name	Lysimachia pendens	Flowering plants	E	US	6/17/2010	O
No common name	Lysimachia scopulensis	Flowering plants	E	US	6/17/2010	O
No common name	Lysimachia venosa	Flowering plants	E	US	6/17/2010	O
No common name	Mariscus fauriei	Flowering plants	E	US	9/26/1996	F
No common name	Mariscus pennatiformis	Flowering plants	E	US	7/10/1999	F
No common name	Mitracarpus maxwelliae	Flowering plants	E	US	10/6/1998	F
No common name	Mitracarpus polycladus	Flowering plants	E	US/foreign	10/6/1998	F
No common name	Munroidendron racemosum	Flowering plants	E	US	8/23/1998	F
No common name	Myrcia paganii	Flowering plants	E	US	9/29/1997	F
No common name	Neraudia angulata	Flowering plants	E	US	8/10/1998	F
No common name	Neraudia ovata	Flowering plants	E	US	9/26/1996	F
No common name	Neraudia sericea	Flowering plants	E	US	12/10/2002	F
No common name	Nesogenes rotensis	Flowering plants	E	US	5/3/2007	F
No common name	Osmoxylon mariannense	Flowering plants	E	US	5/3/2007	F
No common name	Phyllostegia glabra var. lanaiensis	Flowering plants	E	US	9/29/1995	F
No common name	Phyllostegia hirsuta	Flowering plants	E	US	8/10/1998	F
No common name	Phyllostegia hispida	Flowering plants	E	US	6/2/2011	F
No common name	Phyllostegia kaalaensis	Flowering plants	E	US	8/10/1998	F
No common name	Phyllostegia knudsenii	Flowering plants	E	US	9/20/1995	F

TABLE 10.1
Endangered and threatened plant species, November 2011 [CONTINUED]

Common name	Scientific name	Species group	Listing status[a]	U.S. or U.S./foreign listed	Recovery plan date	Recovery plan stage[b]
No common name	Phyllostegia mannii	Flowering plants	E	US	6/2/2011	F
No common name	Phyllostegia mollis	Flowering plants	E	US	8/10/1998	F
No common name	Phyllostegia parviflora	Flowering plants	E	US	7/10/1999	F
No common name	Phyllostegia renovans	Flowering plants	E	US	6/17/2010	O
No common name	Phyllostegia velutina	Flowering plants	E	US	5/11/1998	F
No common name	Phyllostegia waimeae	Flowering plants	E	US	9/20/1995	F
No common name	Phyllostegia warshaueri	Flowering plants	E	US	5/11/1998	F
No common name	Phyllostegia wawrana	Flowering plants	E	US	8/23/1998	F
No common name	Platanthera holochila	Flowering plants	E	US	7/10/1999	F
No common name	Poa siphonoglossa	Flowering plants	E	US	9/20/1995	F
No common name	Polystichum calderonense	Ferns and Allies	E	US	1/17/1995	F
No common name	Pteris lidgatei	Ferns and Allies	E	US	4/10/1998	F
No common name	Remya kauaiensis	Flowering plants	E	US	9/20/1995	F
No common name	Remya montgomeryi	Flowering plants	E	US	9/20/1995	F
No common name	Sanicula mariversa	Flowering plants	E	US	8/10/1998	F
No common name	Sanicula purpurea	Flowering plants	E	US	7/10/1999	F
No common name	Schiedea attenuata	Flowering plants	E	US	6/17/2010	O
No common name	Schiedea haleakalensis	Flowering plants	E	US	7/29/1997	F
No common name	Schiedea helleri	Flowering plants	E	US	9/20/1995	F
No common name	Schiedea hookeri	Flowering plants	E	US	7/10/1999	F
No common name	Schiedea kaalae	Flowering plants	E	US	8/10/1998	F
No common name	Schiedea kauaiensis	Flowering plants	E	US	8/23/1998	F
No common name	Schiedea lydgatei	Flowering plants	E	US	9/26/1996	F
No common name	Schiedea membranacea	Flowering plants	E	US	9/20/1995	F
No common name	Schiedea nuttallii	Flowering plants	E	US	12/10/2002	F
No common name	Schiedea sarmentosa	Flowering plants	E	US	None	—
No common name	Schiedea spergulina var. leiopoda	Flowering plants	E	US	8/23/1998	F
No common name	Schiedea spergulina var. spergulina	Flowering plants	T	US	9/20/1995	F
No common name	Schiedea verticillata	Flowering plants	E	US	3/31/1998	F
No common name	Schoepfia arenaria	Flowering plants	T	US	1/10/1992	F
No common name	Silene alexandri	Flowering plants	E	US	9/26/1996	F
No common name	Silene hawaiiensis	Flowering plants	T	US	9/26/1996	F
No common name	Silene lanceolata	Flowering plants	E	US	9/26/1996	F
No common name	Silene perlmanii	Flowering plants	E	US	8/10/1998	F
No common name	Spermolepis hawaiiensis	Flowering plants	E	US	12/10/2002	F
No common name	Stenogyne angustifolia angustifolia	Flowering plants	E	US	9/20/1993	D
No common name	Stenogyne bifida	Flowering plants	E	US	9/26/1996	F
No common name	Stenogyne campanulata	Flowering plants	E	US	9/20/1995	F
No common name	Stenogyne kanehoana	Flowering plants	E	US	8/10/1998	F
No common name	Stenogyne kealiae	Flowering plants	E	US	6/17/2010	O
No common name	Tectaria estremerana	Ferns and Allies	E	US	1/17/1995	F
No common name	Ternstroemia subsessilis	Flowering plants	E	US	7/31/1995	F
No common name	Tetramolopium arenarium	Flowering plants	E	US	9/26/1996	F
No common name	Tetramolopium filiforme	Flowering plants	E	US	8/10/1998	F
No common name	Tetramolopium lepidotum ssp. lepidotum	Flowering plants	E	US	8/10/1998	F
No common name	Tetramolopium remyi	Flowering plants	E	US	9/29/1995	F
No common name	Tetramolopium rockii	Flowering plants	T	US	9/26/1996	F
No common name	Tetraplasandra bisattenuata	Flowering plants	E	US	6/17/2010	O
No common name	Tetraplasandra flynnii	Flowering plants	E	US	6/17/2010	O
No common name	Thelypteris inabonensis	Ferns and Allies	E	US	1/17/1995	F
No common name	Thelypteris verecunda	Ferns and Allies	E	US	1/17/1995	F
No common name	Thelypteris yaucoensis	Ferns and Allies	E	US	1/17/1995	F

TABLE 10.1
Endangered and threatened plant species, November 2011 [CONTINUED]

Common name	Scientific name	Species group	Listing status[a]	U.S. or U.S./foreign listed	Recovery plan date	Recovery plan stage[b]
No common name	Trematolobelia singularis	Flowering plants	E	US	8/10/1998	F
No common name	Vernonia proctorii	Flowering plants	E	US	7/31/1995	F
No common name	Vigna o-wahuensis	Flowering plants	E	US	7/10/1999	F
No common name	Viola helenae	Flowering plants	E	US	5/31/1994	F
No common name	Viola lanaiensis	Flowering plants	E	US	9/29/1995	F
No common name	Viola oahuensis	Flowering plants	E	US	8/10/1998	F
No common name	Xylosma crenatum	Flowering plants	E	US	9/20/1995	F
Nohoanu	Geranium kauaiense	Flowering plants	E	US	6/17/2010	O
Nohoanu	Geranium multiflorum	Flowering plants	E	US	7/29/1997	F
Northeastern bulrush	Scirpus ancistrochaetus	Flowering plants	E	US	8/25/1993	F
Northern wild monkshood	Aconitum noveboracense	Flowering plants	T	US	9/23/1983	F
North Park phacelia	Phacelia formosula	Flowering plants	E	US	3/21/1986	F
Oha	Delissea rivularis	Flowering plants	E	US	9/20/1995	F
Oha	Delissea subcordata	Flowering plants	E	US	8/10/1998	F
Ohai	Sesbania tomentosa	Flowering plants	E	US	12/10/2002	F
`Oha wai	Clermontia drepanomorpha	Flowering plants	E	US	9/26/1996	F
`Oha wai	Clermontia lindseyana	Flowering plants	E	US	9/26/1996	F
`Oha wai	Clermontia oblongifolia ssp. brevipes	Flowering plants	E	US	6/2/2011	F
`Oha wai	Clermontia oblongifolia ssp. mauiensis	Flowering plants	E	US	7/29/1997	F
`Oha wai	Clermontia peleana	Flowering plants	E	US	5/11/1998	F
`Oha wai	Clermontia pyrularia	Flowering plants	E	US	9/26/1996	F
`Oha wai	Clermontia samuelii	Flowering plants	E	US	7/10/1999	F
`Ohe`ohe	Tetraplasandra gymnocarpa	Flowering plants	E	US	8/10/1998	F
Okeechobee gourd	Cucurbita okeechobeensis ssp. okeechobeensis	Flowering plants	E	US	5/18/1999	F
Olulu	Brighamia insignis	Flowering plants	E	US	8/23/1998	F
Opuhe	Urera kaalae	Flowering plants	E	US	8/10/1998	F
Orcutt's spineflower	Chorizanthe orcuttiana	Flowering plants	E	US	None	—
Osterhout milk-vetch	Astragalus osterhoutii	Flowering plants	E	US	9/30/1992	F
Otay mesa-mint	Pogogyne nudiuscula	Flowering plants	E	US/foreign	9/3/1998	F
Otay tarplant	Deinandra (=Hemizonia) conjugens	Flowering plants	T	US/foreign	12/28/2004	F
Pagosa skyrocket	Ipomopsis polyantha	Flowering plants	E	US	None (Listed in 2011)	—
Pa`iniu	Astelia waialealae	Flowering plants	E	US	6/17/2010	O
Palapalai aumakua	Dryopteris crinalis var. podosorus	Ferns and Allies	E	US	6/17/2010	O
Pallid manzanita	Arctostaphylos pallida	Flowering plants	T	US	4/7/2003	D
Palma de manaca	Calyptronoma rivalis	Flowering plants	T	US	6/25/1992	F
Palmate-bracted bird's beak	Cordylanthus palmatus	Flowering plants	E	US	9/30/1998	F
Palo colorado	Ternstroemia luquillensis	Flowering plants	E	US	7/31/1995	F
Palo de jazmin	Styrax portoricensis	Flowering plants	E	US	7/31/1995	F
Palo de nigua	Cornutia obovata	Flowering plants	E	US	8/7/1992	F
Palo de ramon	Banara vanderbiltii	Flowering plants	E	US	3/15/1991	F
Palo de rosa	Ottoschulzia rhodoxylon	Flowering plants	E	US	9/20/1994	F
Pamakani	Tetramolopium capillare	Flowering plants	E	US	7/29/1997	F
Pamakani	Viola chamissoniana ssp. chamissoniana	Flowering plants	E	US	8/10/1998	F
Papala	Charpentiera densiflora	Flowering plants	E	US	6/17/2010	O
Papery whitlow-wort	Paronychia chartacea	Flowering plants	T	US	5/18/1999	F
Parachute beardtongue	Penstemon debilis	Flowering plants	T	US	None (Listed in 2011)	—
Pariette cactus	Sclerocactus brevispinus	Flowering plants	T	US	4/14/2010	O
Parish's daisy	Erigeron parishii	Flowering plants	T	US	9/30/1997	D
Pauoa	Ctenitis squamigera	Ferns and Allies	E	US	4/10/1998	F

TABLE 10.1

Endangered and threatened plant species, November 2011 [CONTINUED]

Common name	Scientific name	Species group	Listing status[a]	U.S. or U.S./foreign listed	Recovery plan date	Recovery plan stage[b]
Pecos (=puzzle, =paradox) sunflower	Helianthus paradoxus	Flowering plants	T	US	9/15/2005	F
Pedate checker-mallow	Sidalcea pedata	Flowering plants	E	US	7/31/1998	F
Peebles Navajo cactus	Pediocactus peeblesianus var. peeblesianus	Flowering plants	E	US	3/30/1984	F
Peirson's milk-vetch	Astragalus magdalenae var. peirsonii	Flowering plants	T	US	None	—
Pelos del diablo	Aristida portoricensis	Flowering plants	E	US	5/16/1994	F
Pendant kihi fern	Adenophorus periens	Ferns and Allies	E	US	12/10/2002	F
Penland alpine fen mustard	Eutrema penlandii	Flowering plants	T	US	8/31/1993	O
Penland beardtongue	Penstemon penlandii	Flowering plants	E	US	9/30/1992	F
Pennell's bird's-beak	Cordylanthus tenuis ssp. capillaris	Flowering plants	E	US	9/30/1998	F
Persistent trillium	Trillium persistens	Flowering plants	E	US	3/27/1984	F
Peter's Mountain mallow	Iliamna corei	Flowering plants	E	US	9/28/1990	F
Pigeon wings	Clitoria fragrans	Flowering plants	T	US	5/18/1999	F
Pilo	Hedyotis mannii	Flowering plants	E	US	6/2/2011	F
Pilo kea lau li`i	Platydesma rostrata	Flowering plants	E	US	6/17/2010	O
Pima pineapple cactus	Coryphantha scheeri var. robustispina	Flowering plants	E	US/foreign	None	—
Pine Hill ceanothus	Ceanothus roderickii	Flowering plants	E	US	8/30/2002	F
Pine Hill flannelbush	Fremontodendron californicum ssp. decumbens	Flowering plants	E	US	8/30/2002	F
Pismo clarkia	Clarkia speciosa ssp. immaculata	Flowering plants	E	US	9/28/1998	F
Pitcher's thistle	Cirsium pitcheri	Flowering plants	T	US/foreign	9/20/2002	F
Pitkin Marsh lily	Lilium pardalinum ssp. pitkinense	Flowering plants	E	US	None	—
Po`e	Portulaca sclerocarpa	Flowering plants	E	US	5/11/1998	F
Pondberry	Lindera melissifolia	Flowering plants	E	US	9/23/1993	F
Popolo ku mai	Solanum incompletum	Flowering plants	E	US	12/10/2002	F
Prairie bush-clover	Lespedeza leptostachya	Flowering plants	T	US	10/6/1988	F
Presidio clarkia	Clarkia franciscana	Flowering plants	E	US	9/30/1998	F
Presidio Manzanita	Arctostaphylos hookeri var. ravenii	Flowering plants	E	US	10/6/2003	F
Price's potato-bean	Apios priceana	Flowering plants	T	US	2/10/1993	F
Pua `ala	Brighamia rockii	Flowering plants	E	US	9/26/1996	F
Purple amole	Chlorogalum purpureum	Flowering plants	T	US	None	—
Pu`uka`a	Cyperus trachysanthos	Flowering plants	E	US	7/10/1999	F
Pygmy fringe-tree	Chionanthus pygmaeus	Flowering plants	E	US	5/18/1999	F
Red Hills vervain	Verbena californica	Flowering plants	T	US	None	—
Relict trillium	Trillium reliquum	Flowering plants	E	US	1/31/1991	F
Roan Mountain bluet	Hedyotis purpurea var. montana	Flowering plants	E	US	5/13/1996	F
Robust (incl. Scotts Valley) spineflower	Chorizanthe robusta (incl. vars. robusta and hartwegii)	Flowering plants	E	US	12/20/2004	F
Rock gnome lichen	Gymnoderma lineare	Lichens	E	US	9/30/1997	F
Rough-leaved loosestrife	Lysimachia asperulaefolia	Flowering plants	E	US	4/19/1995	F
Rough popcornflower	Plagiobothrys hirtus	Flowering plants	E	US	9/25/2003	F
Round-leaved chaff-flower	Achyranthes splendens var. rotundata	Flowering plants	E	US	10/5/1993	D
Rugel's pawpaw	Deeringothamnus rugelii	Flowering plants	E	US	4/5/1988	F
Running buffalo clover	Trifolium stoloniferum	Flowering plants	E	US	6/27/2007	RF(1)
Ruth's golden aster	Pityopsis ruthii	Flowering plants	E	US	6/11/1992	F
Sacramento Mountains thistle	Cirsium vinaceum	Flowering plants	T	US	9/27/1993	F
Sacramento Orcutt grass	Orcuttia viscida	Flowering plants	E	US	12/15/2005	F
Sacramento prickly poppy	Argemone pleiacantha ssp. pinnatisecta	Flowering plants	E	US	8/31/1994	F
Salt marsh bird's-beak	Cordylanthus maritimus ssp. maritimus	Flowering plants	E	US/foreign	12/6/1985	F
San Benito evening-primrose	Camissonia benitensis	Flowering plants	T	US	9/21/2006	F
San Bernardino bluegrass	Poa atropurpurea	Flowering plants	E	US	None	—
San Bernardino Mountains bladderpod	Lesquerella kingii ssp. bernardina	Flowering plants	E	US	9/30/1997	D

TABLE 10.1

Endangered and threatened plant species, November 2011 [CONTINUED]

Common name	Scientific name	Species group	Listing status[a]	U.S. or U.S./foreign listed	Recovery plan date	Recovery plan stage[b]
San Clemente Island broom	Lotus dendroideus ssp. traskiae	Flowering plants	E	US	1/26/1984	F
San Clemente Island bush-mallow	Malacothamnus clementinus	Flowering plants	E	US	1/26/1984	F
San Clemente Island indian paintbrush	Castilleja grisea	Flowering plants	E	US	1/26/1984	F
San Clemente Island larkspur	Delphinium variegatum ssp. kinkiense	Flowering plants	E	US	1/26/1984	F
San Clemente Island woodland-star	Lithophragma maximum	Flowering plants	E	US	1/26/1984	F
San Diego ambrosia	Ambrosia pumila	Flowering plants	E	US/foreign	None	—
San Diego button-celery	Eryngium aristulatum var. parishii	Flowering plants	E	US	9/3/1998	F
San Diego mesa-mint	Pogogyne abramsii	Flowering plants	E	US	9/3/1998	F
San Diego thornmint	Acanthomintha ilicifolia	Flowering plants	T	US/foreign	None	—
Sandlace	Polygonella myriophylla	Flowering plants	E	US	5/18/1999	F
Sandplain gerardia	Agalinis acuta	Flowering plants	E	US	9/20/1989	F
San Francisco lessingia	Lessingia germanorum (=L.g. var. germanorum)	Flowering plants	E	US	10/6/2003	F
San Francisco Peaks groundsel	Senecio franciscanus	Flowering plants	T	US	7/21/1987	F
San Jacinto Valley crownscale	Atriplex coronata var. notatior	Flowering plants	E	US	None	—
San Joaquin adobe sunburst	Pseudobahia peirsonii	Flowering plants	T	US	None	—
San Joaquin Orcutt grass	Orcuttia inaequalis	Flowering plants	T	US	12/15/2005	F
San Joaquin wooly-threads	Monolopia (=Lembertia) congdonii	Flowering plants	E	US	9/30/1998	F
San Mateo thornmint	Acanthomintha obovata ssp. duttonii	Flowering plants	E	US	9/30/1998	F
San Mateo woolly sunflower	Eriophyllum latilobum	Flowering plants	E	US	9/30/1998	F
San Rafael cactus	Pediocactus despainii	Flowering plants	E	US	10/2/1995	D
Santa Ana River woolly-star	Eriastrum densifolium ssp. sanctorum	Flowering plants	E	US	None	—
Santa Barbara Island liveforever	Dudleya traskiae	Flowering plants	E	US	6/27/1985	F
Santa Clara Valley dudleya	Dudleya setchellii	Flowering plants	E	US	9/30/1998	F
Santa Cruz cypress	Cupressus abramsiana	Conifers and cycads	E	US	9/26/1998	F
Santa Cruz Island bush-mallow	Malacothamnus fasciculatus var. nesioticus	Flowering plants	E	US	9/26/2000	F
Santa Cruz Island dudleya	Dudleya nesiotica	Flowering plants	T	US	9/26/2000	F
Santa Cruz Island fringepod	Thysanocarpus conchuliferus	Flowering plants	E	US	9/26/2000	F
Santa Cruz Island malacothrix	Malacothrix indecora	Flowering plants	E	US	9/26/2000	F
Santa Cruz Island rockcress	Sibara filifolia	Flowering plants	E	US	None	—
Santa Cruz tarplant	Holocarpha macradenia	Flowering plants	T	US	None	—
Santa Monica Mountains dudleyea	Dudleya cymosa ssp. ovatifolia	Flowering plants	T	US	9/30/1999	F
Santa Rosa Island manzanita	Arctostaphylos confertiflora	Flowering plants	E	US	9/26/2000	F
Schweinitz's sunflower	Helianthus schweinitzii	Flowering plants	E	US	4/22/1994	F
Scotts Valley Polygonum	Polygonum hickmanii	Flowering plants	E	US	9/28/1998	F
Scrub blazingstar	Liatris ohlingerae	Flowering plants	E	US	5/18/1999	F
Scrub buckwheat	Eriogonum longifolium var. gnaphalifolium	Flowering plants	T	US	6/20/1996	RF(1)
Scrub lupine	Lupinus aridorum	Flowering plants	E	US	6/20/1996	RF(1)
Scrub mint	Dicerandra frutescens	Flowering plants	E	US	5/18/1999	F
Scrub plum	Prunus geniculata	Flowering plants	E	US	6/20/1996	RF(1)
Seabeach amaranth	Amaranthus pumilus	Flowering plants	T	US	11/12/1996	F
Sebastopol meadowfoam	Limnanthes vinculans	Flowering plants	E	US	None	—
Sensitive joint-vetch	Aeschynomene virginica	Flowering plants	T	US	9/29/1995	F
Sentry milk-vetch	Astragalus cremnophylax var. cremnophylax	Flowering plants	E	US	9/28/2006	F
shale barren rock cress	Arabis serotina	Flowering plants	E	US	8/15/1991	F
Shivwits milk-vetch	Astragalus ampullarioides	Flowering plants	E	US	9/29/2006	F
Short-leaved rosemary	Conradina brevifolia	Flowering plants	E	US	5/18/1999	F
Short's goldenrod	Solidago shortii	Flowering plants	E	US	5/25/1988	F

TABLE 10.1

Endangered and threatened plant species, November 2011 [CONTINUED]

Common name	Scientific name	Species group	Listing status[a]	U.S. or U.S./foreign listed	Recovery plan date	Recovery plan stage[b]
Showy Indian clover	Trifolium amoenum	Flowering plants	E	US	None	—
Showy stickseed	Hackelia venusta	Flowering plants	E	US	12/12/2007	F
Shrubby reed-mustard	Schoenocrambe suffrutescens	Flowering plants	E	US	9/14/1994	F
Siler pincushion cactus	Pediocactus (=Echinocactus, =Utahia) sileri	Flowering plants	T	US	4/14/1986	F
Slender-horned spineflower	Dodecahema leptoceras	Flowering plants	E	US	None	—
Slender Orcutt grass	Orcuttia tenuis	Flowering plants	T	US	12/15/2005	F
Slender-petaled mustard	Thelypodium stenopetalum	Flowering plants	E	US	7/31/1998	F
Slender rush-pea	Hoffmannseggia tenella	Flowering plants	E	US	9/13/1988	F
Slickspot peppergrass	Lepidium papilliferum	Flowering plants	T	US	None	—
Small-anthered bittercress	Cardamine micranthera	Flowering plants	E	US	7/10/1991	F
Small's milkpea	Galactia smallii	Flowering plants	E	US	5/18/1999	F
Small whorled pogonia	Isotria medeoloides	Flowering plants	T	US/foreign	11/13/1992	RF(1)
Smooth coneflower	Echinacea laevigata	Flowering plants	E	US	4/18/1995	F
Snakeroot	Eryngium cuneifolium	Flowering plants	E	US	5/18/1999	F
Sneed pincushion cactus	Coryphantha sneedii var. sneedii	Flowering plants	E	US	3/21/1986	F
Soft bird's-beak	Cordylanthus mollis ssp. mollis	Flowering plants	E	US	2/10/2010	D
Soft-leaved paintbrush	Castilleja mollis	Flowering plants	E	US	9/26/2000	F
Solano grass	Tuctoria mucronata	Flowering plants	E	US	12/15/2005	F
Sonoma alopecurus	Alopecurus aequalis var. sonomensis	Flowering plants	E	US	None	—
Sonoma spineflower	Chorizanthe valida	Flowering plants	E	US	9/29/1998	F
Sonoma sunshine	Blennosperma bakeri	Flowering plants	E	US	None	—
Southern mountain wild-buckwheat	Eriogonum kennedyi var. austromontanum	Flowering plants	T	US	None	—
South Texas ambrosia	Ambrosia cheiranthifolia	Flowering plants	E	US	None	—
Spalding's Catchfly	Silene spaldingii	Flowering plants	T	US	10/12/2007	F
Spreading avens	Geum radiatum	Flowering plants	E	US	4/28/1993	F
Spreading navarretia	Navarretia fossalis	Flowering plants	T	US/foreign	9/3/1998	F
Spring Creek bladderpod	Lesquerella perforata	Flowering plants	E	US	9/8/2006	F
Spring-loving centaury	Centaurium namophilum	Flowering plants	T	US	9/28/1990	F
Springville clarkia	Clarkia springvillensis	Flowering plants	T	US	None	—
Star cactus	Astrophytum asterias	Flowering plants	E	US/foreign	11/6/2003	F
Steamboat buckwheat	Eriogonum ovalifolium var. williamsiae	Flowering plants	E	US	9/20/1995	F
Stebbins' morning-glory	Calystegia stebbinsii	Flowering plants	E	US	8/30/2002	F
St. Thomas prickly-ash	Zanthoxylum thomasianum	Flowering plants	E	US	4/5/1988	F
Suisun thistle	Cirsium hydrophilum var. hydrophilum	Flowering plants	E	US	2/10/2010	D
Swamp pink	Helonias bullata	Flowering plants	T	US	9/30/1991	F
Telephus spurge	Euphorbia telephioides	Flowering plants	T	US	6/22/1994	F
Tennessee yellow-eyed grass	Xyris tennesseensis	Flowering plants	E	US	6/24/1994	F
Terlingua Creek cat's-eye	Cryptantha crassipes	Flowering plants	E	US	4/5/1994	F
Texas ayenia	Ayenia limitaris	Flowering plants	E	US/foreign	None	—
Texas poppy-mallow	Callirhoe scabriuscula	Flowering plants	E	US	3/29/1985	F
Texas prairie dawn-flower	Hymenoxys texana	Flowering plants	E	US	4/13/1990	F
Texas snowbells	Styrax texanus	Flowering plants	E	US	7/31/1987	F
Texas trailing phlox	Phlox nivalis ssp. texensis	Flowering plants	E	US	3/28/1995	F
Texas wild-rice	Zizania texana	Flowering plants	E	US	2/14/1996	RF(1)
Thread-leaved brodiaea	Brodiaea filifolia	Flowering plants	T	US	None	—
Tiburon jewelflower	Streptanthus niger	Flowering plants	E	US	9/30/1998	F
Tiburon mariposa lily	Calochortus tiburonensis	Flowering plants	T	US	9/30/1998	F
Tiburon paintbrush	Castilleja affinis ssp. neglecta	Flowering plants	E	US	9/30/1998	F
Tiny polygala	Polygala smallii	Flowering plants	E	US	5/18/1999	F
Tobusch fishhook cactus	Ancistrocactus tobuschii	Flowering plants	E	US	3/18/1987	F
Todsen's pennyroyal	Hedeoma todsenii	Flowering plants	E	US	1/31/2001	RF(2)
Triple-ribbed milk-vetch	Astragalus tricarinatus	Flowering plants	E	US	None	—

TABLE 10.1
Endangered and threatened plant species, November 2011 (CONTINUED)

Common name	Scientific name	Species group	Listing status[a]	U.S. or U.S./foreign listed	Recovery plan date	Recovery plan stage[b]
Uhiuhi	Caesalpinia kavaiense	Flowering plants	E	US	5/6/1994	F
Uinta Basin hookless cactus	Sclerocactus wetlandicus	Flowering plants	T	US	4/14/2010	O
Ute ladies'-tresses	Spiranthes diluvialis	Flowering plants	T	US	9/21/1995	D
Uvillo	Eugenia haematocarpa	Flowering plants	E	US	9/11/1998	F
Vahl's boxwood	Buxus vahlii	Flowering plants	E	US	4/28/1987	F
Vail Lake ceanothus	Ceanothus ophiochilus	Flowering plants	T	US	None	—
Ventura Marsh Milk-vetch	Astragalus pycnostachyus var. lanosissimus	Flowering plants	E	US	None	—
Verity's dudleya	Dudleya verityi	Flowering plants	T	US	9/30/1999	F
Vine Hill clarkia	Clarkia imbricata	Flowering plants	E	US	None	—
Virginia round-leaf birch	Betula uber	Flowering plants	T	US	9/24/1990	RF(2)
Virginia sneezeweed	Helenium virginicum	Flowering plants	T	US	10/2/2000	D
Virginia spiraea	Spiraea virginiana	Flowering plants	T	US	11/13/1992	F
Wahane	Pritchardia aylmer-robinsonii	Flowering plants	E	US	1/0/1900	O
Walker's manioc	Manihot walkerae	Flowering plants	E	US/foreign	12/12/1993	F
Water howellia	Howellia aquatilis	Flowering plants	T	US	9/24/1996	D
Wawae`iole	Huperzia mannii	Ferns and Allies	E	US	7/29/1997	F
Wawae`iole	Lycopodium (=Phlegmariurus) nutans	Ferns and Allies	E	US	8/10/1998	F
Welsh's milkweed	Asclepias welshii	Flowering plants	T	US	9/30/1992	F
Wenatchee Mountains checkermallow	Sidalcea oregana var. calva	Flowering plants	E	US	9/30/2004	F
Western lily	Lilium occidentale	Flowering plants	E	US	3/31/1998	F
Western prairie fringed Orchid	Platanthera praeclara	Flowering plants	T	US/foreign	9/30/1996	F
West Indian Walnut (=Nogal)	Juglans jamaicensis	Flowering plants	E	US/foreign	12/9/1999	F
Wheeler's peperomia	Peperomia wheeleri	Flowering plants	E	US	11/26/1990	F
White birds-in-a-nest	Macbridea alba	Flowering plants	T	US	6/22/1994	F
White bladderpod	Lesquerella pallida	Flowering plants	E	US	10/16/1992	F
White-haired goldenrod	Solidago albopilosa	Flowering plants	T	US	9/28/1993	F
White irisette	Sisyrinchium dichotomum	Flowering plants	E	US	4/10/1995	F
White-rayed pentachaeta	Pentachaeta bellidiflora	Flowering plants	E	US	9/30/1998	F
White sedge	Carex albida	Flowering plants	E	US	None	—
Wide-leaf warea	Warea amplexifolia	Flowering plants	E	US	2/17/1993	F
Willamette daisy	Erigeron decumbens var. decumbens	Flowering plants	E	US	6/29/2010	F
Willowy monardella	Monardella linoides ssp. viminea	Flowering plants	E	US/foreign	None	—
Winkler cactus	Pediocactus winkleri	Flowering plants	T	US	10/2/1995 and 12/6/2007	D and O
Wireweed	Polygonella basiramia	Flowering plants	E	US	5/18/1999	F
Wright fishhook cactus	Sclerocactus wrightiae	Flowering plants	E	US	12/24/1985	F
Yadon's piperia	Piperia yadonii	Flowering plants	E	US	6/17/2005	F
Yellow larkspur	Delphinium luteum	Flowering plants	E	US	None	—
Yreka phlox	Phlox hirsuta	Flowering plants	E	US	9/21/2006	F
Zapata bladderpod	Lesquerella thamnophila	Flowering plants	E	US	8/25/2004	F
Zuni fleabane	Erigeron rhizomatus	Flowering plants	T	US	9/30/1988	F

[a]E = endangered; T = threatened.
[b]F = final; D = draft; RD = draft revision; RF = final revision; O = other.

SOURCE: Adapted from "Generate Species List," in *Species Reports*, U.S. Department of the Interior, U.S. Fish & Wildlife Service, November 2011, http://ecos.fws.gov/tess_public/pub/adHocSpeciesForm.jsp (accessed November 8, 2011), and "Listed FWS/Joint FWS and NMFS Species and Populations with Recovery Plans (Sorted by Listed Entity)," in *Recovery Plans Search*, U.S. Department of the Interior, U.S. Fish & Wildlife Service, November 2011, http://ecos.fws.gov/tess_public/pub/speciesRecovery.jsp?sort=1 (accessed November 8, 2011)

TABLE 10.2

Plant species with the highest expenditures under the Endangered Species Act, fiscal year 2010

Ranking	Species	Expenditure
1	Ladies' = -tresses, Ute (Spiranthes diluvialis)	$2,789,999
2	Goldfields, Contra Costa (Lasthenia conjugens)	$1,855,592
3	Seagrass, Johnson's (Halophila johnsonii)	$1,443,585
4	Amaranth, seabeach (Amaranthus pumilus)	$1,035,827
5	Cactus, Siler pincushion (Pediocactus (=Echinocactus,=Utahia) sileri)	$928,400
6	Water-umbel, Huachuca (Lilaeopsis schaffneriana var. recurva)	$808,674
7	Peppergrass, Slickspot (Lepidium papilliferum)	$674,205
8	Paintbrush, golden (Castilleja levisecta)	$639,981
9	Brodiaea, thread-leaved (Brodiaea filifolia)	$526,858
10	Geocarpon minimum (=Sci name)	$514,400

SOURCE: Adapted from "Table 2. Species Ranked in Descending Order of Total FY 2010 Reported Expenditures, Not Including Land Acquisition Costs," in *Federal and State Endangered and Threatened Species Expenditures: Fiscal Year 2010*, U.S. Department of the Interior, U.S. Fish and Wildlife Service, 2010, http://www.fws.gov/endangered/esa-library/pdf/2010.EXP.FINAL.pdf (accessed October 25, 2011)

FIGURE 10.3

Ute Ladies'-tresses

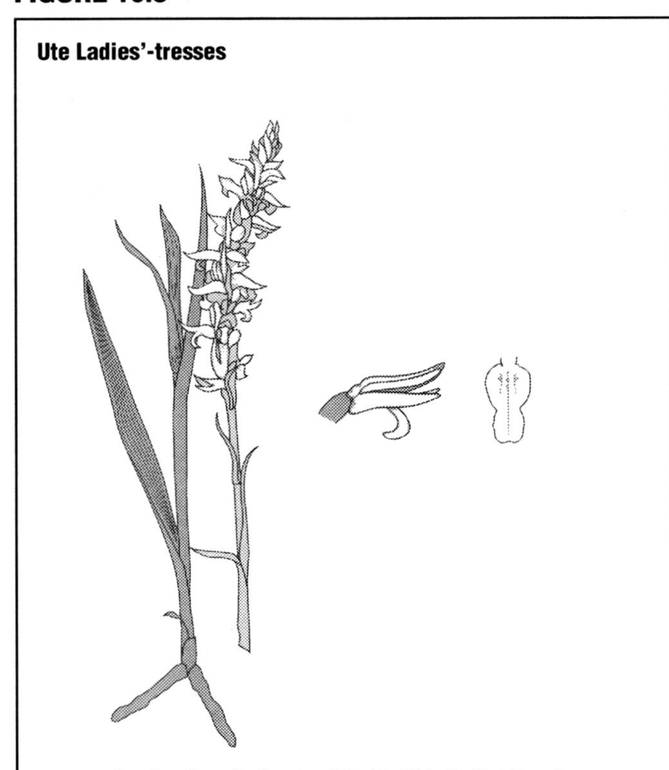

SOURCE: Carolyn Crawford, artist, "Untitled," in B. Heidl et al., *Wyoming's Threatened and Endangered Plant Species: Ute Ladies'-Tresses Orchid*, Wyoming Bureau of Land Management in collaboration with Wyoming Natural Diversity Database, 2008, http://www.blm.gov/pgdata/etc/medialib/blm/wy/programs/botany/docs.Par.33555.File.dat/UteLadTrOrchid.pdf (accessed November 12, 2011)

plant group and are associated mostly with tropical and subtropical regions. In general, these plants are characterized by stems with long protruding leaves called fronds. Fern allies are plants with similar life cycles to ferns, but without their stem or leaf structure. Examples of fern allies include club mosses and horsetails.

As of November 2011, there were 29 fern and allied species listed under the ESA. (See Table 10.1.) Nearly all were endangered.

Lichens

Lichens are not truly plants. Scientists place them in the fungi kingdom, instead of the plant kingdom. Lichens are plantlike life forms composed of two separate organisms: a fungus and an alga. In *Red List of Threatened Species Version 2011.2*, the IUCN indicates that 17,000 described lichens are known to science. Biologists believe there are up to 4,000 lichen species in the United States. They are found in many different habitats and grow extremely slowly. Some lichens look like moss, whereas others appear more like traditional plants with a leafy or blade structure. Lichens do not have a "skin" to protect them from the atmosphere. As a result, they are highly sensitive to air contaminants and have disappeared from many urban areas, presumably because of air pollution. Lichens are most predominant in undisturbed forests, bogs, and wetlands, particularly in California, Florida, Hawaii, the Appalachians, and the Pacific Northwest. They are commonly found on rocky outcroppings. Lichens provide a foodstuff for some animals and are used by some bird species in nest building.

As of November 2011, there were two lichen species listed under the ESA: rock gnome lichen (*Gymnoderma lineare*) and Florida perforate cladonia (*Cladonia perforata*). (See Table 10.1.) Both were classified as endangered. During the 1970s rock gnome lichen was virtually wiped out in the Great Smoky Mountains National Park in Tennessee due to zealous collecting by scientists. Florida perforate cladonia is found only in rosemary scrub habitats in portions of Florida. It is endangered due to loss or degradation of those habitats.

Flowering Plants

Flowering plants are vascular plants with flowers—clusters of specialized leaves that participate in reproduction. The flowers of some species are large and colorful, whereas others are extremely small and barely noticeable to humans. In *Red List of Threatened Species Version 2011.2*, the IUCN indicates that there are 268,000 described flowering plants. Biologists estimate that 80% to 90% of all plants on the earth are flowering plants.

As of November 2011, there were 760 flowering plants listed as endangered or threatened under the ESA. (See Table 10.1.) They come from a wide variety of taxonomic groups and are found in many different habitats. Some of the largest families represented are:

FIGURE 10.4

Vernal pool regions of California and southern Oregon

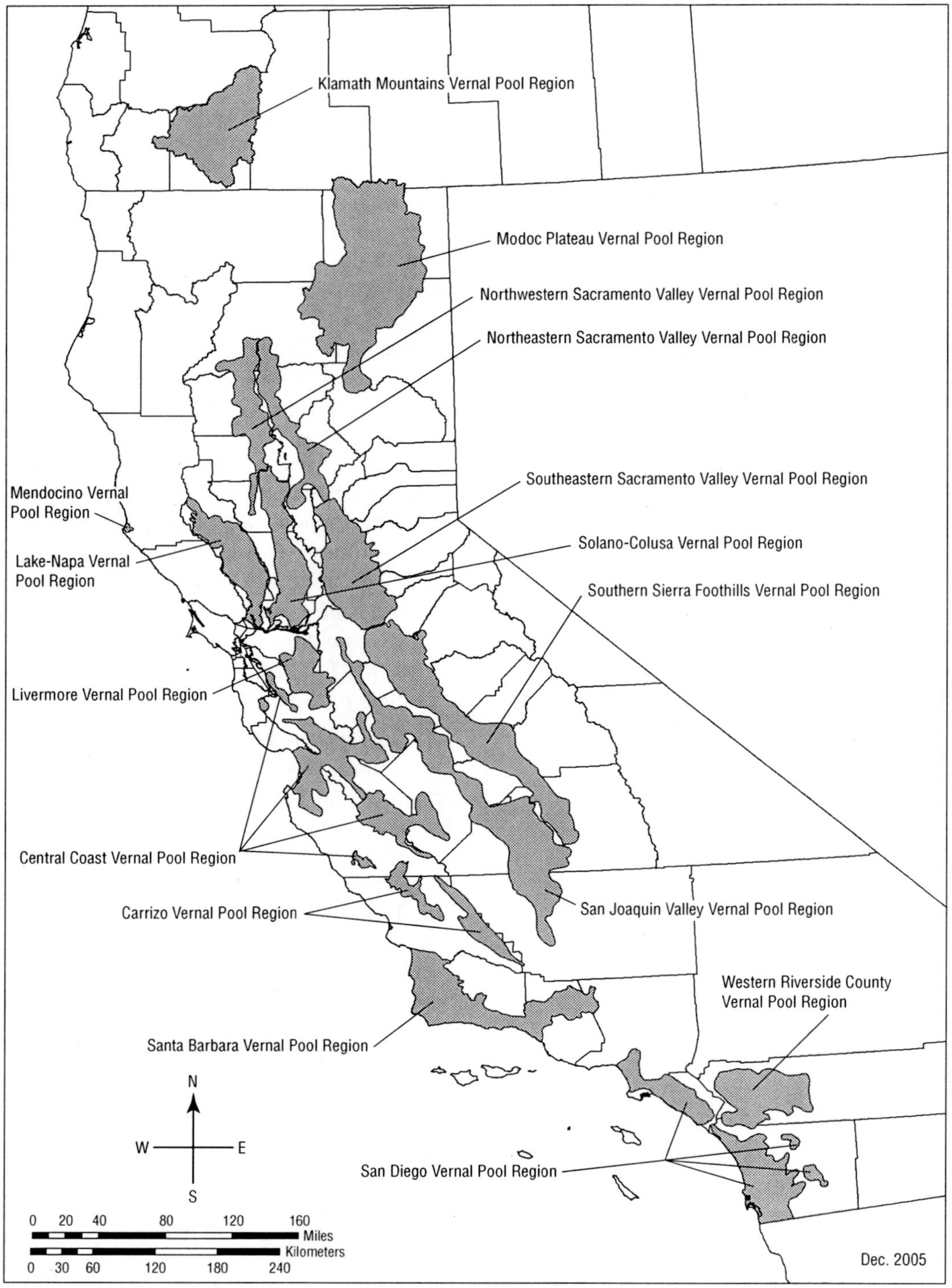

SOURCE: "Figure I-1. Map of Recovery Plan Area Showing Location of Vernal Pool Regions," in *Recovery Plan for Vernal Pool Ecosystems of California and Southern Oregon*, U.S. Department of the Interior, U.S. Fish and Wildlife Service, December 2005, http://ecos.fws.gov/docs/recovery_plan/060614.pdf (accessed November 12, 2011)

FIGURE 10.5

Range of Johnson's seagrass

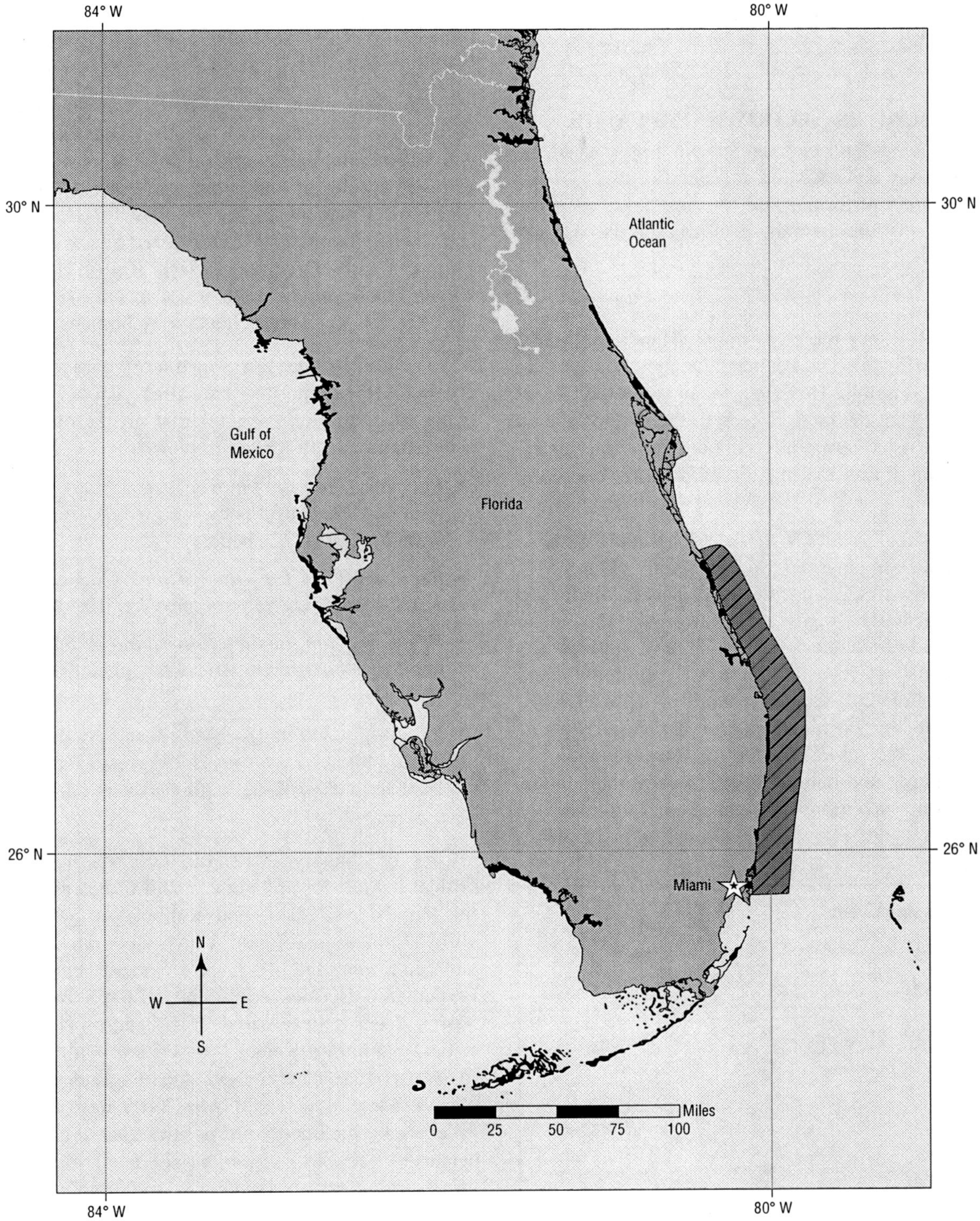

Note: Map represents approximate range of species. Offshore distances are approximate.

SOURCE: "Johnson's Seagrass Range," in *Johnson's Seagrass (Halophila Johnsonii)*, U.S. Department of Commerce, National Oceanic and Atmospheric Administration, National Marine Fisheries Service, December 2007, http://www.nmfs.noaa.gov/pr/pdfs/rangemaps/johnsonsseagrass.pdf (accessed November 12, 2011)

- Asteraceae (asters, daisies, and sunflowers)
- Brassicaceae (mustard and cabbage)
- Campanulaceae (bellflowers)
- Fabaceae (legumes and pulses)
- Lamiaceae (mints)

GEOGRAPHICAL BREAKDOWN OF PLANTS

Most threatened and endangered plant species in the United States are concentrated in specific areas of the country. Hawaii, California, and the Southeast (particularly Florida) are home to the majority of listed plant species.

Hawaiian Plants

Figure 10.6 shows the eight major islands that make up the state of Hawaii. The island of Oahu is home to the state's capital Honolulu. However, Oahu is not the largest of the islands. That distinction goes to the island labeled "Hawaii," which is commonly called "the Big Island." In the following discussion, the term *Hawaii* refers to the entire state.

According to the USFWS, in "48 Kaua'i Species Protected under Endangered Species Act" (March 10, 2010, http://www.fws.gov/pacificislands/news%20releases/Kauai48ListingNR031010.pdf), Hawaii has the most endangered and threatened species of any state in the country. Because of its isolation from continental land masses, many of the species found in Hawaii exist nowhere else in the world. In fact, an estimated 90% of Hawaiian plant species are endemic. Because of large-scale deforestation and habitat destruction on the islands, Hawaii is home to more threatened and endangered plants than any other state in the nation. The USFWS (http://ecos.fws.gov/tess_public/pub/adHocSpeciesForm.jsp) reports that as of November 2011, there were 319 listed plant species in Hawaii. Nearly all are flowering plants and have an endangered listing. The most recent additions to the list were 45 plants found on Kauai; they were listed in May 2010.

FIGURE 10.6

The main Hawaiian Islands

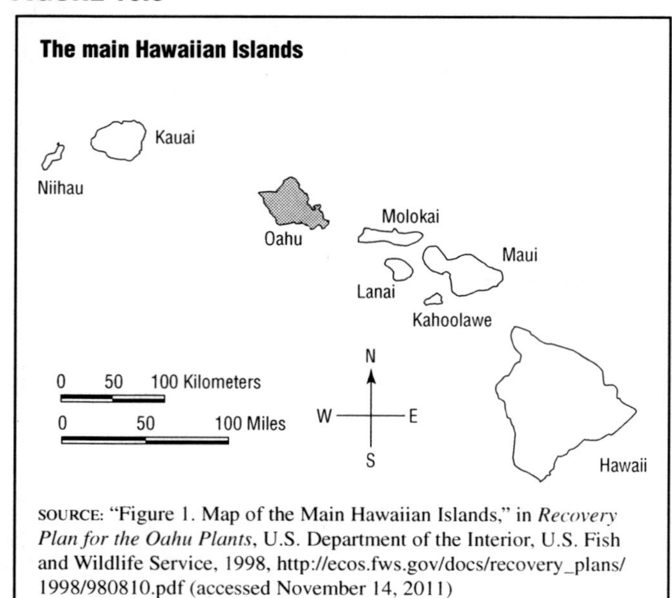

SOURCE: "Figure 1. Map of the Main Hawaiian Islands," in *Recovery Plan for the Oahu Plants*, U.S. Department of the Interior, U.S. Fish and Wildlife Service, 1998, http://ecos.fws.gov/docs/recovery_plans/1998/980810.pdf (accessed November 14, 2011)

Hawaiian plants have suffered from the introduction of invasive predators such as cows, pigs, and insects, as well as from the loss of critical pollinators with the decline of many species of native birds and insects. According to Marie M. Bruegemann, in "A Plan for Hawaiian Plants and Their Ecosystems" (*Endangered Species Bulletin*, vol. 28, no. 4, July–December 2003), 100 of Hawaii's 1,500 known plant species are believed to have become extinct since the islands were colonized by humans.

The USFWS has developed more than a dozen recovery plans for imperiled Hawaiian plants. Many of the plans cover multiple species that are found in the same ecosystem or habitat type, such as:

- *Recovery Plan for the Big Island Plant Cluster* (September 1996, http://ecos.fws.gov/docs/recovery_plan/960926a.pdf)—22 species
- *Recovery Plan for Oahu Plants* (August 1998, http://ecos.fws.gov/docs/recovery_plan/980810.pdf)—66 species
- *Recovery Plan for the Multi-island Plants* (July 1999, http://ecos.fws.gov/docs/recovery_plan/990710.pdf)—26 species
- *Recovery Outline for the Kauai Ecosystem* (June 2010, http://ecos.fws.gov/docs/recovery_plan/Recovery%20Outline%20Kauai%20Ecosystem.pdf)—45 plant species

As of January 2012, the USFWS had designated critical habitat for endangered and threatened species on the Big Island, Maui, and Kahoolawe (which together with Molokai and Lanai were once one large island) and on Kauai. In "48 Kaua'i Species Protected under Endangered Species Act," the USFWS reports that in March 2010 it designated 26,582 acres (10,757 ha) of critical habitat on Kauai. Most of the land (21,666 acres [8,768 ha]) is state owned, the remainder is privately owned. More than 1,000 acres (405 ha) originally proposed to be included were not designated critical habitat because "the designation would have had a negative effect on the private landowner's voluntary ongoing and future conservation activities." The land is owned by a corporation, but is managed by the Nature Conservancy, a private nonprofit organization.

Californian Plants

The USFWS (http://ecos.fws.gov/tess_public/pub/adHocSpeciesForm.jsp) reports that as of November 2011

California was home to more than 180 imperiled plants, including several types of checker-mallow, dudleya, evening primrose, grass, jewelflower, larkspur, manzanita, milk-vetch, paintbrush, rock-cress, spineflower, and thistle.

MILK-VETCH. Milk-vetch is an herbaceous perennial flowering plant that is found in various parts of the world. It received its common name during the 1500s because of a belief among European farmers that the plant increased the milk yield of goats. As of November 2011, there were 11 species of milk-vetch listed as threatened or endangered in California:

- Ash meadows milk-vetch (*Astragalus phoenix*)
- Braunton's milk-vetch (*Astragalus brauntonii*)
- Clara Hunt's milk-vetch (*Astragalus clarianus*)
- Coachella Valley milk-vetch (*Astragalus lentiginosus* var. *coachellae*)
- Coastal dunes milk-vetch (*Astragalus tener* var. *titi*)
- Cushenbury milk-vetch (*Astragalus albens*)
- Fish Slough milk-vetch (*Astragalus lentiginosus* var. *piscinensis*)
- Lane Mountain milk-vetch (*Astragalus jaegerianus*)
- Peirson's milk-vetch (*Astragalus magdalenae* var. *peirsonii*)
- Triple-ribbed milk-vetch (*Astragalus tricarinatus*)
- Ventura Marsh milk-vetch (*Astragalus pycnostachyus* var. *lanosissimus*)

Peirson's milk-vetch is a plant with a long history of litigation and controversy in California. It is found in only one small area of Imperial County in the southern part of the state. This area is the Imperial Sand Dunes Recreation Area (ISDRA), which is managed by the Bureau of Land Management (BLM). ISDRA has a remote and barren landscape that is dominated by huge rolling sand dunes—the Algodones Dunes, the largest sand dune fields in North America. ISDRA covers 185,000 acres (75,000 ha) and is a popular destination for off-highway vehicle (OHV) riders, receiving more than 1 million visitors annually.

In 1998 Peirson's milk-vetch was designated a threatened species by the USFWS because of the threat of destruction by OHVs and other recreational activities at ISDRA. The agency decided not to designate critical habitat at that time, fearing the remaining plants would be subject to deliberate vandalism. The BLM was sued by conservation groups and accused of not consulting with the USFWS about the threats to Peirson's milk-vetch before establishing a management plan for ISDRA. In 2000, in response to that lawsuit, the BLM closed more than a third of ISDRA to OHV use.

In October 2001 a petition to delist the species was submitted on behalf of the American Sand Association, the San Diego Off-Road Coalition, and the Off-Road Business Association. A month later two lawsuits were filed against the USFWS by conservation organizations challenging the agency's decision not to designate critical habitat for the species. Under court order, the USFWS proposed critical habitat in 2003.

In August 2004 the USFWS (http://www.fws.gov/policy/library/2004/04-17575.html) issued a final designation of critical habitat for Peirson's milk-vetch that encompassed 21,836 acres (8,837 ha) of ISDRA. This was less than half of the acreage originally proposed. The reduction was made after an economic analysis revealed that closure of ISDRA areas to OHV use would have a negative impact on local businesses. In February 2008 the USFWS (http://edocket.access.gpo.gov/2008/pdf/08-545.pdf) issued a new final designation of critical habitat that included 12,105 acres (4,899 ha), substantially less area than was designated in 2004. The new designation excluded acreage due to the "disproportionately high economic and social impacts" of including it in the critical habitat.

In June 2004 the USFWS completed a status review that was triggered by the 2001 delisting proposal and found that the species should remain listed as threatened. In 2005 the original petitioners and additional OHV and motorcycle associations submitted a new petition to delist Peirson's milk-vetch. In a 90-day finding the USFWS announced that the petition did present adequate information indicating that delisting might be warranted. However, in July 2008 the agency issued a 12-month finding that delisting was not warranted.

In September 2008 the USFWS (http://ecos.fws.gov/docs/five_year_review/doc1995.pdf) completed a five-year status review of Peirson's milk-vetch in which the agency recommended no changes to the species' threatened listing under the ESA. The USFWS noted that OHV activity remained the "primary threat" to the plant's survival.

In April 2010 the BLM published *Draft Imperial Sand Dunes Recreation Area Management Plan* (http://www.blm.gov/pgdata/etc/medialib/blm/ca/pdf/elcentro/planning/2010.Par.32016.File.dat/ISDRAMP_PublicMtg_pp0410.pdf) based on the 2008 USFWS designation of critical habitat. The plan includes several alternatives for limiting OHV use in various areas to protect Peirson's milk-vetch habitat. As of January 2012, the plan had not been finalized.

LOS ANGELES BASIN MOUNTAIN PLANTS. Many species of threatened and endangered plants have reached their precarious state due to urbanization and other

FIGURE 10.7

Distribution of six endangered and threatened plant species in the mountains surrounding the Los Angeles basin

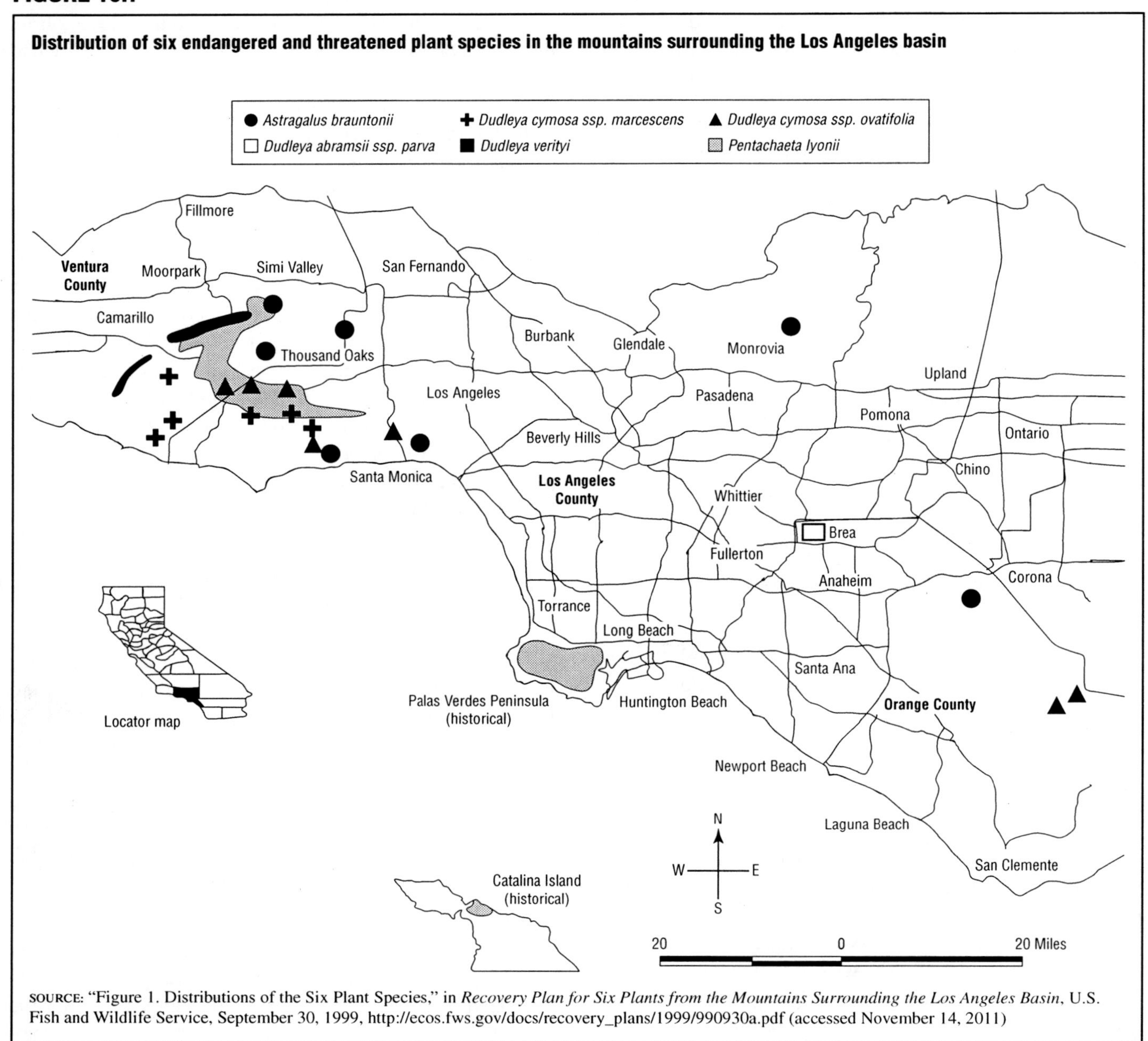

SOURCE: "Figure 1. Distributions of the Six Plant Species," in *Recovery Plan for Six Plants from the Mountains Surrounding the Los Angeles Basin*, U.S. Fish and Wildlife Service, September 30, 1999, http://ecos.fws.gov/docs/recovery_plans/1999/990930a.pdf (accessed November 14, 2011)

human activity. Figure 10.7 shows the species distribution of six threatened and endangered plant species that are found in the mountains surrounding the Los Angeles basin:

- Braunton's milk-vetch (*Astragalus brauntonii*)
- Conejo dudleya (*Dudleya abramsii* ssp. *parva*)
- Lyon's pentachaeta (*Pentachaeta lyonii*)
- Marcescent dudleya (*Dudleya cymosa* ssp. *marcescens*)
- Santa Monica Mountains dudleya (*Dudleya cymosa* ssp. *ovatifolia*)
- Verity's dudleya (*Dudleya verityi*)

In September 1999 the USFWS published *Recovery Plan for Six Plants from the Mountains Surrounding the Los Angeles Basin* (http://ecos.fws.gov/docs/recovery_plans/1999/990930a.pdf). The agency cited threats including "urban development, recreational activities, alteration of fire cycles, fire suppression and pre-suppression (fuel modification) activities, over collecting, habitat fragmentation and degradation, and competition from invasive weeds. Several of the plants are also threatened with extinction from random events because their numbers and ranges are so limited."

In May 2011 the USFWS (http://www.gpo.gov/fdsys/pkg/FR-2010-05-21/pdf/2010-12170.pdf#page=1) published notice that it had completed five-year status reviews for 96 California and Nevada species, including dozens of California plants. No listing status changes were recommended for any of the plant species.

Southeastern Plants

As of November 2011, the USFWS (http://ecos.fws.gov/tess_public/pub/adHocSpeciesForm.jsp) indicated that the southeastern states contained more than 100 threatened and endangered plant species listed under the ESA. The most common families are the Asteraceae (asters, daisies, and sunflowers), Fabaceae (legumes and pulses), and Lamiaceae (mints).

FLORIDIAN PLANTS. Approximately 50 of the threatened and endangered plant species listed under the ESA are found only in Florida. They include multiple species of mint, pawpaw, rosemary, and spurge.

Many of Florida's imperiled plants are found in the southern part of the state—the only subtropical ecological habitat in the continental United States. The majority of native plant species located in the bottom half of the southern Florida ecosystem originated from the tropics.

In May 1999 the USFWS published *South Florida Multi-species Recovery Plan* (http://ecos.fws.gov/docs/recovery_plan/990518_1.pdf), which covered 68 species, including 35 plant species. In 2004 an implementation schedule for many, but not all, of the species in the plan was issued, and it divided the ecosystem into ecological communities. The following ecological communities include plant species:

- Florida scrub/scrubby flatwoods/scrubby high pine—19 plant species
- Pine rocklands—five plant species
- Beach dune/coastal strand—one plant species
- Freshwater marsh/wet prairie—one plant species
- Mesic and hydric pine flatwoods—one plant species
- Tropical hardwood hammock—one plant species

The recovery and restoration tasks outlined in the recovery plan are to be implemented through the creation of a team of federal, state, and local governmental agencies; Native American tribal governments; academic representatives; industry representatives; and members of the private sector. The schedule prioritizes the plan's recovery actions and estimates the annual costs for implementing the actions in each ecological community.

South Florida Multi-species Recovery Plan is considered a landmark plan, because it was one of the first recovery plans to focus on an ecosystem approach to recovery, rather than on a species-by-species approach.

PONDBERRY. Another southeastern imperiled plant is the pondberry. (See Figure 10.8.) The plant is a member of the Lauraceae family, which includes dozens of trees and shrubs. The pondberry is a shrub that grows to be around 6 feet (1.8 m) tall. It is deciduous (meaning it loses its leaves during some part of the year) and produces tiny red berries during late summer or early fall. It prefers lowland habitats with moist soil. In "Pondberry" (October 11, 2006, http://www.dnr.sc.gov/marine/mrri/acechar/specgal/berrypon.htm), the South Carolina Department of Natural Resources notes that "pondberry does not have any particular aesthetic value, nor are any horticultural, medicinal, or other economic uses known."

The pondberry has been listed under the ESA as endangered since 1986. A recovery plan for the species was issued in 1993. The elimination and development of lowland forests across the South contributed to its imperiled condition. According to Zoë Hoyle, in "Pondberry: Modest but Mysterious" (*Compass*, no. 6, July 2006),

FIGURE 10.8

Pondberry

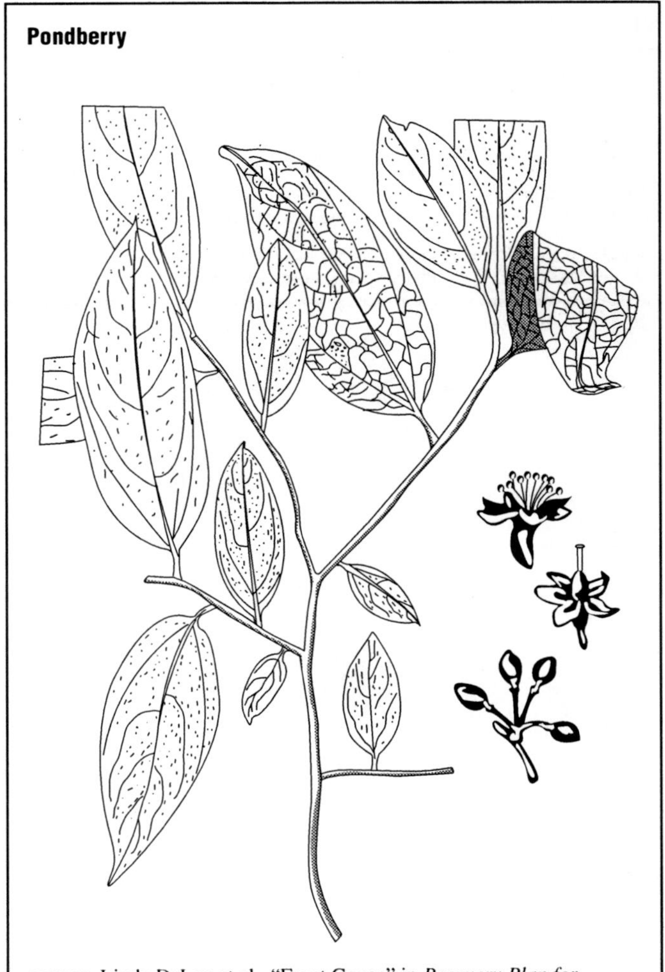

SOURCE: Linda DeLay et al., "Front Cover," in *Recovery Plan for Pondberry (Lindera melissifolia)*, U.S. Department of the Interior, U.S. Fish and Wildlife Service, 1993, http://ecos.fws.gov/docs/recovery_plan/930923a.pdf (accessed November 18, 2011)

scientists believe that the pondberry has always been rare. Approximately three dozen separate populations of the species are known to exist. They are so scattered that scientists believe the plant's seeds must have been spread by floods in the distant past. Extensive flood control measures implemented across the South in recent decades have likely limited the further spread of the species. Hoyle notes that approximately $5 million was spent by various federal agencies between 2002 and 2006 to study factors that affect the plant's survival. In April 2010 the USFWS (http://www.gpo.gov/fdsys/search/citation.result.FR.action?federalRegister.volume=2010&federalRegister.page=18233&publication=FR) initiated a five-year status review for 10 southeastern species, including the pondberry. As of January 2012, the results of that review had not been published.

THREATENED AND ENDANGERED FOREIGN SPECIES OF PLANTS

In *Red List of Threatened Species Version 2011.2*, the IUCN indicates that there were 9,158 species of threatened plants (including flowering plants, gymnosperms, ferns and allies, mosses, and lichens) in 2011. This was 63% of the 14,498 evaluated plant species. However, less than 3% of described plant species have been studied in sufficient detail to assess their status, and the actual number of threatened species is likely to be much higher. The IUCN notes that 324,674 flowering plants, gymnosperms, ferns and allies, mosses, and lichens are known and described around the world.

The majority (8,527) of IUCN-listed species are flowering plants, a diverse and well-studied group. Other IUCN-listed species include 377 gymnosperms (conifers, cycads, ginkgos, and gnetophytes), 163 ferns and allies, 80 true mosses, and two lichens. Habitat loss is the primary reason for the threatened status of the vast majority of IUCN-listed plants.

As of November 2011, the USFWS listed only three totally foreign species of plants as threatened or endangered under the ESA. (See Table 10.3.) All three are found in Mexico or in Central or South America.

TABLE 10.3

Foreign endangered and threatened plant species, November 2011

Common name	Scientific name	Species group	Listing status	Historic range
Chilean false larch	Fitzroya cupressoides	Conifers and cycads	Threatened	Chile, Argentina
Costa Rican jatropha	Jatropha costaricensis	Flowering plants	Endangered	Costa Rica
Guatemalan fir (=pinabete)	Abies guatemalensis	Conifers and cycads	Threatened	Mexico, Guatemala, Honduras, El Salvador

SOURCE: Adapted from "Generate Species List," in *Species Reports*, U.S. Department of the Interior, U.S. Fish & Wildlife Service, November 2011, http://ecos.fws.gov/tess_public/pub/adHocSpeciesForm.jsp (accessed November 8, 2011)

IMPORTANT NAMES AND ADDRESSES

Alaska Fisheries Science Center
National Oceanic and Atmospheric Administration
7600 Sand Point Way NE, Bldg. 4
Seattle, WA 98115
(206) 526-4000
FAX: (206) 526-4004
E-mail: afsc.webmaster@noaa.gov
URL: http://www.afsc.noaa.gov/

AmphibiaWeb
Museum of Vertebrate Zoology
University of California
3101 Valley Life Sciences
Berkeley, CA 94720-3160
(510) 642-3567
E-mail: supportamphibiaweb@berkeley.edu
URL: http://amphibiaweb.org/

Bureau of Land Management
1849 C St., Rm. 5665
Washington, DC 20240
(202) 208-3801
FAX: (202) 208-5242
URL: http://www.blm.gov/

Bureau of Reclamation
1849 C St. NW
Washington, DC 20240-0001
(202) 513-0501
FAX: (202) 513-0309
URL: http://www.usbr.gov/

Center for Biological Diversity
PO Box 710
Tucson, AZ 85702-0710
(520) 623-5252
1-866-357-3349
FAX: (520) 623-9797
E-mail: center@biologicaldiversity.org
URL: http://www.biologicaldiversity.org/

Center for Plant Conservation
PO Box 299
St. Louis, MO 63166-0299
(314) 577-9450
FAX: (314) 577-9465
E-mail: cpc@mobot.org
URL: http://www.centerforplantconservation.org/

Congressional Research Service
Library of Congress
101 Independence Ave. SE
Washington, DC 20540
URL: http://www.loc.gov/crsinfo/

Convention on International Trade in Endangered Species of Wild Fauna and Flora Secretariat
International Environment House
11 Chemin des Anémones
Geneva, CH-1219, Switzerland
(011-41-22) 917-81-39/40
FAX: (011-41-22) 797-34-17
E-mail: info@cites.org
URL: http://www.cites.org/

Defenders of Wildlife
1130 17th St. NW
Washington, DC 20036
(202) 682-9400
1-800-385-9712
E-mail: defenders@mail.defenders.org
URL: http://www.defenders.org/index.php/

Earth System Research Laboratory
National Oceanic and Atmospheric Administration
325 Broadway
Boulder, CO 80305-3337
URL: http://www.esrl.noaa.gov/gmd/

Ecosystems—Wildlife:
Terrestrial and Endangered Resources Program
U.S. Geological Survey
12201 Sunrise Valley Dr.
Reston, VA 20192
(703) 648-4019
1-888-275-8747
FAX: (703) 648-4238
URL: http://ecosystems.usgs.gov/wter/

Environmental Defense Fund
257 Park Ave. South
New York, NY 10010
1-800-684-3322
URL: http://www.edf.org/

Goddard Institute for Space Studies
National Aeronautics and Space Administration
2880 Broadway
New York, NY 10025
URL: http://www.giss.nasa.gov/

Greenpeace U.S.A.
702 H St. NW, Ste. 300
Washington, DC 20001
(202) 462-1177
1-800-722-6995
FAX: (202) 462-4507
E-mail: info@wdc.greenpeace.org
URL: http://www.greenpeace.org/usa/

Heritage Foundation
214 Massachusetts Ave. NE
Washington, DC 20002-4999
(202) 546-4400
E-mail: info@heritage.org
URL: http://www.heritage.org/

International Union for Conservation of Nature
Rue Mauverney 28
Gland, 1196, Switzerland
(011-41-22) 999-0000
FAX: (011-41-22) 999-0002
E-mail: mail@iucn.org
URL: http://www.iucn.org/

International Whaling Commission
The Red House
135 Station Rd.
Impington, CB24 9NP, United Kingdom
(011-44-0) 1223-233-971
FAX: (011-44-0) 1223-232-876
E-mail: secretariat@iwcoffice.org
URL: http://www.iwcoffice.org/index.htm

Land Trust Alliance
1660 L St. NW, Ste. 1100
Washington, DC 20036
(202) 638-4725
FAX: (202) 638-4730
E-mail: info@lta.org
URL: http://www.lta.org/

National Association of Home Builders
1201 15th St. NW
Washington, DC 20005
(202) 266-8200
1-800-368-5242
FAX: (202) 266-8400
URL: http://www.nahb.com/

National Audubon Society
225 Varick St.
New York, NY 10014
(212) 979-3000
URL: http://www.audubon.org/

National Forest Service
U.S. Department of Agriculture
1400 Independence Ave. SW
Washington, DC 20250-0003
1-800-832-1355
E-mail: info@fs.fed.us
URL: http://www.fs.fed.us

National Marine Fisheries Service
National Oceanic and Atmospheric Administration
1315 East-West Hwy.
Silver Spring, MD 20910
(301) 713-2379
FAX: (301) 713-2384
URL: http://www.nmfs.noaa.gov/

National Marine Mammal Laboratory
National Oceanic and Atmospheric Administration
7600 Sand Point Way NE, Bldg. 4
Seattle, WA 98115
(206) 526-4000
FAX: (206) 526-4004
URL: http://www.afsc.noaa.gov/nmml/

National Ocean Service
National Oceanic and Atmospheric Administration
1305 East-West Hwy.
Silver Spring, MD 20910
(301) 713-3010
E-mail: nos.info@noaa.gov
URL: http://oceanservice.noaa.gov/

National Park Service
1849 C St. NW
Washington, DC 20240
(202) 208-3818
URL: http://www.nps.gov/

National Research Council
National Academies
500 Fifth St. NW
Washington, DC 20001
(202) 334-2000
URL: http://www.nas.edu/nrc/

National Wildlife Federation
11100 Wildlife Center Dr.
Reston, VA 20190-5362
(703) 438-6000
1-800-822-9919
URL: http://www.nwf.org/

National Wildlife Health Center
U.S. Geological Survey
6006 Schroeder Rd.
Madison, WI 53711-6223
(608) 270-2400
FAX: (608) 270-2415
E-mail: AskNWHC@usgs.gov
URL: http://www.nwhc.usgs.gov/

National Wildlife Refuge System
U.S. Fish and Wildlife Service
1-800-344-9453
URL: http://www.fws.gov/refuges/

Natural Resources Defense Council
40 W. 20th St.
New York, NY 10011
(212) 727-2700
FAX: (212) 727-1773
E-mail: nrdcinfo@nrdc.org
URL: http://www.nrdc.org/

Nature Conservancy
4245 N. Fairfax Dr., Ste. 100
Arlington, VA 22203-1606
(703) 841-5300
1-800-628-6860
E-mail: member@tnc.org
URL: http://nature.org/

Northeast Fisheries Science Center
National Oceanic and Atmospheric Administration
166 Water St.
Woods Hole, MA 02543-1026
(508) 495-2000
URL: http://www.nefsc.noaa.gov/

Northern Prairie Wildlife Research Center
U.S. Geological Survey
8711 37th St. SE
Jamestown, ND 58401
(701) 253-5500
FAX: (701) 253-5553
E-mail: npwrc@usgs.gov
URL: http://www.npwrc.usgs.gov/

Northwest Fisheries Science Center
National Oceanic and Atmospheric Administration
2725 Montlake Blvd. East
Seattle, WA 98112-2097
(206) 860-3200
FAX: (206) 860-3217
URL: http://www.nwfsc.noaa.gov/

Pacific Legal Foundation
National Litigation Center
930 G St.
Sacramento, CA 95814
(916) 419-7111
FAX: (916) 419-7747
URL: http://www.pacificlegal.org/

Sierra Club
85 Second St., Second Floor
San Francisco, CA 94105
(415) 977-5500
FAX: (415) 977-5797
E-mail: information@sierraclub.org
URL: http://www.sierraclub.org/

TRAFFIC North America—Regional Office
1250 24th St. NW
Washington, DC 20037
(202) 293-4800
FAX: (202) 775-8287
E-mail: tna@wwfus.org
URL: http://www.traffic.org/

United Nations Environment Programme
United Nations Ave., Gigiri
PO Box 30552
Nairobi, 00100, Kenya
(011-254-20) 7621234
FAX: (011-254-20) 7624489/90
E-mail: unepinfo@unep.org
URL: http://www.unep.org/

U.S. Army Corps of Engineers
441 G St. NW
Washington, DC 20314-1000
(202) 761-0011
E-mail: hq-publicaffairs@usace.army.mil
URL: http://www.usace.army.mil/

U.S. Department of Justice
Environment and Natural Resources Division
Law and Policy Section
PO Box 7415
Ben Franklin Station
Washington, DC 20044
(202) 514-2701
E-mail: webcontentmgr.enrd@usdoj.gov
URL: http://www.justice.gov/enrd/

U.S. Fish and Wildlife Service
Endangered Species Program
4401 N. Fairfax Dr., Rm. 420
Arlington, VA 22203
URL: http://www.fws.gov/endangered/

U.S. Government Accountability Office
441 G St. NW
Washington, DC 20548
(202) 512-3000
E-mail: contact@gao.gov
URL: http://www.gao.gov/

Water Resource Center
U.S. Environmental Protection Agency
1200 Pennsylvania Ave. NW
Washington, DC 20460
(202) 566-1729

1-800-832-7828
E-mail: center.water-resource@epa.gov
URL: http://water.epa.gov/aboutow/ownews/wrc/

Western Ecological Research Center
U.S. Geological Survey
3020 State University Dr. East
Modoc Hall, Rm. 3006
Sacramento, CA 95819
(916) 278-9492
FAX: (916) 278-9475
URL: http://www.werc.usgs.gov/

WildEarth Guardians
516 Alto St.
Santa Fe, NM 87501
(505) 988-9126
FAX: (505) 213-1895
E-mail: info@wildearthguardians.org
URL: http://www.wildearthguardians.org/site/PageServer

Wilderness Society
1615 M St. NW
Washington, DC 20036
1-800-843-9453
URL: http://www.wilderness.org/

World Wildlife Fund
1250 24th St. NW
Washington, DC 20037-1193
(202) 293-4800
URL: http://www.worldwildlife.org/

RESOURCES

A leading source of information on endangered species is the U.S. Fish and Wildlife Service (USFWS), an agency of the U.S. Department of the Interior. The USFWS oversees the Endangered Species List. Its comprehensive website (http://www.fws.gov/endangered/) includes news stories on threatened and endangered species, information about laws protecting endangered species, regional contacts for endangered species programs, and a searchable database called the Threatened and Endangered Species System (http://ecos.fws.gov/tess_public/) with information on all listed species. Each listed species has an information page that provides details regarding the status of the species (whether it is listed as threatened or endangered and in what geographic area), *Federal Register* documents pertaining to listing, information on habitat conservation plans and national wildlife refuges pertinent to the species, and, for many species, links to descriptions of biology and natural history. Particularly informative are the recovery plans published for many listed species. These recovery plans detail the background research on the natural history of endangered species and list measures that should be adopted to aid in conservation.

The USFWS also maintains updated tables of the number of threatened and endangered species by taxonomic group, as well as lists of threatened and endangered U.S. species. The agency publishes the bimonthly *Endangered Species Bulletin*, which provides information on new listings, delistings, and reclassifications, besides news articles on endangered species. Finally, the USFWS prints an annual report on expenditures that have been made under the Endangered Species Act.

The National Marine Fisheries Service (NMFS) is responsible for the oversight of threatened and endangered marine animals and anadromous fish. The NMFS is a division of the National Oceanic and Atmospheric Administration (NOAA), which also operates the National Marine Mammal Laboratory, the Northeast Fisheries Science Center, and the Northwest Fisheries Science Center. NOAA's National Ocean Service is an excellent source of information about marine creatures.

The U.S. Geological Survey (USGS) performs research on many imperiled animal species. Data sheets on individual species are available from the USGS National Wildlife Health Center. Other important USGS centers include the Western Ecological Research Center, the Northern Prairie Wildlife Research Center, the Nonindigenous Aquatic Species information resource at the Center for Aquatic Resource Studies (http://nas.er.usgs.gov/), and the Ecosystems—Wildlife: Terrestrial and Endangered Resources Program.

Information on federal lands and endangered species management can be found via the National Wildlife Refuge System, the National Park Service, and the National Forest Service. Information on water quality in the United States is available via the U.S. Environmental Protection Agency (http://www.epa.gov/gateway/science/water.html). Information on wetlands can be found at the USFWS's National Wetlands Inventory (http://www.fws.gov/wetlands/).

Data and information related to global warming are available from the National Aeronautics and Space Administration's Goddard Institute for Space Studies and from NOAA's Earth System Research Laboratory, Global Monitoring Division.

Other federal agencies that proved useful for this book were the U.S. Department of Justice's Environment and Natural Resources Division, which defends the federal government in court cases involving environmental laws, such as the Endangered Species Act, and the U.S. Army Corps of Engineers, which maintains the National Inventory of Dams (http://geo.usace.army.mil/pgis/f?p=397:5:554178539665691::NO). The Department of the

Interior operates the National Atlas of the United States (http://nationalatlas.gov/), an online mapping tool. The U.S. Government Accountability Office and the Congressional Research Service publish a number of reports that assess the policies and effectiveness of the Endangered Species Program.

The International Union for Conservation of Nature (IUCN) provides news articles on a wide array of worldwide conservation issues and maintains the *Red List of Threatened Species Version 2011.2* (http://www.iucnredlist.org/). This site includes an extensive database of information on IUCN-listed threatened species. Species information that is available includes the Red List endangerment category, the year the species was assessed, the countries in which the species is found, a list of the habitat types the species occupies, major threats to continued existence, and current population trends. Brief descriptions of ecology and natural history and of conservation measures for protecting listed species are also available. Searches can also be performed by taxonomic group, Red List categories, country, region, or habitat.

The Convention on International Trade in Endangered Species of Wild Fauna and Flora (CITES) provides information on international trade in endangered species. It includes a species database of protected flora and fauna in the three CITES appendixes, as well as information on the history and aims of the convention and its current programs.

Many private organizations are dedicated to the conservation of listed species and their ecosystems. Readers with interest in a particular endangered species are advised to conduct Internet searches to locate these groups. The Save the Manatee Club, which focuses on West Indian manatees, and the Save Our Springs Alliance, which focuses on protection of the endangered Barton Springs salamander, are only two of many examples.

BirdLife International provides diverse resources on global bird conservation. It is an association of non-governmental conservation organizations that has over 2.5 million members worldwide.

AmphibiaWeb (http://amphibiaweb.org/) provides detailed information on global amphibian declines. It maintains a watch list of recently extinct and declining species, discusses potential causes of amphibian declines and deformities, and provides detailed information on amphibian biology and conservation. AmphibiaWeb also sponsors a discussion board where readers can submit questions regarding amphibians.

TRAFFIC (http://www.traffic.org/) was originally founded to help implement the CITES treaty, but it now addresses diverse issues in the wildlife trade. It is a joint wildlife trade monitoring organization of the World Wildlife Fund and the IUCN. TRAFFIC publishes several periodicals and report series on the wildlife trade, including the *TRAFFIC Bulletin*, the TRAFFIC Online Report Series, and the *Species in Danger Series*.

The International Whaling Commission provides information on whaling regulations, whale sanctuaries, and other issues that are associated with whales and whaling. The Center for Biological Diversity plays a major role in the petition and listing process under the Endangered Species Act and conducts scientific investigations regarding imperiled species.

A useful historical resource was *The Endangered Species Act at Thirty: Renewing the Conservation Promise* (Dale D. Goble, J. Michael Scott, and Frank W. Davis, eds., 2006). The Gallup Organization provided information related to public polls that have been conducted in recent years concerning environmental issues.

Information Plus sincerely thanks all the previously mentioned organizations for the valuable information they provided.

INDEX

Page references in italics refer to photographs. References with the letter t *following them indicate the presence of a table. The letter* f *indicates a figure. If more than one table or figure appears on a particular page, the exact item number for the table or figure being referenced is provided.*

A

B

J

K

L

M

N

O

P

Q

R

S

T

U

V

W

Y

Z